Agricultural Field Experiments

BOOKS IN SOILS, PLANTS, AND THE ENVIRONMENT

Soil Biochemistry, Volume 1, edited by A. D. McLaren and G. H. Peterson
Soil Biochemistry, Volume 2, edited by A. D. McLaren and J. Skujiņš
Soil Biochemistry, Volume 3, edited by E. A. Paul and A. D. McLaren
Soil Biochemistry, Volume 4, edited by E. A. Paul and A. D. McLaren
Soil Biochemistry, Volume 5, edited by E. A. Paul and J. N. Ladd
Soil Biochemistry, Volume 6, edited by Jean-Marc Bollag and G. Stotzky
Soil Biochemistry, Volume 7, edited by G. Stotzky and Jean-Marc Bollag
Soil Biochemistry, Volume 8, edited by Jean-Marc Bollag and G. Stotzky

Organic Chemicals in the Soil Environment, Volumes 1 and 2, edited by C. A. I. Goring and J. W. Hamaker
Humic Substances in the Environment, M. Schnitzer and S. U. Khan
Microbial Life in the Soil: An Introduction, T. Hattori
Principles of Soil Chemistry, Kim H. Tan
Soil Analysis: Instrumental Techniques and Related Procedures, edited by Keith A. Smith
Soil Reclamation Processes: Microbiological Analyses and Applications, edited by Robert L. Tate III and Donald A. Klein
Symbiotic Nitrogen Fixation Technology, edited by Gerald H. Elkan
Soil-Water Interactions: Mechanisms and Applications, edited by Shingo Iwata, Toshio Tabuchi, and Benno P. Warkentin
Soil Analysis: Modern Instrumental Techniques, Second Edition, edited by Keith A. Smith
Soil Analysis: Physical Methods, edited by Keith A. Smith and Chris E. Mullins
Growth and Mineral Nutrition of Field Crops, N. K. Fageria, V. C. Baligar, and Charles Allan Jones
Semiarid Lands and Deserts: Soil Resource and Reclamation, edited by J. Skujiņš
Plant Roots: The Hidden Half, edited by Yoav Waisel, Amram Eshel, and Uzi Kafkafi
Plant Biochemical Regulators, edited by Harold W. Gausman
Maximizing Crop Yields, N. K. Fageria
Transgenic Plants: Fundamentals and Applications, edited by Andrew Hiatt
Soil Microbial Ecology: Applications in Agricultural and Environmental Management, edited by F. Blaine Metting, Jr.
Principles of Soil Chemistry: Second Edition, Kim H. Tan

Water Flow in Soils, edited by Tsuyoshi Miyazaki
Handbook of Plant and Crop Stress, edited by Mohammad Pessarakli
Genetic Improvement of Field Crops, edited by Gustavo A. Slafer
Agricultural Field Experiments: Design and Analysis, Roger G. Petersen
Environmental Soil Science, Kim H. Tan

Additional Volumes in Preparation

Mechanisms of Plant Growth and Improved Productivity: Modern Approaches, edited by Amarjit S. Basra

Selenium in the Environment, edited by W. T. Frankenberger and Sally Benson

Plant–Environment Interactions, edited by Robert E. Wilkinson

Agricultural Field Experiments

Design and Analysis

Roger G. Petersen

Oregon State University
Corvallis, Oregon

MARCEL DEKKER

MARCEL DEKKER, INC. NEW YORK • BASEL

Library of Congress Cataloging-in-Publication Data

Petersen, Roger G.
Agricultural field experiments: design and analysis / Roger G. Petersen.
p. cm. -- (Books in soils, plants, and the environment)
Includes bibliographical references (p.) and index.
ISBN 0-8247-8912-1
1. Field experiments. 2. Field experiments--Statistical methods. 3. Experimental design. I. Title. II. Series.
S540.F5P47 1994
630'.724--dc20 93-43399
CIP

The publisher offers discounts on this book when ordered in bulk quantities. For more information, write to Special Sales/Professional Marketing at the address below.

This book is printed on acid-free paper.

Marcel Dekker, Inc.
270 Madison Avenue, New York, New York 10016

Current printing (last digit):
10 9 8 7 6 5 4

PRINTED IN THE UNITED STATES OF AMERICA

Preface

This book is written for the agricultural research scientist, particularly the scientist involved with field experiments. It is also aimed at the student preparing for a career in agricultural research. The intent is to assist the scientist in the design, conduct, analysis, and interpretation of field experiments, and to serve as a source of reference for the scientist and student.

It is presumed that readers have been exposed to an introduction to statistical methods. In particular, it is assumed that readers are familiar with the ideas of interval estimation and significance testing. It is further assumed that they have an acquaintance with the basics of the analysis of variance and simple linear regression and correlation. It is expected that readers are familiar with such common statistical tables as Student's t, F, and chi-square. The necessary mathematical background is minimal; at most, a knowledge of college algebra is required.

I have attempted to put together a book that is somewhat different from the usual text on experimental designs. While standard designs are discussed, other topics of particular interest to the field scientist are included. For example, there is an expanded discussion, in Chapter 6, on the combined analysis of data from several experiments, a necessary, but often neglected, part of any research program. Some topics, such as stability analysis (Section 5.6), have not previously been brought together in a book of this type. Further, the combined analysis of on-station yield trials (Section 6.7) has not, to my knowledge, appeared before in print. I have also included sections on other areas of research that have been addressed in journal articles but that have not been included in books devoted to agricultural field research. These topics include: pasture research (Chapter 8), farming systems research (Chapter 9), and intercropping research (Chapter 10).

I have tried to organize the material into sections directed toward particular areas of agricultural research, such as agronomy (Chapter 4) and plant breeding (Chapter 5), rather than into sections devoted to particular classes of experimental designs. I have attempted to include topics that I have found to be particularly valuable to me when working with agricultural scientists during more than 30 years as an agricultural experiment station statistician and consulting biometrician.

It is my belief that the only purpose for statistical and biometrical procedures is to aid the research scientist in obtaining and interpreting research data. For this reason I have tried to illustrate the types of detailed data analysis that I believe to be necessary to obtain the maximum amount of information from a set of research data. I have included a section entitled "Report of Statistical Analysis" with each numerical example. The purpose of this section is to illustrate the presentation of the results of a statistical analysis in terms that can be understood by the agricultural scientist.

The material in the first five chapters forms the basis of a one-term course entitled "Field Plot Technique" presented in the Crop Science Department at Oregon State University. The students are graduate students with majors in the applied agricultural sciences. For many of the students, this course serves as an alternative to the usual course in experimental design.

Some of the tables in the Appendix are reproduced with the kind permission of the original authors and publishers. I am grateful to the literary executor of the late Sir Ronald A. Fisher, F.R.S., to Dr. Frank Yates, F.R.S., and to the Longman Group Ltd., London, for permission to extract Table III from their book *Statistical Tables for Biological, Agricultural and Medical Research* (6th edition, 1974), which appears as Table A2 in this book. The table of percentage points of the F distribution (Table A3), the table of percentage points of the ratio $s^2(max)/s^2(min)$ (Table A4), the table of percentage points of the chi-square distribution (Table A5), and the table of percentage points of the Studentized range (Table A6) appear as Tables 18, 31, 8, and 29, respectively, in *Biometrika Tables for Statisticians*, Vol. 1 (3rd edition, 1966). These tables are reproduced with the kind permission of the Biometrika Trustees. The table of minimum-average-risk t values (Table A7) is reproduced with the permission of the American Statistical Association.

I am indebted to a number of people who contributed in no small way to the ideas and material in this book. The material in the chapter on pasture trials is largely the work of the late H. L. "Curly" Lucas, with whom I had the pleasure of working for a number of years. I would also like to express my appreciation to the scientists at the agricultural experiment stations in North Carolina and Oregon, and those at the International Center for Agricultural Research in the Dry Areas (ICARDA) in Aleppo, Syria, and in the National Agricultural Research Centre (NARC) in Islamabad, Pakistan, for providing many interesting problems and for many fruitful discussions on field research techniques. I would also like to thank the many students whose comments helped to clarify the class notes upon which the first part of this book is based. I am grateful to my wife, Jean, who provided encouragement and support throughout this project. I would also like to show my appreciation to Ms. Genevieve Downing, who served as my secretary and who was largely responsible for putting the manuscript into final shape.

Roger G. Petersen

Contents

Agricultural Field Experiments

1

Basic Principles

1.1 INTRODUCTION

For many branches of agricultural science, plant science in particular, the small-plot field experiment is the fundamental tool for obtaining new information in a research program. Research programs in crop science can have a number of objectives:

1. To develop new varieties that are superior to existing varieties in yielding ability, disease resistance, quality, and so on.
2. To study the effects of changes in the factors of production, and to identify the levels of these factors that optimize yield, quality, and other characteristics.
3. To explore new technologies, new crops, and new areas for potential agricultural production.
4. To demonstrate new knowledge to growers, who must ultimately decide whether or not to use it.
5. To develop a basic understanding of the factors that control the production of crops and how these factors operate.

To meet these objectives, crop scientists conduct experiments of one type or another. In many cases these are small-plot field experiments. Because the resources that can be devoted to research in any program are necessarily limited, it is essential that the research be conducted in such a way as to minimize cost and maximize the information generated for the resources used. In this regard statistical procedures, particularly those of experimental design, can be of assistance. In this book we attempt to look at research techniques and, in particular, the concepts of experimental design as they apply to field experiments.

1.2 TYPES OF FIELD EXPERIMENTS

Plant scientists are involved in a number of types of experiments. The most basic classification is into those involving applied research and those involving basic research. Sometimes an experiment will contain elements of both. More often than not, field experiments are oriented toward applied research.

For the most part, field experiments are one of two types: variety trials and agronomy trials (trials concerned with the factors of production). There are, however, other types of field experiments: pasture trials, those involving perennial crops, farming systems research, and intercropping trials, for example. Each type of experiment has its own special features, and we shall examine some of them briefly.

1.2.1 Variety Trials

Variety trials form the backbone of a plant breeding program. They range from the initial studies involving small plots (single plants or single rows) to replicated yield trials involving fairly large plots (2.5 × 5.0 m, 8 × 16 ft). In the initial studies, plant material is selected by a subjective judgment based on the experience of the plant breeder.

In the replicated yield trials one or more check varieties are included to serve as standards against which the performance of the new selections can be compared. Such trials should also include a long-term check, a variety that stays the same from year to year, to serve as a baseline against which to measure the progress of the breeding program.

1.2.2 Agronomy Trials

Agronomy trials include fertilizer studies, time, rate, and density of planting, tillage studies, and pest and weed control studies, among others. Because the response to changes in the level of one factor is often conditioned by the levels of other factors it is almost essential that the treatments in agronomy trials include combinations of multiple levels of two or more production factors. In general, the plot size for agronomy trials is larger than that for variety trials. This is because of the size of the machinery involved in applying the treatments and because of the border effects introduced, in part, by the use of machinery.

1.2.3 Experiments with Perennial Crops

Experiments with perennial crops involve special problems because the same crop occupies the same plot for 2 or more years. Differences in performance from plot to plot are likely to persist, and the effects of treatments are likely to accumulate. Because the treatments are not randomly assigned to the plots each year, it is not possible to use years as a replication of the experiment. This means that extra care must be used in setting up the experiment. Consideration should also be given to using large plots in the initial years of the trial to permit the introduction of new treatments into the trial after it has been running for some time.

1.2.4 Pasture Experiments

Pasture experiments are different from other types because the response to treatments is measured by grazing animals. Since animals are used, the plots must be fairly large. Plots must be big enough to support one or more animals for an extended period of time. (*Note*: The pasture and not the animal is the experimental unit in a pasture trial.) Because of the costs involved in conducting a full-scale pasture trial, the initial evaluation of prospective treatments is often done on small plots with clipping being used to simulate grazing. This procedure may be useful in the initial screening, but the ultimate evaluation must be done by the grazing animal.

1.2.5 Farming Systems Research

Farming systems research is an attempt to move new agricultural technologies from the research center to the farm. In farming systems research an improved technology, or package of technologies, is selected on the basis of the results of on-station research. The assumption is that the new technology is in some way superior to the practice presently being used. An area called the *target area* is identified. The target area contains farms on which a common production system is presently in use and for which the new technology is expected to lead to improved production.

A number of farms in the target area are chosen. Two large plots are then laid out on each farm. On one of these plots the new package of technologies is used, while on the other plot the farmer's usual practice is used. On each plot the yield, as well as the costs of inputs and the revenue from outputs, is measured. The aim of the research is to provide a measure of gain in net revenue from the new system as compared to that of the old.

Because farming systems research also provides a way of demonstrating new technology to farmers in the target area, the plots must be large and located close to each other so that side-by-side comparisons can be made by the local farmers.

1.2.6 Intercropping Experiments

Intercropping experiments involve growing two or more crops at the same time on the same piece of land. The crops need not be planted nor harvested at exactly the same time, but they are usually grown together for a significant part of the growing season. Intercropping is used extensively in the tropics and subtropics. It is used by subsistence farmers who are trying to ensure themselves against total crop failure should one of the components of the intercropping system fail. The aim is to provide a measure of yield stability. In other instances it is hoped that the yield would be greater from the combination of crops than it would from any of the component crops grown alone.

Experiments designed to study intercropping face a number of problems. The treatments in the experiment must include the crops by themselves as well as the intercrop combinations. If there is a need to determine the optimum ratio of land to be devoted to the component crops, several ratios and several planting configurations should be included in the trial. In any case the number of treatments to be included may become quite large. Further, if the aim is to examine the stability of the system, the experiment must be conducted for several years.

1.3 GENERAL CONSIDERATIONS

Suppose that a wheat breeder wants to compare the yield of a newly developed variety with that of a well-known standard variety. He has two basic objectives: The first is to answer the question: Is the yield of the new variety superior to that of the old? The second question, related to the first, is: How large is the true difference in yields?

Breeders will usually run field experiments to obtain data on which to base their answers. The answer to the first question involves a test of the hypothesis that there is no difference in yield against the alternative that a difference exists. A breeder answers the second question by using the data to compute an estimate, preferably an interval estimate, of the true difference in yield.

Almost all experimentation is done to answer one or both of these questions. An additional purpose is to obtain information on why treatments affect the experimental material as they do. It is the job of the crop scientist to conduct research in such a way as to provide precise information to meet these objectives, and to conduct this research in an efficient way.

Suppose that a wheat breeder decided to conduct an experiment to compare the yield of the new variety with that of the old. The breeder could plant the new variety on one plot of ground and the old variety on another. At harvest the yield of the new variety could be compared with that of the old. This procedure would appear to offer a simple solution to the problem. It does not, however, permit the breeder to draw a valid conclusion. The difference observed may be due to a real difference between varieties or it may be due to other factors.

1.4 EXPERIMENTAL ERROR

To understand why the comparison of yields from only two plots does not support a valid conclusion, assume that the breeder selects two plots which are as nearly alike as possible in all respects. Then, suppose that the same variety is planted on each plot. No matter how careful the breeder is in land preparation, seeding, cultivation, and harvesting, the plots will differ to some extent in fertility, drainage, soil moisture, and the like. As a result, the yields will differ even though every effort has been made to assure that the plots are as alike as possible. This variation between plots treated alike is always present. The breeder must be able to separate this variation from any real varietal difference if a valid comparison of one variety with another is to be made.

To make valid comparisons the breeder the research must be conducted in such a way that the variation among plots treated alike can be measured. That is, the breeder must be able to measure the variation that would exist if all the plots were planted to the same variety. It can be concluded that two varieties differ only if the observed yield difference is sufficiently larger than what would have been found if all the plots had been planted to the same variety.

Variation among experimental units (plots) treated alike is called *experimental error.* This is not to imply that mistakes have been made in running the experiment. Rather, it is a term used to describe the variation among plots subjected to the same treatment. This variation may be natural plant-to-plant variation or slight variation in the conduct of the experiment, applying the treatments, weighing errors, and so on. Although experimental error can never be eliminated completely, it can be reduced

One of the goals of modern experimental design is to provide a measure of experimental error. Another goal is to reduce experimental error as much as possible within the limitations of the resources that can be committed to the experiment.

1.4.1 Sources of Experimental Error

There are a number of sources of variation in field experiments, and these sources all contribute to experimental error. Some of these variations can be reduced by the use of an appropriate experimental design, some cannot. Plot size and shape, experimental design, layout of the experiment, and other factors have an effect on experimental error. Other factors contribute to error as well. In general, the sources of natural variability in field experiments are (1) plant variability, (2) seasonal variability, and (3) soil variability.

Plant variations include type of plant, competition among plants within plots, and variability between plots. Ordinarily, there is larger variation among larger plants (maize) than among smaller plants (cereals). Similarly, the variation among closely spaced plants is smaller than it is among widely spaced plants, and variation is lower for plots with a uniform stand than it is for plots with a patchy stand.

Plot-to-plot variation is another factor to be considered. Plots that are surrounded by similar plots are usually less variable than plots next to alleys, borders, plots with plants of different height, and so on. This variability is due to differential competition between plants at the edges of the plots and is sometimes called a *border effect*. We examine border effects in more detail later.

Seasonal variability is another contributor to experimental error. Climatic factors vary differently from year to year. Similarly, rodents, insects, and disease cause more damage in some years than in others. Some crops, and some crop varieties, are less affected by seasonal variation than are others. In any case, it is well to sample a range of seasonal variation by conducting tests for several (3 or more) years before drawing firm conclusions.

Soil variability is present almost everywhere, even over a small area in a single field. This variability is the result of differences in texture, depth, moisture-holding capacity, drainage, available nutrients, and other factors. Further, these differences tend to persist from year to year. If the pattern of soil variability is known, its effect on experimental error can be reduced by such factors as proper choice of experimental design; size, shape, and orientation of plots; and planned grouping of plots. If time and resources are available and if a particular field is to be used for several years as an experimental site, the pattern of variability can be mapped with a uniformity trial.

1.4.2 Uniformity Trials

To conduct a uniformity trial the area is planted uniformly to a single crop. Cultural practices are conducted throughout the trial in as uniform a way as possible. The trial is then partitioned into small units, or plots. These are harvested individually and the individual yields, together with the location of the plot from which they came, are recorded. These can then be used to construct a yield map of the site, such as that shown in Figure 1.1.

The raw yield map may be smoothed to remove minor irregularities and some of the normal plant-to-plant variation. This smoothing can be done in a number of ways, of which one of the easiest to use is running medians of three plots. For this purpose the yield of a plot is replaced by the median of three adjacent plots in the same row. This procedure is called a *3M smooth* in the row. The 3M row smooth for the yields in Figure 1.1 is shown in Figure 1.2. The next step is to do a 3M smooth in the columns of the map that has been smoothed by rows. The yield map has now been smoothed in two directions. The resulting smooth map is presented in Figure 1.3. The last step is to delineate areas of equal yield. This can be done by using different colors, shading, or crosshatching to develop a smooth yield contour map of the area. The final map of the sample data is presented in Figure 1.4.

Such a map can be of help in deciding on the layout of the field plan. It can also be used in determining the shape and orientation of the plots and in grouping the plots into blocks of similar plots. Further, information from a uniformity trial can be used to determine the optimum size of plot as well. We consider this aspect further when we look at plot size.

1.5 EXPERIMENTAL ERROR MEASUREMENT AND CONTROL

Modern experimental procedures include a number of features that permit the crop scientist to measure and control experimental error. These features include (1) replication, (2) randomization, and (3) blocking.

1.5.1 Replication

It is sometimes possible for a scientist to obtain an estimate of experimental error from a previous experiment involving similar treatments on similar plots. The preferred procedure, however, is to obtain an estimate of error from the experiment itself. To do this the scientist must repeat, or *replicate*, each treatment in the experiment. That is, each treatment must appear on more than one plot in the trial.

There are a number of reasons for replication in an experiment:

1. It provides an estimate of *experimental error* because it provides observations on several plots receiving the same treatment. In this way the variation among plots treated alike can be measured.

488	406	491	479	438	367
488	392	457	438	296	510
457	359	436	396	503	337
440	403	388	450	420	384
325	365	369	343	361	376
347	462	393	390	368	437
424	387	408	354	386	402
426	384	346	389	375	375
448	454	347	374	321	403
476	430	490	428	435	452
416	445	440	415	449	442
387	382	319	393	438	455
470	418	353	402	470	410
446	426	352	394	427	435
410	393	385	411	446	441
388	420	389	419	414	348
303	333	389	396	473	371
315	328	365	426	455	375

Figure 1.1 Site map of yield data (g/plot) of wheat from a uniformity trial.

488	488	479	479	438	367
488	457	438	438	438	510
457	436	396	436	396	337
440	403	403	420	420	384
325	365	365	361	361	376
347	393	393	390	390	437
424	408	387	386	386	402
426	384	384	375	375	375
448	448	374	347	374	403
476	476	430	435	435	452
416	440	440	440	422	422
387	382	382	393	438	455
470	418	402	402	410	410
446	426	394	394	427	435
410	393	393	411	441	441
388	389	419	414	414	348
303	333	389	396	396	371
314	328	365	426	426	375

Figure 1.2 Yield data from Figure 1.1 smoothed by running medians of 3 in rows (3M smooth in rows).

488	488	479	479	438	367
488	457	438	438	438	367
457	436	396	420	396	376
440	403	396	420	396	376
347	393	393	390	390	384
347	393	387	386	386	402
424	393	387	386	386	402
426	408	384	375	375	402
448	448	384	375	375	403
448	448	430	435	435	422
416	440	430	435	435	452
416	418	402	402	422	422
446	418	394	394	427	435
446	418	394	402	427	435
410	393	393	411	427	435
388	389	393	411	414	371
315	333	365	414	414	371
315	328	365	426	426	375

Figure 1.3 Data from Figure 1.2 smoothed further by running medians of 3 in columns (3M smooth in columns).

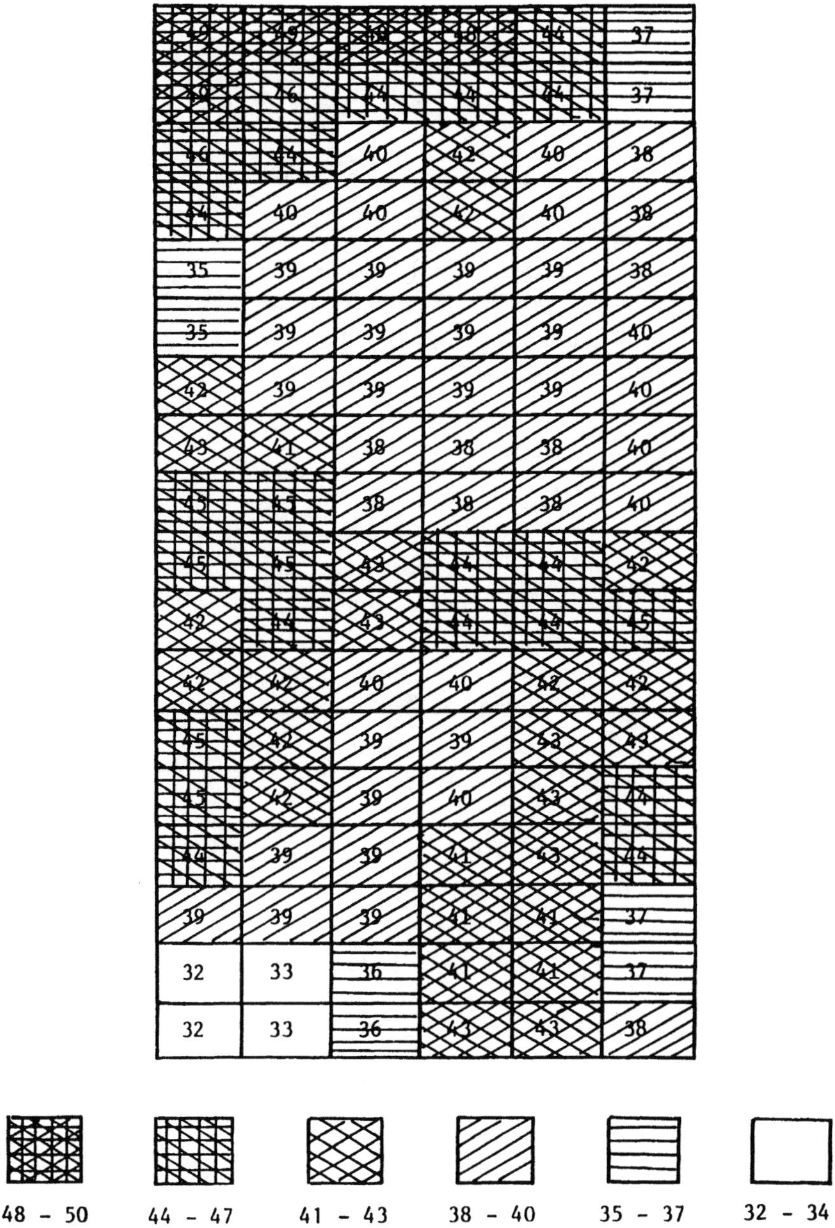

Figure 1.4 Smoothed variability map from a uniformity trial with wheat (yield in dg/plot).

2. It *increases precision* by reducing standard errors. Recall that the standard error of a mean, $s_{\bar{y}}$, is given by

$$s_{\bar{y}} = \sqrt{\frac{s^2}{n}}$$

where s^2 is the sample variance and n is the number of observations (replications) on which the mean is based. Clearly, as the number of replications, n, is increased, the standard error, $s_{\bar{y}}$, decreases.
3. It can *broaden the base for making inferences*. As replication is increased a wider variety of plots can be brought into the experiment. This can lead to a greater range of conditions over which the results of the experiment can be applied.

Before considering the factors that determine the number of replications, we need to recall two relations:

1. $s_{\bar{y}} = \sqrt{s^2/n}$, as given previously.
2. $L(\mu_1 - \mu_2) = (\bar{y}_1 - \bar{y}_2) \pm t_\alpha\sqrt{2s^2/n}$, where
   $L(\mu_1 - \mu_2)$ = $(1 - \alpha)100\%$ confidence interval estimate of the true difference between the means of treatments 1, $\mu_1$and 2, μ_2
   $\bar{y}_1, \bar{y}_2$ = sample means for treatments 1 and 2
   t_α = two-tailed t at the α probability level
   s^2 = sample variance per plot
   n = number of replications

Thus the number of replications determine, in part, the precision with which the true means of the treatments can be estimated and the magnitude of the differences that can be detected. With this background we see that some of the things that determine the number of replications are:

1. Pattern and magnitude of variability in the soils of the experimental site
2. Size of the difference to be detected
3. Required significance level
4. Number of treatments
5. Amount of resources that can be devoted to the experiment
6. Limitations in cost, labor, time, and so on.

1.5.2 Randomization

In addition to replication there is another factor that a plant scientist must consider. Suppose that a wheat breeder had decided to use five plots of a new variety and five plots of an old one to test the difference between the two varieties. Suppose, further, that he had laid out a strip of 10 plots and assigned varieties as shown in Figure 1.5.

Now, suppose that unknown to the breeder, there is a fertility gradient ranging from low at one end of the strip to high at the other. In each pair of plots in the field plan the new variety has been assigned to the plot with the higher fertility level. An observed difference between varieties will reflect a fertility difference in favor of the new variety as well as a varietal difference if it exists. To avoid this problem the treatments should be assigned to the plots in such a way that no treatment is favored by being assigned consistently to the better plots. To do this the breeder should resort to a process called *randomization*.

With proper randomization any plot is equally likely to be assigned to any treatment. In some experimental designs randomization is restricted in certain ways, but in no design is it completely eliminated. We shall look at the details of the randomization procedure for each type of experimental design when we consider the individual designs.

There are at least two reasons for randomization:

1. To *eliminate bias*. Randomization assures that no treatment is favored or discriminated against by the way it is assigned to the plots.
2. To *ensure independence* among observations. This is required for valid significance tests and interval estimates.

1.5.3 Blocking

Another feature of good experimental technique is blocking. The procedure here is to group the plots, or experimental units, into blocks, or groups, of homogeneous units. Treatments are then assigned at random to the plots within the blocks. The basic goal is to remove block-to-block variation from experimental error. Treatment comparisons are then made within the blocks on plots that are as nearly alike as possible.

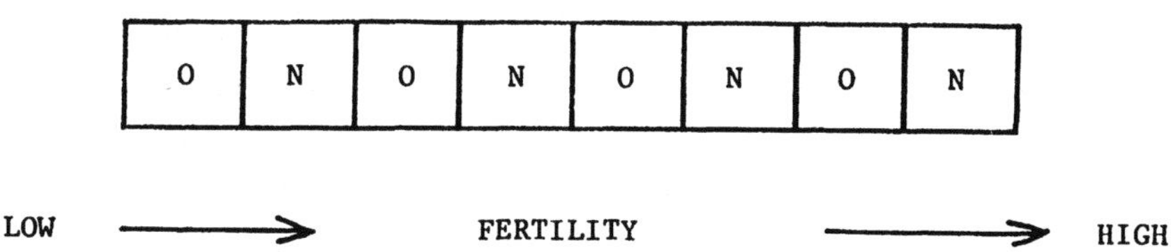

Figure 1.5 Plot layout for new (N) and old (O) varieties assigned systematically to a strip of plots with a fertility gradient.

For example, the field under consideration for an experiment might contain two types of soil. We could select plots on each soil type and randomly assign treatments to plots within soil types. In this case soil type would be a blocking factor and the difference between soil types would be removed from experimental error in the analysis of the data. Of course, criteria other than soil type can be used as blocking factors. The key to successful blocking is to maximize the variation among blocks and to minimize the variation among plots within blocks. We shall consider the concept of blocking in more detail as we look at specific experimental designs.

There are a number of reasons for blocking:

1. It can *increase precision*. Differences among blocks are removed from experimental error in the analysis of the experimental data.
2. Treatment *comparisons are more uniform*. Because comparisons are made within blocks of uniform plots the observed differences more nearly reflect true differences among treatments.
3. It can sometimes *increase the information* from an experiment. Blocking permits the scientist to sample a wider range of conditions and hence can broaden the scope of inferences.

Note: Contrary to usage in some areas of agricultural research, blocks and replications are not the same thing. A *block* is simply a group of plots, whereas a *replication* is a repetition of a treatment in an experiment.

1.5.4. Reduction of Experimental Error

We have indicated earlier that one of the goals of sound research procedure is to reduce experimental error and thus increase precision. Recall that the standard error, $s_{\bar{y}}$, of a mean is a measure of the precision with which the mean is estimated. The smaller is the standard error the more precise is the estimate. Recall, again, that

$$s_{\bar{y}} = \sqrt{\frac{s^2}{n}}$$

where s^2 is the error variance per plot and n is the number of observations (replications) per treatment. We can reduce $s_{\bar{y}}$ in the following ways:

1. *Increase the size of the experiment.* Increase the size by increasing the replication or by increasing the number of treatments. We are attempting to reduce $s_{\bar{y}}$ by increasing n or by increasing the degrees of freedom with which s^2 is computed, in this way, reducing $s_{\bar{y}}$. As the size of the experiment is increased, however, more heterogeneous

material must be used, with consequent increase in s^2. Thus if carried too far, increasing size might be self-defeating.

2. *Proper selection of treatments.* Factorial combinations of treatments (discussed later) result in hidden replication for some comparisons, an effective increase in n.
3. *Refine the experimental technique.* Here the attempt is to reduce s^2.
4. *Blocking.* To the extent that block differences are removed from experimental error, s^2 will be reduced.
5. *Measure a concomitant variable.* Covariance analysis with the concomitant variable as a covariable can sometimes reduce s^2.

1.6 PLANNING CONSIDERATIONS

In addition to replication, randomization, and blocking, a few other points should be given consideration in planning field experiments. These include, but are not limited to, a statement of the problem, a statement of the objectives, the choice of treatments to be used, the choice of experimental site, the measurements to be made, and the statistical analysis of the resulting data. Other topics, such as the choice of experimental design and the field plot itself, are deferred for more detailed discussion later.

1.6.1 Problem and Objectives

A statement of the problem and the objectives is essential. Unless this can be done, there is little hope that the experiment will be a success. It is common in the initial planning stage to attempt to do too much in a single experiment. This is usually the result of too little thought about the objectives of the experiment, about the resources that can be devoted to the experiment, and about the complexity of the analysis of the resulting data. The preparation of a written statement of the objectives, listed in order of their priority, will often result in a more concise and realistic experiment.

1.6.2 Treatments

The treatments to be used in an experiment are determined by the objectives. They are selected to answer the questions raised by the problem and the specific objectives of the experiment. If the objective is to evaluate new selections in a breeding program, the treatments are the new selections along with one or more standard varieties against which the new selections can be compared.

On the other hand, agronomy trials are often designed to answer such questions as: What rate of seeding should be used? Is fertilizer necessary?

If so, what kind and how much? What row spacing will produce maximum yield? Treatments for such trials should be selected to answer these specific questions. Often, the response to changes in one production factor is conditioned by the level of one or more other factors. For this reason, treatments that consist of combinations of factors at different levels are often useful. We consider multifactor treatment combinations when we look at factorial experiments.

1.6.3 Choice of Experimental Site

The choice of experimental site affects the conduct of the experiment and the interpretation of the results in several ways. It determines the population to which the results of the experiment will apply. Thus the population about which information is wanted must be defined carefully and the site must be one that represents the population. If the results are to be used by grass seed growers in the Willamette valley, for example, the trial must be conducted in the valley and not in central Oregon. In theory the site should be selected at random from among all possible sites in the population. In practice, however, a site is usually selected because it is accessible, available, and so on. As long as the site is representative, however, the results are usually acceptable.

Often, the research is conducted on an experiment station belonging to the research organization. Equally often the area within the station has been altered so much by previous research that the population it represents is restricted to the station itself. This causes little difficulty if the problem is one of basic research where the aim is to find out something about the mechanism of treatment action. For applied research, on the other hand, it must be assumed that the results are "like" what would have occurred in the area of interest, with the realization that recommendations based on such results are uncertain.

The characteristics of the site also determine, to some extent, the type of experimental design to be used, the placement of the trial, the necessity for and the formation of blocks, and the size, shape, and orientation of plots. All sites have variation in such characteristics as soil, topography, and physical features. The experiment should be planned in such a way that the effects of these sources of variability are minimized. Differences in topography and physical features that are easy to recognize should be indicated on a sketch map of the site to assist in planning. The pattern of soil variability, on the other hand, is usually not obvious. Sometimes it can be guessed by noting such features as hills, slopes, low spots, and drainage patterns. Such guessed patterns might be included on the sketch map. If the site is to be used for several years and if sufficient resources are available, the soil pattern can be mapped using the uniformity trial, described earlier.

1.6.4 Statistical Analysis

When the experiment has been completed, the data should be subjected to statistical analysis. The usual first step is to conduct an analysis of variance of the data. The analysis of variance (ANOVA) does the following in a systematic way:

1. Partitions the total variation in the data into components associated with such sources of variation as error, treatments, and grouping categories (blocks).
2. In particular, it provides an estimate, s^2, of experimental error that can be used to construct interval estimates and significance tests.
3. Provides a format for testing the significance of the several variance sources in the partition.

Once the ANOVA has been completed, the data are subjected to additional analysis to provide specific information on the questions raised by the objectives of the experiment. This may take the form of comparing various groups of means or of attempting to select treatments that are "best" in some sense. We outline and illustrate the ANOVA of each design we consider. We also look at methods of separating means once the preliminary analysis has been completed.

1.7 DEFINITIONS

An *experiment* is a planned inquiry to discover new facts, or to confirm or deny the results of previous investigations.

A *treatment* is a procedure whose effect on the experimental material is to be measured.

A particular class of related treatments is often called a *factor.*

The states of a factor (i.e., the treatments within the class) are called the *levels* of the factor.

An *experimental unit* is the piece of experimental material to which one trial of a single treatment is applied.

A *sampling unit* is that fraction of the experimental unit on which the effect of the treatment is measured.

A group of homogeneous experimental units is called a *block.*

The quantity that is measured on the experimental material is often called the *yield.*

An *experimental design* is a set of rules by which the treatments are assigned to the experimental units.

Experimental error is the variation among experimental units that have been treated alike.

If treatments are assigned to a set of units in such a way that every unit is equally likely to receive any treatment, the assignment is said to be *random.*

When a treatment appears more than once in an experiment, the treatment is said to be *replicated.*

FURTHER READING

Complete citations for the books listed for further reading are given in the bibliography at the end of the book.

1. Cochran and Cox, Chaps. 1 and 2.
2. Cox, Chaps. 1–6.
3. Le Clerg, Leonard, and Clark, Chaps. 1–3 and 10.
4. Little and Hills, Chaps. 1–3.
5. Petersen, Chap. 1.
6. Steel and Torrie, Chap. 6.

2

The Field Plot

2.1 INTRODUCTION

The fundamental experimental unit in field experiments is the field plot. This is the piece of experimental material to which one treatment is applied, and it serves as the vehicle for evaluating the response of the material to the treatment. Depending on the material involved, the plot can take a number of different forms. In the early stages of a plant breeding program, for example, the plot might be a single plant or a short row of plants. With field crops, such as cereals or forages, the plot may be a small area of land and the crop growing on it. With horticultural crops the plot may also be an area of land upon which rows of crops are growing. In other cases the plot might be one or more bushes of ornamental shrubs or fruits. In range or pasture evaluation the plot might be a line transect of a given length across a vegetated area. With orchard crops the plot may be one or more trees, or in some cases, single limbs on a tree.

Although rectangular plots are most common plots of other shapes are also used. When studying the effect of different plant densities, for instance, fan-shaped plots with high densities at the narrow end and low densities at the wide end of the fan are often used. Sometimes the shape of the plot is determined by the machinery used to apply the treatments to conduct the experiment. Sprinklers rotating around a central pivot might call for circular plots. We discuss the factors affecting plots as if we were dealing with rectangular plots. For the most part, however, the same principles apply regardless of the shape of the plot.

2.2 PLOT SIZE

There are a number of factors that affect plot size. These include, but are not limited to:

1. Type of crop
2. Type of experiment
3. Phase of the research program
4. Variability of the experimental site
5. Presence and nature of border effects
6. Type of machinery to be used
7. Number and type of treatments
8. Land area available
9. Cost

2.2.1 Effect on Variability

Studies carried out with uniformity trials have shown that variability per plot tends to decrease as plot size is increased up to a size of about 0.05 ha (0.1 acre). However, there is also a tendency toward higher experimental error as plot size increases because larger plots mean that a bigger, more variable area must be brought into the experiment. On the other hand, very small plots tend to have high variability because:

1. Losses at harvest or errors in measurement have relatively greater effect.
2. Variability with row crops in small plots may be large because of reduced plant numbers.
3. Competition and border effects are likely to be relatively greater in small plots.

2.2.2 Lower Limits

There are some factors that set a lower limit on plot size. Plots should be large enough to permit the removal of borders, where necessary, with a large enough remainder to provide a reasonable yield sample. Plots must also be large enough to accommodate the machinery needed to seed, fertilize, cultivate, and harvest the plots. Once the plots are large enough to be handled conveniently, little is gained by making them larger. For a given total area in the experiment, precision is increased faster by increasing the number of replications than by increasing plot size.

2.2.3 Type of Crop

Plot size is determined, in part, by the type of crop being grown. Smaller plots are required for cereals and other closely seeded crops than for row crops. Larger plots are required for fruit, nut, and other tree crops. Finally, because they are evaluated by grazing animals, pasture experiments require even larger plots.

Plot sizes for agronomy trials (fertilizer, tillage, etc.) are usually larger than for breeding trials. This is because the plots must be large enough to permit the operation of machinery in them and to adjust for "machinery drag" from one plot to another.

2.2.4 Phase of the Research Program

Plot sizes tend to increase as the research program progresses. In a cereal breeding program, for example, the plot may be as small as a single plant in the initial stages of the program. In later generations single-row plots may be used, then two-row plots. Preliminary selection for yield is often made using unreplicated two- or four-row plots. Advanced yield trials use replicated four- or six-row plots. Finally, farmer's field verification trials often contain 0.1-ha or 0.1-acre plots.

2.2.5 Summary

We summarize the effects of several factors on plot size in Table 2.1.

Table 2.1 **Factors Affecting Plot Size**

Factor	Increasing plot size Small plots ——————→ Large Plots
Soil variability	Uniform ——————→ Heterogeneous
Crop	Turf →Cereals →Row crops→Trees →Pasture
Research phase	Early ——————→ Late
Experiment type	Breeding ->Fertilizer ->Tillage->Irrigation
Machinery	None ———→ Research ———→ Farm scale

2.3 SOIL VARIABILITY INDEX

The optimum size of plot, that which will minimize variance for a fixed cost or minimize cost for a fixed variance, depends on the costs involved in conducting the trial and on the soil variability at the experimental site. Cost data are not too difficult to obtain, but a measurement of soil variability is not too easy to come by.

The basic work on the relationship between plot size and variance per plot was developed by Smith (1938), who proposed the following relation:

$$V_x = \frac{V}{x^b}$$

where

V = variance of a unit plot
V_x = variance, on a per unit basis, of plots formed from x adjacent units
x = plot size in multiples of adjacent unit plots
b = index of soil variability

The index can vary from zero to 1. When $b = 1$,

$$V_x = \frac{V}{x}$$

and the units that go to make up the plots are independent (uncorrelated) of each other. This is the situation that exists if the soil is highly heterogeneous. If adjacent plots are perfectly correlated (no variability), then $b = 0$ and $V_x = V$. In this case there is no relation between variance and plot size. This condition exists if the soil is highly uniform.

2.3.1 Variability Index Calculation

The calculation of Smith's index of soil variability may be illustrated using the wheat uniformity data presented in Figure 1.1. These data are from plots 5 m long and 2 m wide, an area of 10 m^2. These constitute the basic plots for constructing the variability index. The data can be combined in various ways to give yields of plots of different sizes. Examples of such combinations are presented in Figures 2.1 to 2.3. Figure 2.1, for example, gives a map of yields on plots 5 m long and 4 m wide, twice the size of the basic unit, formed by combining plots from adjacent rows of Figure 1.1. Similarly, Figure 2.2 gives a map of yields from plots twice the size of the basic unit plot formed by combining plots from adjacent columns of Figure 1.1. Finally, Figure 2.3 shows a map of yields on plots 10 m

976	798	948	917	734	877
877	762	824	846	923	721
672	827	762	733	729	813
850	771	844	743	761	777
924	884	837	802	756	855
803	827	759	808	887	877
916	844	705	796	897	845
798	813	774	830	860	789
618	661	754	822	928	746

Figure 2.1 Yield map (g/plot) for plots 4 m × 5 m, constructed by combining yields of plots from adjacent rows of Figure 1.1.

long by 4 m wide, four times the basic unit plot size, formed by combining adjacent rows and adjacent columns of Figure 1.1.

By taking different combinations of rows and columns 19 plot yield maps were constructed from the basic yield data of Figure 1.1. Characteristics of these yield maps are given in Table 2.2. The data for units per plot, x, and variance per unit, V_x, are transformed to a logarithmic scale as shown in Table 2.3. Using linear regression procedures the following equation is fitted to the transformed data:

$$\log V_x = \log V - b \log x$$

where

V_x = variance, per unit, of plots of x units
V = variance of a unit plot
x = number of unit plots
b = index of soil variability

894	970	805
880	895	806
796	832	840
843	838	804
690	712	737
809	783	805
811	762	788
810	825	750
902	721	724
906	918	887
861	855	871
769	712	893
888	755	880
872	746	862
803	796	887
808	808	762
636	785	844
643	791	830

Figure 2.2 Yield map (g/plot) for plots 2 m × 10 m, constructed by combining yields of plots from adjacent columns of Figure 1.1.

1774	1865	1611
1639	1670	1644
1499	1495	1542
1621	1587	1538
1808	1639	1611
1630	1567	1764
1760	1501	1742
1611	1604	1649
1279	1576	1674

Figure 2.3 Yield map (g/plot) for plots 4 m × 10 m, constructed by combining yields of plots from adjacent rows and columns of Figure 1.1.

For the present data we find that

$$\log V_x = 3.3205 - 0.7394 \log x$$

Thus the index of variability for this set of wheat yield data is

$$b = 0.7394$$

A plot of the transformed data and the fitted equation is shown in Figure 2.4.

Koch and Rigney (1951) have shown that the index of soil variability, b, can be estimated using data from certain types of experimental designs. Further, Hathaway and Williams (1958) have presented a weighting procedure for obtaining a minimum-variance unbiased estimate of b from plots of different sizes. Examination of a large number of sets of data indicate that a value of b = 0.5 will serve as a reasonable approximation if data from which a more accurate estimate can be made are not available.

2.4 OPTIMUM PLOT SIZE

Once an estimate of b is available, the optimum plot size—that which will minimize cost for a fixed variance or minimize variance for a fixed cost—

Table 2.2 **Summary of Characteristics of Yield Maps of Plot of Various Sizes Formed by Combining Yields (g/plot) of Adjacent Plots in a Uniformity Trial on Wheat (Figure 1.1)**

Combination		Number of plots	Area per plot	Units per plot	Variance per plot	Variance per unit	CV
Rows	Columns	n	m.^{2}	x	s^2 *	V_x**	%
1	1	108	10	1	2,126.66	2,126.66	11.35
1	2	54	20	2	4,820.68	1,205.17	8.54
1	3	36	30	3	8,425.34	936.15	7.53
1	6	18	60	6	21,231.16	589.75	5.97
2	1	54	20	2	5,631.70	1,407.92	9.23
2	2	27	40	4	13,879.30	867.46	7.25
2	3	18	60	6	24,810.93	689.19	6.46
2	6	9	120	12	56,913.45	395.23	4.89
3	1	36	30	3	9,069.74	1,007.75	7.81
3	2	18	60	6	24,926.81	692.41	6.47
3	3	12	90	9	46,168.06	569.98	5.87
3	6	6	180	18	124,460.26	384.14	4.82
4	1	24	40	4	9,766.78	610.42	6.02
4	2	12	80	8	25,347.72	396.06	4.85
4	3	8	120	12	40,210.84	279.24	4.07
4	6	4	240	24	158,624.23	275.39	4.05
6	1	18	60	6	11,352.10	315.34	4.37
6	2	9	120	12	24,724.20	171.70	3.22
6	3	6	180	18	55,942.66	172.66	3.23

*$s^2 = \Sigma_i(y_i - \bar{y})^2/(n - 1$, where
s^2 = variance per plot
y_i = yield on the i-th plot of the yield map, $i = 1,2, \ldots ,n$
$\bar{y}$ = mean yield $= \Sigma_i\, y_i/n$
n = number of plots on the yield map.
**$V_x = s^2/x^2$, where
V_x = variance per unit plot
x = units per plot

Table 2.3 **Basic Data for Computing Smith's Index of Variability**

Combination	x	Log(x)	V_x	Log(V_x)
1X1	1	0.0000	2,126.66	3.3327
1X2	2	0.3010	1,205.17	3.0810
1X3	3	0.4771	936.15	2.9713
1X6	6	0.7782	589.75	2.7707
2X1	2	0.3010	1,407.92	3.1486
2X2	4	0.6021	867.46	2.9382
2X3	6	0.7782	689.19	2.8383
2X6	12	1.0792	395.23	2.5969
3X1	3	0.4771	1,007.75	3.0034
3X2	6	0.7782	692.41	2.8404
3X3	9	0.9542	569.98	2.7759
3X6	18	1.2553	384.14	2.5845
4X1	4	0.6021	610.42	2.7856
4X2	8	0.9031	396.06	2.5987
4X3	12	1.0792	279.24	2.4460
4X6	24	1.3802	275.39	2.4399
6X1	6	0.7782	315.34	2.4988
6X2	12	1.0792	171.70	2.2348
6X3	18	1.2553	172.66	2.2327

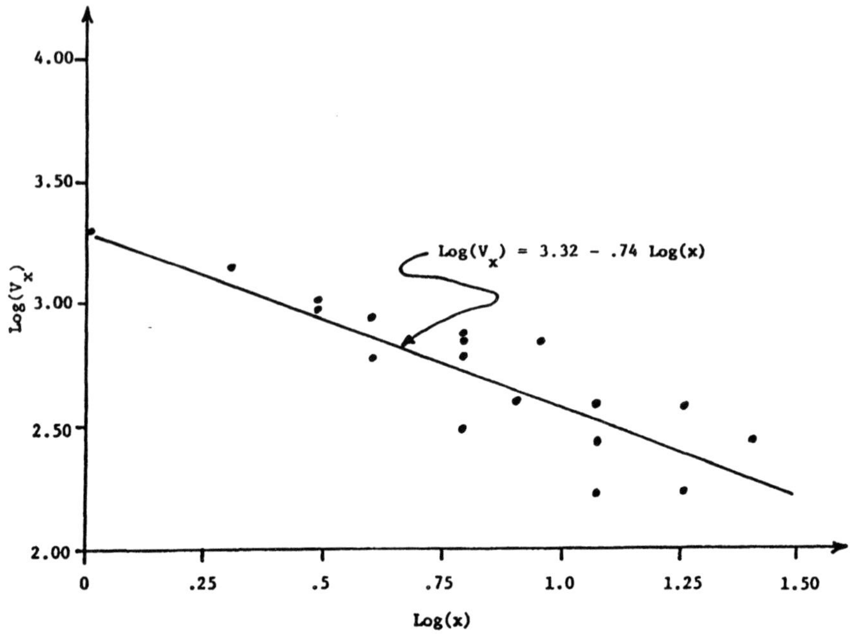

Figure 2.4 Relation between log(plot size) and log(unit variance) and the determination of Smith's index of variability.

can be determined provided that certain costs are known. Suppose that

$$T = K_1 + K_2x$$

where

T = total cost per plot (measured in dollars or minutes)
K_1 = a constant (measured in dollars or minutes) that is independent of plot size
K_2 = a constant (measured in dollars or minutes) that depends on plot size
x = number of unit plots

Then the optimum plot size, x_{opt}, was found by Smith (1938) to be

$$x_{opt} = \frac{bK_1}{(1 - b)K_2}$$

2.4.1 Example

For the wheat data we found a variability index of 0.74. In this data set the unit plot was 5 m long and 2 m wide, an area of 10 m^2. Suppose that

$K_1 = \$3.00 =$ cost associated with each plot regardless of size
$K_2 = \$5.00 =$ cost per 10 m^2 of plot area

Under these conditions the optimum size of plot, x_{opt}, would be

$$x_{opt} = \frac{bK_1}{(1 - b)K_2}$$

$$= \frac{(0.74)(3.00)}{(1 - 0.74)(5.00)} = \frac{2.22}{1.30}$$

$$= 1.71$$

Hence the optimum plot would have an area, A, of

$$A = (1.71)(10) = 17.10 \text{ m}^2$$

This could be achieved with a plot 2 m wide and 8.55 m long.

2.5 CONVENIENT PLOT SIZE

While Smith's optimum plot size is based on costs, Hathaway (1963) has developed a formula for estimating plot size based on soil variability, size of difference to be detected, and significance level of the test. Basically, Hathaway incorporated Smith's b into the formula given by Cochran and Cox (1957) for calculating numbers of replications. The Hathaway equation is

$$x^b = \frac{2(t_1 + t_2)^2C^2}{nd^2}$$

where

x^b = number of units of plots of size x = plot size
b = Smith's coefficient of heterogeneity
t_1 = t at the significance level for the test
t_2 = t in the t table at 2(1 − P), where P is the selected probability of obtaining a significant result
C^2 = $(CV)^2$ = square of the coefficient of variation from a previous experiment
n = number of replications
d = difference to be detected (as a percentage of the mean)

Alternatively, it is possible to compute the detectable difference for a given number of replications and plot size as

$$d^2 = \frac{2(t_1 + t_2)^2C^2}{nx^b}$$

where the terms are as defined before.

2.5.1 Example

Suppose that we want to run a variety trial for 50 selections using a randomized block design with three blocks. What size of plot must be used to detect a difference of 25% of the mean with probability 0.8 using a 5% significance level test?

For this problem,

$$\begin{aligned}
\text{error df} &= (3 - 1)(50 - 1) = 98 \\
t_1 &= t_{0.05(98)} = 1.984 \\
t_2 &= t_{0.40(98)} = 0.845 \\
b &= 0.74 \\
C &= 11\%
\end{aligned}$$

Then

$$\begin{aligned}
x^b &= \frac{2(t_1 + t_2)^2 C^2}{nd^2} = \frac{2(1.984 + 0.845)^2(11)^2}{(3)(25)^2} \\
&= \frac{1936.78}{1875} \\
x^b &= 1.0330 \\
b \log x &= \log(1.0330) \\
\log x &= \frac{\log(1.0330)}{0.74} = \frac{0.0141}{0.74} = 0.0190 \\
x &= 1.0447 \text{ basic units}
\end{aligned}$$

Since the basic plot has an area of 10.00 m^2, the required plot size for this example would be

$$(10.00)(1.0447) = 10.45 \text{ m}^2$$

This could be done using a plot 2 m wide and 5.23 = 5.25 m long.

An alternative example would be to consider the size of the difference that can be detected for a given plot size and number of replications. Suppose that we want to find the difference that can be detected four times out of five ($P = 0.80$) using a 5% significance level test. Suppose, further, that we want to use a plot 2 m wide and 7 m long replicated four times. Recall that the basic information for this example is

$$\begin{aligned}
A &= 2 \text{ m} \times 5 \text{ m} = 10 \text{ m}^2 \\
b &= 0.74 \\
C &= 11\%
\end{aligned}$$

$$t_1 = t_{0.05(98)} = 1.984$$
$$t_2 = t_{0.40(98)} = 0.845$$

Then

$$d^2 = \frac{2(t_1 + t_2)^2C^2}{nx^b}$$

and

$$x^b = \frac{2 \text{ m} \times 7 \text{ m}}{10 \text{ m}^2} = 1.40 \text{ basic units}$$

Hence

$$d^2 = \frac{2(1.984 + 0.845)^2(11)^2}{(4)(1.40^{0.74})}$$
$$= \frac{(2)(8.0032)(121)}{(4)(1.2827)} = \frac{1936.77}{5.1308}$$
$$= 377.48$$
$$d = 19.43$$

With a plot size of 2 m × 7 m using four replications, a difference of 19.43% of the mean could be detected four times out of five with a 5% significance level test.

We can use the equation of this example to determine the detectable difference for various numbers of replications and various sizes of plot. The results of such a set of calculations are presented in Figure 2.5. This figure can be used to decide on the number of replications and size of plot required to detect a given difference. For example, suppose that it is desired to detect a difference of 15% of the mean. Consideration of Figure 2.5 shows that this would require a plot of about 1.125 units (11.25 m^2) with eight replications, a plot of about 1.675 units (16.75 m^2) with six replications, or a plot of about 2.80 units (28.0 m^2) with four replications. A difference of 15% could not be detected if only two replications were used.

2.6 PLOT SHAPE AND ORIENTATION

Many studies with plots of different shape have indicated that long, narrow plots are to be preferred. Plots of this shape are usually more economical for such field operations as seeding, tillage, cultivation, and harvest. In general, if there is a fertility or other gradient in the field, the plots should be oriented with the longest plot dimension in the direction of greatest variability. This is illustrated in Figure 2.6.

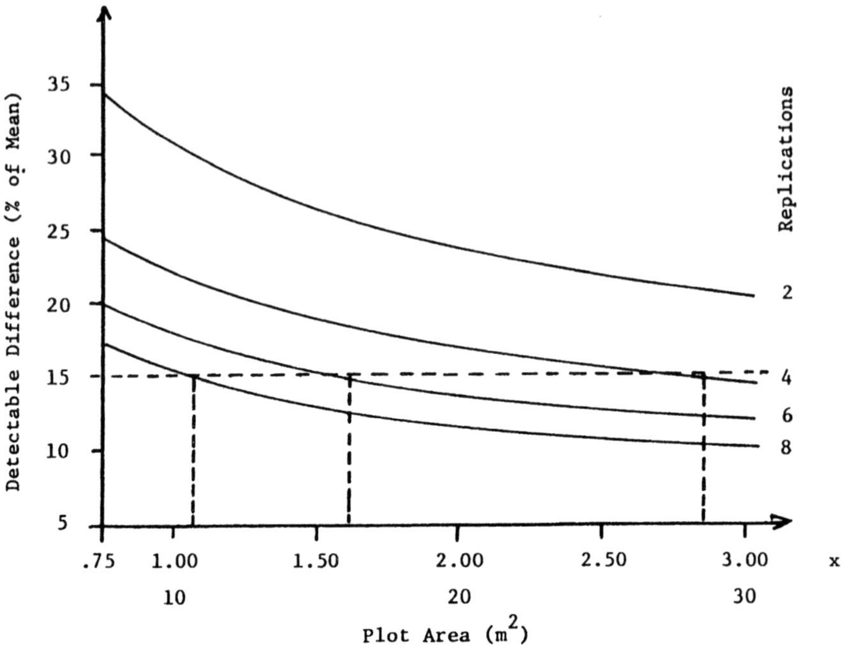

Figure 2.5 Difference (percent of mean) detected four times out of five with a 5% significance level test at various plot sizes and numbers of replications. (b = 0.74.)

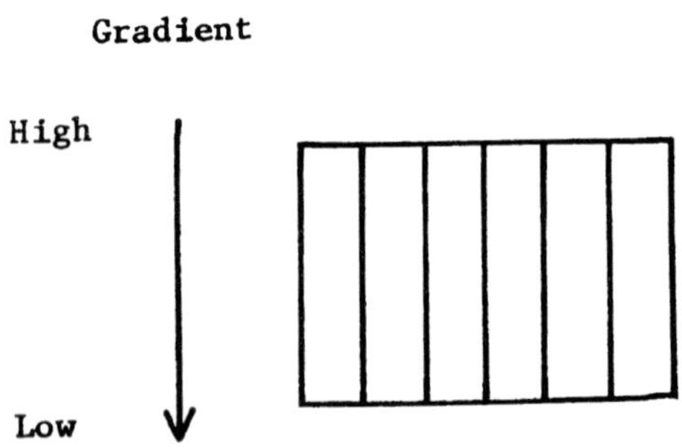

Figure 2.6 Orientation of plots with respect to a gradient.

2.7 BORDER EFFECTS

It is a common observation that plants along the edges of plots often perform differently than those in the center of the plot. This edge, or border, effect can be of considerable significance, particularly with very small plots. There are a number of reasons for this effect. Border plants at the outer boundary of an experiment or next to an unplanted alley have an advantage over interior plants as regards competition for air, sunlight, nutrients, and moisture. Similarly, plants with a tall-growth habit have a competitive advantage when grown next to plants with a low-growth habit. With fertilizer or tillage studies it is not uncommon for the effects of one treatment to be carried into an adjacent plot because of the difficulty of maintaining precise control of machinery changes from one plot to another ("machinery drag").

Border effects are so common that almost all experiments are conducted in such a way as to minimize them. The usual practice is to remove the ends of the plots at harvest and then to harvest only the center of the plot. This practice is shown in Figure 2.7. In some trials it is the practice to group plots containing plants of similar growth patterns together to minimize differential competition effects. When this is done it may not be necessary to eliminate material from the plot borders when measuring yield.

Alleys are usually left between plots in fertilizer and tillage experiments. The aim is to eliminate fertilizer drag from one plot to another or to provide space to adjust the machinery between treatments. These alleys may be seeded or left bare. In any case, provision should be made to discard material from the plot borders before measuring yield.

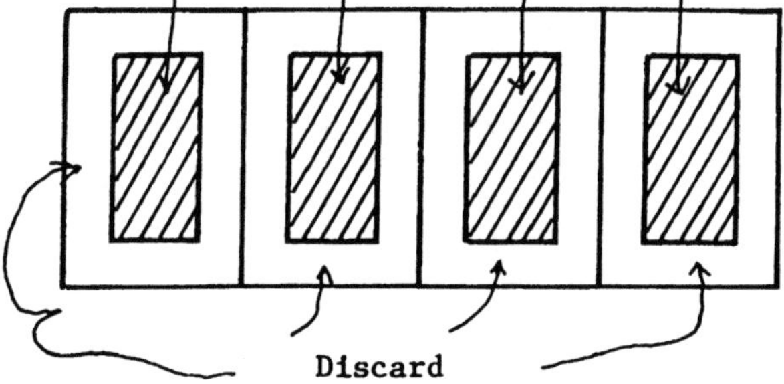

Figure 2.7 Harvesting procedure for eliminating border effects.

REFERENCES

Cochran, W. G., and G. M. Cox, 1957. *Experimental Designs*, 2nd ed. Wiley, New York, pp. 20–23.

Hathaway, W. H., 1963. Convenient plot size. *Agron. J.* 53:279–280.

Hathaway, W. H., and E. J. Williams, 1958. Efficient estimation of the relationship between plot size and variability of crop yields. *Biometrics* 14:207–222.

Koch, E. J., and J. A. Rigney, 1951. A method of estimating plot size from experimental data. *Agron. J.* 43:17–21.

Smith, H. F., 1938. An empirical law describing heterogeneity in the yields of agricultural crops. *J. Agric. Sci.* 28:1–23.

FURTHER READING

Complete citations for books for further reading are given in the bibliography at the end of the book.

1. LeClerg, Leonard, and Clark, Chap. 9.
2. Little and Hills, Chap. 13.
3. Snedecor and Cochran, Chap. 9.
4. Steel and Torrie, Chap. 10.

3

Basic Experimental Designs

3.1 INTRODUCTION

With the preceding discussion as a background, we now turn our attention to experimental designs themselves. As we stated previously, an *experimental design* is a rule, or formal plan, for the assignment of the treatments to the plots, or experimental units, in the experiment. Designs differ from each other primarily in the way the plots are grouped before the treatments are applied. In effect, designs differ in the amount of restriction they impose on the random assignment of treatments to the plots.

3.1.1 Reasons for Designed Experiments

Experiments are designed for a number of purposes:

1. To provide an estimate of experimental error.
2. To increase precision by planned grouping of plots and the elimination of group differences from experimental error.
3. To provide information needed to perform tests of significance and to construct interval estimates.
4. To facilitate the application of treatments, cultural operations, and harvest of the experimental plots.

3.1.2 Factors to Be Considered

There are a number of factors that affect the choice of an experimental design. These include but are not restricted to:

1. Physical and topographic features of the experimental site
2. Amount and pattern of soil variability
3. Number and nature of the treatments
4. Crop to be used
5. Duration of the experiment
6. Nature of the machinery to be used
7. Size of the difference to be detected
8. Significance level to be used
9. Amount of experimental resources (land, seed, fertilizer) that can be devoted to the study
10. Cost (money, time, personnel) required to complete the experiment

As a cardinal rule: *Choose the simplest experimental design that will give the required precision within the limits of the available resources.* In this section we look at three basic experimental designs, beginning with the simplest and moving to designs of increasing complexity.

3.2 COMPLETELY RANDOMIZED DESIGN

The simplest, least restrictive experimental design is the *completely randomized design* (CRD). In a CRD the treatments are assigned to the plots without restriction. That is, with the CRD every plot is equally likely to be assigned to any treatment.

3.2.1 Advantages, Disadvantages, and Uses

The CRD has a number of advantages:

1. *Flexibility.* Any number of treatments and any number of replications may be used. Further, the number of replications per treatment need not be the same for all treatments, although the design is most efficient if the treatments are equally replicated.
2. *Simple statistical analysis.* The analysis is not complicated by unequal replication of treatments. This is not the case with more complex designs.
3. *Missing plots.* The analysis is not complicated by missing data.
4. *Maximum error degrees of freedom* (df). No other design with the same total number of plots and treatments provides more error df than does the CRD.

The CRD has one primary disadvantage:

1. Low precision if the plots are not uniform. Often, it is possible to group the plots into blocks of uniform plots, with a resulting reduction in experimental error.

Although other designs may be more precise, there are a number of uses for the CRD:

1. Where the experimental site is relatively uniform or where a recognizable basis for grouping does not exist. (In this regard I believe that the CRD should be used more frequently than it is.)
2. Where a large fraction of the plots may not respond or may be lost during the course of the experiment.
3. Where the number of plots is limited. The CRD provides maximum df for error in this case.

3.2.2 Design Construction

In the completely randomized design there is no restriction on the assignment of treatments to the plots. The only stipulation is that before the assignment of treatments begins, any treatment is equally likely to be assigned to any plot. Suppose that we have p treatments and $n > p$ plots. To construct a CRD, we randomly assign treatment 1 to r_1 plots, treatment 2 to r_2 plots, . . . , and treatment p to r_p plots, so that $\sum_i r_i = n$. The key to construction is the random assignment of treatments to the plots. There are a number of ways that this can be done, but in any case, some sort of mechanical procedure should be used because it is nearly impossible to prevent personal bias otherwise. We consider two randomization procedures: (1) by lot, and (2) using a random number table.

To illustrate the randomization methods, suppose that we want to conduct an experiment with three treatments, $p = 3$, each replicated four times, $r = 4$. We will require $n = 3 \times 4 = 12$ plots. We locate the plots on a map of the experimental site and number them as shown in Figure 3.1.

1	2	3	4
5	6	7	8
9	10	11	12

Figure 3.1 Site map with 12 numbered plots.

Randomization *by lot*:

1. Take 12 slips of paper and number them from 1 to 12.
2. Place them in a box or hat and mix them thoroughly.
3. Draw the slips one at a time without replacement.
4. Assign the first treatment to the plots whose numbers were on the first four slips drawn, the second treatment to the second four, and the third treatment to the last four slips drawn.

Another randomization procedure makes use of a *random number table*:

1. Select a starting point somewhere in a table of random numbers.
2. By moving up or down, right or left, from the starting point, write down the first 12 three-digit numbers you come to. (Three-digit numbers are used because they are less likely to contain the same number than are lists of numbers with fewer digits.)
3. Rank the numbers from smallest to largest. These ranks correspond to plot numbers on the field plan.
4. Assign the first treatment to the first four plots in sequence, the second treatment to the second four plots in sequence, and the third treatment to the remaining plots.

To illustrate the procedure, suppose that the drawing resulted in the numbers displayed, in the sequence in which they were drawn, in Table 3.1. The resulting field plan would be as presented in Figure 3.2.

3.2.3 Analysis

In the analysis of data from a completely randomized design we want to do three things:

1. Estimate the treatment means or treatment effects.
2. Estimate the standard error of a treatment mean or the difference between two means. With this we can compute interval estimates.
3. Test the significance of differences among the treatment means.

We can do these things in a systematic way using a procedure called the analysis of variance (ANOVA).

Suppose that we have data from an experiment run as a CRD with p treatments. Suppose, in general, that there are r_j replications of the j-th treatment so that there are a total of $n = \Sigma_j r_j$ plots (observations). Let y_{ij} be the i-th observation on the j-th treatment. The steps in the ANOVA are

Table 3.1 **Assignment of Treatments Using a Random Number Table**

Sequence	Random Number	Rank (plot number)	Treatment
1	481	5	A
2	516	6	A
3	991	12	A
4	062	1	A
5	804	11	B
6	675	9	B
7	154	2	B
8	437	4	B
9	571	7	C
10	769	10	C
11	639	8	C
12	428	3	C

1 A	2 B	3 C	4 B
5 A	6 A	7 C	8 C
9 B	10 C	11 B	12 A

Figure 3.2 Field plan with treatments randomly assigned to plots.

as follows:

1. Set up a table of observations and compute the treatment totals and means. For this purpose, let

   $T_j = \Sigma_i\, y_{ij}$ = sum of the observations of the j-th treatment
   $\bar{y} = T_j/r_j$ = mean of the j-th treatment
   $G = \Sigma_j\, T_j$ = grand total of all observations
   $\bar{\bar{y}} = G/n = G/\Sigma_j\, r_j$ = grand mean

   Then the data are summarized as in Table 3.2.
2. Set up an analysis-of-variance (ANOVA) table, Table 3.3.

The entries in the ANOVA table are obtained in the following way:

1. The total variation among observations is partitioned into components attributable to two sources of variation: (1) variation among treatment means (treatment), and (2) variation among plots within treatments (experimental error or, simply, error).
2. The degrees of freedom, df, on each line are one less than the number of observations in the variance source on that line. There are $n = \Sigma_j\, r_j$ total observations, giving $n - 1$ df on the total line.

Table 3.2 **Table of Observations, Totals, and Means for a Completely Randomized Design**

	Treatment				
	1	2	...	p	
	y_{11}	y_{12}	...	y_{1p}	
	y_{21}	y_{22}	...	y_{2p}	
	.	.		.	
	.	.		.	
	.	.		.	
	y_{r11}	y_{r22}	...	y_{rpp}	Total
Sum	T_1	T_2	...	T_p	G
Mean	$\bar{y}_1$	$\bar{y}_2$	...	$\bar{y}_p$	$\bar{\bar{y}}$

Table 3.3 **Analysis of Variance of Data from a CRD**

ANOVA

Source of Variation (Source)	Degrees of Freedom (df)	Sum of Squares (SS)	Mean Square (MS)	F
Total	n−1	SSTot		
Treatment	p−1	SST	MST	F_T
Error	n−p	SSE	MSE	

There are p treatments, giving p − 1 df on the treatment line. The df for error are n − p, which is obtained by difference: (n − 1) − (p − 1) = n − p.

3. The entries in the "sum of squares" column are computed in the following way:
   a. First, compute a correction term, C:
   $$C = \frac{G^2}{n}$$
   Then
   b. $SSTot = \Sigma_j \Sigma_i y_{ij}^2 - C.$
   c. $SST = \Sigma_j (T_j^2/r_j) - C.$
   d. $SSE = SSTot - SST.$
4. The entries in the "mean square" column are obtained by dividing the sums of squares by their corresponding degrees of freedom. (*Note*: A mean square is not computed on the total line.)
   a. $MST = SST/(p - 1).$
   b. $MSE = SSE/(n - p).$
5. Finally,
   $$F_T = \frac{MST}{MSE}$$

The computations are simplified to some extent if the treatments are equally replicated. In this case we have

$$r_1 = r_2 = \cdots = r_p = r$$

and the experiment contains a total of n = rp plots. The ANOVA table is as shown in Table 3.4.

Table 3.4 **Analysis of Variance of a CRD with Equal Replication**

ANOVA

Source	df	SS	MS	F
Total	rp−1	SSTot		
Treatment	p−1	SST	MST	F_T
Error	p(r−1)	SSE	MSE	

The computations involved are

$$C = \frac{G^2}{rp}$$

$$SST = \frac{1}{r} \sum_j T_j^2 - C$$

Everything else is computed in the same way under both conditions.

Having completed the ANOVA table, we can use the entries for a couple of things:

1. F_T = MST/MSE, with (p − 1), p(r − 1) df, can be used to judge the significance of differences among the treatment means. That is, F is a test statistic for

   $$H_0: \mu_1 = \mu_2 = \cdots = \mu_p$$

   (the true means of all treatments are the same) against

   $$H_a: \mu_1 \neq \mu_2 \neq \cdots \neq \mu_p$$

   (at least one mean differs from the others). We compare F_T with F in the F table at p − 1, n − p df. If F_T is greater than F in the 1% table, we say that the differences are highly significant. If F_T is greater than F in the 5% table but less than the F in the 1% table, we say that the differences are significant. If F_T is smaller than the 5% F, we conclude that the means are not significantly different.
2. MSE = s^2 is the sample estimate of the experimental error. We can use it to compute standard errors and interval estimates.

3.2.4 Means and Standard Errors

The true means, μ_j, of the treatments are estimated by the sample means, $\bar{y}_j$. We have

$$\mu_j \doteq \bar{y}_j = \frac{T_j}{r_j}$$

where "$\doteq$" means "is estimated by." The sample standard error of the j-th treatment mean, $s_{\bar{y}j}$, is computed as

$$s_{\bar{y}j} = \sqrt{\frac{s^2}{r}} = \sqrt{\frac{MSE}{r_j}}$$

which, in general, is computed separately for each mean. If

$$r_1 = r_2 = \cdots = r_p = r \qquad \text{(equal replication)}$$

then

$$s_{\bar{y}} = \sqrt{\frac{MSE}{r}}$$

for all means.

The $(1 - \alpha)100\%$ confidence interval estimate, $L(\mu_j)$, of the true mean of the j-th treatment is given by

$$L(\mu_j) = \bar{y}_j \pm t_\alpha\sqrt{\frac{MSE}{r_j}}$$

where t_α is the two-tailed t at the α probability level with $n - p$ df.

Often we are interested in the differences between the means rather than in the means themselves. That is, we want to estimate $\mu_j - \mu_{j'}$ (difference between the true means of treatment j and treatment j′) or to test the significance of the difference between the two. As might be expected, the sample estimate of the difference is simply the difference between the sample means. We have

$$\mu_j - \mu_{j'} \doteq \bar{y}_j - \bar{y}_{j'}$$

If we want an interval estimate or a test of the significance of the difference, we require the sample standard error of the difference. In general, this standard error, $s_{(\bar{y}j - \bar{y}j')}$, is

$$s_{(\bar{y}j - \bar{y}j')} = \sqrt{s^2\left(\frac{1}{r_j} + \frac{1}{r_{j'}}\right)} = \sqrt{MSE\left(\frac{1}{r_j} + \frac{1}{r_{j'}}\right)}$$

where r_j and $r_{j'}$ are the numbers of observations on treatments j and j′. Note that for equal replication

$$r_1 = r_2 = \cdots = r_p = r$$

Then

$$s_{(\bar{y}j - \bar{y}j')} = \sqrt{\frac{2s^2}{r}} = \sqrt{\frac{2MSE}{r}}$$

for all differences.

We can use this standard error to construct an interval estimate of the true difference. The $(1 - \alpha)100\%$ confidence interval estimate, $L(\mu_j - \mu_{j'})$, of the difference between the true means, μ_j and $\mu_{j'}$, of treatments j and j′ is

$$L(\mu_j - \mu_{j'}) = (\bar{y}_j - \bar{y}_{j'}) \pm t_\alpha \sqrt{MSE\left(\frac{1}{r_j} + \frac{1}{r_{j'}}\right)}$$

where t_α is the two-tailed t at the α probability level with $n - p$ degrees of freedom. Again, with equal replication

$$r_1 = r_2 = \cdots r_p = r$$

and

$$L(\mu_j - \mu_{j'}) = (\bar{y}_j - \bar{y}_{j'}) \pm t_\alpha \sqrt{\frac{2MSE}{r}}$$

in which t_α has $p(r - 1)$ df.

Subject to certain restrictions, which we discuss later, we can use the standard error to test the significance of the difference between the treatment means. We want to test

H_0: $\mu_j = \mu_{j'}$, or, alternatively, H_0: $\mu_j - \mu_{j'} = 0$ against the alternative, H_a: $\mu_j \neq \mu_{j'}$, or H_a: $\mu_j - \mu_{j'} \neq 0$. In general, the test statistic for this test is

$$t = \frac{\bar{y}_j - \bar{y}_{j'}}{\sqrt{MSE(1/r_j + 1/r_{j'})}}$$

or, with equal replication,

$$t = \frac{\bar{y}_j - \bar{y}_{j'}}{\sqrt{2MSE/r}}$$

where t is the sample of t and $\bar{y}_j$, $\bar{y}_{j'}$, r_j, $r_{j'}$, and MSE are taken from the analysis of the experimental data. This t has $n - p$ df.

If t is *numerically greater* than the 1% two-tailed t at n − p df, we say that the difference is *highly significant*(**). If 1%t > |t| > 5%t, the difference is *significant*(*), and if |t| < 5%t, the difference is *not significant* (NS). (*Note*: |t| = sample t with the sign ignored.)

3.2.5 Numerical Example

The farming systems program of a major agricultural research center established a set of on-farm demonstration plots throughout an agricultural district near the research center. A single plot was located within a lentil field on each of 20 farms in the district. Each plot was fertilized at a recommended rate, treated for sitona weevil, and treated with a herbicide to control weeds. The goal was to demonstrate the difference between the farmer's usual practice and the production package recommended by the scientists at the research center.

At maturity a portion of each plot was harvested for yield. The farms were then classified by soil type and the yields were analyzed, using a CRD analysis, to see if there were yield differences due to soil type. Seed yields (kg/plot) were as shown in Table 3.5.

Table 3.5 **Lentil Seed Yields (kg/plot) from 20 Demonstration Plots on Various Soil Types**

	Soil Type 1	2	3	4	5	
	42.2	28.4	18.8	41.5	33.0	
	34.9	28.0	19.5	36.3	26.0	
	29.7	22.8	13.1	31.7	30.6	
		18.5	10.1	31.0		
		19.4		28.2		Total
Sum	106.8	117.1	61.5	168.7	89.6	543.7
r_j	3	5	4	5	3	20
$\bar{y}_j$	35.60	23.42	15.38	33.74	29.87	27.18

Data Analysis The analysis of variance of the lentil seed yield data is given in Table 3.6. The entries in the ANOVA table are computed as follows:

1. *Degrees of freedom*:

   Total df $= \sum_j r_j - 1 = 20 - 1 = 19$

   Soil type df $= p - 1 = 5 - 1 = 4$

   Error df $=$ total df $-$ soil type df $= 19 - 4 = 15$

2. *Sums of squares*:

$$C = \frac{G^2}{\sum_j r_j} = \frac{543.7^2}{20} = 14{,}780.4845$$

$$\text{SSTot} = \sum_i \sum_j y_{ij}^2 - C = 42.2^2 + 34.9^2 + \cdots + 30.6^2 - C$$

$$= 16{,}219.6900 - 14{,}780.4845 = 1439.2055$$

$$\text{SST} = \sum_j \frac{T_j^2}{r_j} - C = \frac{(106.8)^2}{3} + \cdots + \frac{(89.6)^2}{3} - C$$

$$= 15{,}858.1158 - 14{,}780.4845 = 1077.6313$$

$$\text{SSE} = \text{SSTot} - \text{SST} = 1439.2055 - 1077.6313 = 361.5742$$

Table 3.6 **Analysis of Seed Yields (kg/plot) from a Lentil Demonstration Trial**

ANOVA

Source	df	SS	MS	F
Total	19	1,439.2055		
Soil type	4	1,077.6313	269.4078	11.18**
Error	15	361.5742	24.1049	

**Significant at the 1% level.

3. *Mean squares*:

$$\text{MST} = \frac{\text{SST}}{p - 1} = \frac{1077.6313}{4} = 269.4078$$

$$\text{MSE} = \frac{\text{SSE}}{n - p} = \frac{361.5742}{15} = 26.1049$$

4. *F*:

$$F_T = \frac{\text{MST}}{\text{MSE}} = \frac{269.4078}{24.1049} = 11.1765$$

$F_{0.01}(4, 15) = 4.8932 < 11.1765$ (the means are significantly different at the 1% level)

5. *Coefficient of variation* (CV):

$$\text{CV} = \left(\frac{\sqrt{\text{MSE}}}{\bar{\bar{y}}}\right)100 = \left(\frac{\sqrt{24.1049}}{27.18}\right)100 = 18.1\%$$

Standard errors: $s_{\bar{y}j} = \text{MSE}/r_j$

1. $s_{\bar{y}1} = \sqrt{24.1049/3} = 2.83.$
2. $s_{\bar{y}2} = \sqrt{24.1049/5} = 2.20.$
3. $s_{\bar{y}3} = \sqrt{24.1049/4} = 2.45.$
4. $s_{\bar{y}4} = \sqrt{24.1049/5} = 2.20.$
5. $s_{\bar{y}5} = \sqrt{24.1049/3} = 2.83.$

95% interval estimate of the true mean of soil type 4, $L(\mu_4)$:

$$L(\mu_4) = \bar{y}_4 \pm t_{0.05}\sqrt{\frac{\text{MSE}}{5}} \qquad \text{t has 15 df}$$

$$= 33.74 \pm 2.131\sqrt{\frac{24.1049}{5}}$$

$$= 33.74 \pm 4.69$$

Standard error of the difference between type 1 and type 2:

$$s_{(\bar{y}1 - \bar{y}2)} = \sqrt{\text{MSE}\left(\frac{1}{r_1} + \frac{1}{r_2}\right)}$$

$$= \sqrt{24.1049\left(\frac{1}{3} + \frac{1}{5}\right)} = 3.58$$

Report of Statistical Analysis

Analysis of yield data from 20 lentil demonstration plots indicates that there are highly significant differences in yield among the five soil types represented in the trial. The results are presented in Table 3.7.

It is apparent that soil type 1 produces the highest yield of lentil seed, although it is doubtful that the difference between types 1 and 4 is significant. Soil type 3 is clearly inferior to the others.

3.3 RANDOMIZED BLOCK DESIGN

The randomized block design (RBD) or, more correctly, the randomized complete block design, is the design used most often in agricultural research. It is nearly as easy to use as the CRD, yet it provides an opportunity for increased precision over the CRD. In the RBD the plots (experimental units) are first classified into groups, or blocks, of plots that are as nearly alike as possible. The treatments are then assigned to the plots within blocks in such a way that each treatment occurs the same number of times, usually once, within each block. The object is to make the variation from plot to plot as small as possible within the blocks while maximizing the variation among blocks. For this reason it is essential that not only should the plots be as alike as possible within the blocks, but during the course of the experiment uniform techniques should be used on all the plots within a block. Changes in technique or in other conditions that might affect the results should be made between blocks.

3.3.1 Blocking Considerations

There are a number of criteria on which plots may be grouped into blocks. In field experiments neighboring plots tend to perform more nearly alike

Table 3.7 **Mean Yields (kg/plot) of Lentils Grown on Five Soil Types**

Soil Type	1	2	3	4	5
Mean yield	35.60	23.42	15.38	33.74	29.87
Replications	3	5	4	5	3
Standard error	2.83	2.20	2.45	2.20	2.83

than do plots that are separated from each other. In the absence of other identifiable sources of variation, then, blocks may be formed from a square, compact group of adjacent plots. Although such a configuration may be pleasing to the eye of the experiment station superintendent, it may not result in increased precision.

The author has examined a large number of sets of data from field experiments using randomized block designs in which the blocks consisted of groups of adjacent plots. These data were from experiments conducted on experiment stations as well as experiments conducted off-station. On average, as compared to a completely randomized design, the efficiency of a randomized block design in these instances was increased by only about 5%. Hence blocking by grouping adjacent plots may not really be worth the effort of delineating the blocks.

If a fertility or other gradient exists, however, blocks should be long and narrow, with the long axis perpendicular to the gradient as illustrated in Figure 3.3. Blocks should be laid out to avoid the effect of such physical features as roads, fences, rows of trees, ridges, water courses, soil-type boundaries, and buildings. To do this it is sometimes necessary to have irregularly shaped blocks which are at some distance from each other. Figure 3.4 illustrates the placement of blocks to avoid the effect of some physical features.

In some instances it may be useful to place each block at a different location. This is permitted as long as the locations are not so different as to result in differential treatment effects. Placing blocks at different locations

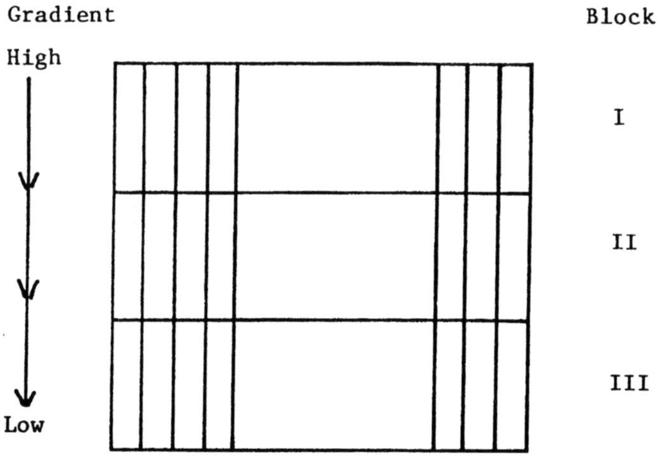

Figure 3.3 Placement of blocks with respect to a gradient.

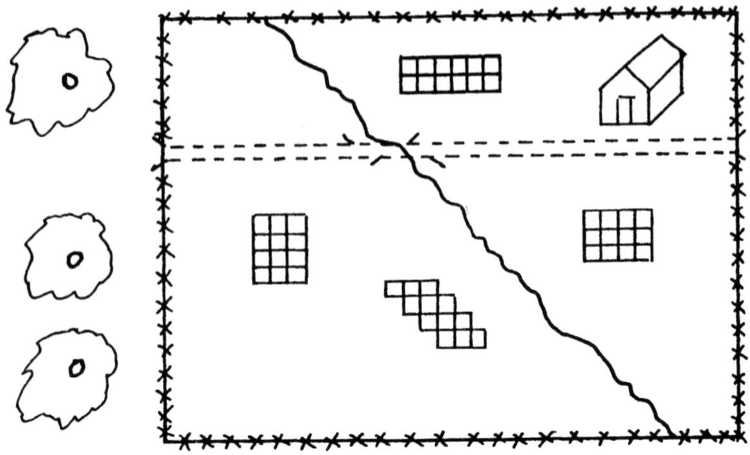

Figure 3.4 Placement of blocks to accommodate physical features of the research site.

may be helpful because it brings a wider range of environmental conditions into the experiment and can broaden the inferences that can be drawn from the results of the trial.

Once the blocks have been formed it is best to handle each block as a unit during the course of the experiment. Field operations, such as seeding, cultivation, and harvest, should be completed in one block before moving to another. Similarly, if operations are to be performed by more than one person, each should be assigned to a different block. To repeat, the *key* to success with a randomized block design is to *minimize the variation within* blocks and to *maximize the variation among* blocks.

3.3.2 Advantages, Disadvantages, and Uses

There are a number of advantages in favor of a randomized block design:

1. It can remove one source of variation from experimental error and thus increase precision.
2. By placing blocks under different conditions it can broaden the scope of the trial.
3. Any number of treatments and any number of blocks may be used. The only restriction is that each treatment must be replicated the same number of times in each block.
4. Statistical analysis of the results is fairly simple.

There are also some disadvantages to this design:

1. Missing data can cause some difficulty in the analysis. One or two missing plots can be handled fairly easily but numerous missing data can cause real problems.
2. Assignment of treatments by mistake to plots in the wrong block can lead to problems in the analysis.
3. The design is less efficient than others in the presence of more than one source of unwanted variation.
4. If the plots are uniform, the RBD is less efficient than the CRD.

The randomized block design has a number of uses:

1. It can be used to eliminate one source of unwanted variation. Often, it provides satisfactory precision without the need for a more complex design.
2. It provides unbiased estimates of the means of the blocking factor. Hence these means can be estimated using the randomized block design.

3.3.3 Randomization

In constructing a randomized block design, the plots are grouped into blocks. Normally, the number of plots in each block is equal to the number of treatments, while the number of blocks is equal to the number of replications per treatment. That is, in the usual randomized block design each treatment occurs once, and only once, in each block. Once the blocks have been formed, the treatments are assigned at random to the plots within the blocks. Randomization may be done either by lot or by using a random number table as described for the CRD. A separate randomization is used in each block.

3.3.4 Analysis

The data analysis proceeds in much the same way as for the completely randomized design:

1. Construct a table of totals and means.
2. Compute the entries in an ANOVA table.
3. Compute a CV.
4. Conduct significance tests.
5. Compute means and standard errors.

To illustrate the analysis suppose that we have a RBD with r blocks and p treatments. Let y_{ij} represent the yield of the j-th treatment in the i-th block.

1. Construct a two-way block by treatment table and compute totals and means as shown in Table 3.8.
2. Construct an ANOVA table. This takes the same general form as the ANOVA table for the CRD except that it includes an additional source of variation for blocks. The ANOVA table is displayed as Table 3.9.

The entries in Table 3.9 are obtained as follows:

1. *Source*: Total variation is partitioned into components due to blocks, treatments, and error.
2. *df*: Degrees of freedom are one less than the number of classes in each variance source. The error df value is obtained by difference: dfE = dfTot − dfR − dfT.

Table 3.8 **Summary Table of Data from a Randomized Block Design**[a]

Block	Treatment 1	2	...	p	Sum	Mean
1	y_{11}	y_{12}	...	y_{1p}	R_1	$\bar{y}_{1.}$
2	y_{21}	y_{22}	...	y_{2p}	R_2	$\bar{y}_{2.}$
.	.	.		.	.	.
.	.	.		.	.	.
.	.	.		.	.	.
r	y_{r1}	y_{r2}	...	y_{rp}	R_r	$\bar{y}$
Sum	T^1	T^2	...	T^p	G	
Mean	$\bar{y}^{.1}$	$\bar{y}^{.2}$		$\bar{y}^{.p}$		$\bar{\bar{y}}$

[a] $R_i = \Sigma_j y_{ij}$ = total yield of the i-th block
$T_j = \Sigma_j y_{ij}$ = total yield of the j-th treatment
$G = \Sigma_j R_i = \Sigma_j T_j$ = grand total of all yields
$\bar{y}^{i\cdot} = R_i/p$ = mean of the i-th block
$\bar{y}^{\cdot j} = T_j/r$ = mean of the j-th treatment
$\bar{y} = G/rp$ = grand mean

Table 3.9 **ANOVA Table for a Randomized Block Design**

ANOVA

Source	df	SS	MS	F
Total	rp - 1	SSTot		
Block	r - 1	SSR	MSR	F_R
Treatment	p - 1	SST	MST	F_T
Error	(r-1)(p-1)	SSE	MSE	

3. *Sums of squares*
   a. First compute a correction term, $C = G^2/rp$. Then
   b. $SSTot = \Sigma_i \Sigma_j y^2_{ij} - C$.
   c. $SSR = (1/p) \Sigma_i R_i^2 - C$.
   d. $SST = (1/r) \Sigma_j T_j^2 - C$.
   e. $SSE = SSTot - SSR - SST$.
4. *Mean squares*: Divide the sums of squares by the degrees of freedom on the same line in the ANOVA.
   a. $MSR = SSR/(r - 1)$.
   b. $MST = SST/(p - 1)$.
   c. $MSE = SSE/(r - 1)(p - 1)$.

   F: MSE is the divisor for all F ratios.
   a. $F_R = MSR/MSE$.
   b. $F_T = MST/MSE$.
5. *CV* (coefficient of variation): $\% \text{ CV} = (\text{MSE}/\bar{\bar{y}})100$.

3.3.5 Significance Tests

F_T, with $(p - 1)$ and $(r - 1)(p - 1)$ df, is a test statistic for the hypothesis that all treatment means are equal against the alternative that at least one mean differs from the others. In other words, F_T tests the significance of the differences among treatment means. If F_T is larger than the 1% F in the table, the differences are said to be highly significant (**). If F_T is greater than the 5% F but smaller than the 1% F, the differences are significant (*). If F_T is smaller than the 5% F, the differences are not significant (NS).

F_R, with (r − 1) and (r − 1)(p − 1) df, is only an approximate test statistic for differences among blocks. If F_R is greater than the F in the F table, it is an indication that blocking has been effective in reducing experimental error. There is no appropriate error term in the ANOVA of an RBD for testing the significance of block differences. F_R is not always computed because it is only approximate and, more to the point, if blocking is done correctly, the differences among blocks are deliberately maximized. It seems pointless in this case to test the hypothesis that the differences among blocks are zero.

3.3.6 Means and Standard Errors

With the randomized block design, as with the CRD, the sample means estimate the true means of the treatments. We have

$$\mu_j \doteq \bar{y}_j = \frac{T_j}{r}$$

Similarly, with this ANOVA, as in the ANOVA for the CRD, MSE = s^2 is an estimate of the variance per plot (variation among plots treated alike). It is used in the computation of standard errors and in the construction of interval estimates.

The standard error of a treatment mean, $s_{\bar{y}}$, is

$$s_{\bar{y}} = \sqrt{\frac{MSE}{r}}$$

Since r is the same for all treatments, $s_{\bar{y}}$ is the same for all. The (1 − α)100% confidence interval estimate of a treatment mean, $L(\mu_j)$, is

$$L(\mu_j) = (\bar{y}_j) \pm t_\alpha\sqrt{\frac{MSE}{r}}$$

where t_α is the two-tailed, α-level t with (r − 1)(p − 1) df.

Again, we are often more interested in the difference between means than in the means themselves. In this case the standard error of a difference, $s_{\bar{d}}$, is

$$s_{\bar{d}} = \sqrt{\frac{2MSE}{r}}$$

The (1 − α)100% confidence interval estimate, $L(\mu_j - \mu_{j'})$, of a difference is given by

$$L(\mu_j - \mu_{j'}) = (\bar{y}_j - \bar{y}_{j'}) \pm t_\alpha\sqrt{\frac{2MSE}{r}}$$

where, again, t_α is the two-tailed t with (r − 1)(p − 1) df.

If we want to test the significance of the difference between two means, we can use t as a test statistic under certain restrictions which we discuss later. We compute

$$t = \frac{\bar{y}_j - \bar{y}_{j'}}{\sqrt{\dfrac{2MSE}{r}}}$$

If |t| (sample t with sign ignored) is greater than the 1% t with (r − 1) (p − 1) df. The difference is said to be highly significant (**). If it is less than the 1% t but greater than the 5% t, the difference is significant (*), and if it is smaller than the 5% t, the difference is not significant (NS).

For a discussion of *missing plots* and *relative efficiency* for the randomized block design, refer to Appendix on p. 76.

3.3.7 Numerical Example

An agronomist wanted to determine the effect of different sources of nitrogen on the yield of barley used as a forage crop. There were five sources to be compared: (1) $(NH_4)_2SO_4$, (2) NH_4NO_3, (3) $CO(NH_2)_2$, (4) $Ca(NO_3)_2$, and (5) $NaNO_3$. He also decided to use a control treatment with no nitrogen.

Because he wanted to apply the results over a fairly wide range of conditions, he decided to conduct the trial on four types of soil. For his experimental design he chose a randomized block design with soil type as a blocking factor. He located six plots on each of the four soil types. He then randomly assigned the treatments to the plots within each soil type. The field plan and yields (kg/plot) were as shown in Figure 3.5. The yield data are summarized in Table 3.10.

Data Analysis The analysis of variance of this data set is given in Table 3.11.

Computations:

$$C = \frac{G^2}{rp} = \frac{(740.5)^2}{24} = 22{,}847.51$$

$$\begin{aligned} SSTot &= \sum_i \sum_j y_{ij}^2 - C = 32.1^2 + 30.1^2 + \cdots + 26.7^2 - C \\ &= 23{,}339.87 - 22{,}847.51 = 492.36 \end{aligned}$$

$$\begin{aligned} SSR &= \frac{1}{p} \sum_i R_i^2 - C = \frac{1}{6}(161.0^2 + \cdots + 187.3^2) - C \\ &= 23{,}040.06 - 22{,}847.51 = 192.56 \end{aligned}$$

$$\begin{aligned} SST &= \frac{1}{r} \sum_j T_j^2 - C = \frac{1}{4}(145.0^2 + \cdots + 101.4^2) - C \\ &= 23{,}102.79 - 22{,}847.51 = 255.28 \end{aligned}$$

I	II	III	IV
(5) 26.1	(1) 35.6	(6) 26.7	(4) 31.4
(3) 25.4	(2) 31.5	(1) 41.9	(2) 30.8
(6) 23.2	(6) 24.8	(5) 33.8	(6) 26.7
(2) 30.1	(5) 31.0	(4) 35.6	(1) 35.4
(4) 24.1	(3) 27.4	(3) 33.8	(5) 31.9
(1) 32.4	(4) 33.0	(2) 37.1	(3) 31.1

Treatments (applied at a rate of 50 kg/ha N):

(1) $(NH_4)_2SO_4$ (4) $Ca(NO_3)_2$

(2) NH_4NO_3 (5) $NaNO_3$

(3) $CO(NH_2)_2$ (6) Control

Figure 3.5 Field plan for a source-of-nitrogen trial on forage barley.

$$SSE = SSTot - SSR - SST = 492.36 - 192.56 - 255.2 = 44.52$$

$$MSR = \frac{SSR}{r-1} = \frac{192.56}{3} = 64.19$$

$$MST = \frac{SST}{p-1} = \frac{255.28}{5} = 51.06$$

$$MSE = \frac{SSE}{(r-1)(p-1)} = \frac{44.52}{15} = 2.97$$

$$F_R = \frac{MSR}{MSE} = \frac{64.19}{2.97} = 21.61 \qquad 2{,}15 \text{ df}$$

$$F_T = \frac{MST}{MSE} = \frac{51.06}{2.97} = 17.19 \qquad 5{,}15 \text{ df}$$

Table 3.10 **Yield (kg/plot) of Forage Barley Under Various Sources of Nitrogen**

Treat.	Soil type I	II	III	IV	Total	Mean
1	32.1	35.6	41.9	35.4	145.0	36.25
2	30.1	31.5	37.1	30.8	129.5	32.38
3	25.4	27.1	33.8	31.1	117.7	29.42
4	24.1	33.0	35.6	31.4	124.1	31.02
5	26.1	31.0	33.8	31.9	122.8	30.70
6	23.2	24.8	26.7	26.7	101.4	25.35
Total	161.0	183.3	208.9	187.3	740.5	
Mean	26.83	30.55	34.82	31.22		30.85

Table 3.11 **Analysis of Variance of Barley Forage Yields Under Various Types of Nitrogen Fertilizer**[a]

ANOVA

Source	df	SS	MS	F
Total	23	492.36		
Soils(block)	3	192.56	64.19	21.61
Fert.(treat.)	5	255.28	51.06	17.19**
Error	15	44.52	2.97	

[a]CV = 5.6%.

**Significant at the 1% level.

Significance tests:

1. $F_T = 17.19$; $F_{0.01(5,15)} = 4.56$. Since $17.19 > 4.56$, we conclude that differences among nitrogen source means are significant at the 1% level (highly significant).
2. $F_R = 21.61$; $F_{0.01(3,15)} = 5.42$. Since $21.61 > 5.42$, we conclude that blocking on soil type was effective in reducing experimental error.

Standard errors:

1. Standard error of a treatment mean, $s_{\bar{y}}$:

$$s_{\bar{y}} = \sqrt{\frac{MSE}{r}} = \sqrt{\frac{2.97}{4}} = 0.86$$

2. Standard error of a difference between two treatment means, $s_{\bar{d}}$:

$$s_{\bar{d}} = \sqrt{\frac{2MSE}{r}} = \sqrt{\frac{(2)(2.97)}{4}} = 1.22$$

Report of Statistical Analysis

Analysis of the results of an experiment designed to determine the effect of different sources of nitrogen on the forage dry matter produced by barley indicates that differences among sources were highly significant. Ammonium sulfate, $(NH_4)_2SO_4$, produced the highest mean yield, 36.25 kg/plot, and $CO(NH_2)_2$ produced the lowest, 29.42 kg/plot, among the nitrogen treatments. When no nitrogen was added the yield was only 25.35 kg/plot. The results are summarized in Table 3.12.

Blocking on soil type was effective in reducing the experimental error in this trial. This is indicated by the large F for soils and by the relatively small coefficient of variation (5.6%) for the trial.

Table 3.12 **Mean Yield (kg/plot) of Dry Matter from Barley Treated with Nitrogen from Various Sources**[a]

Source	$(NH_4)_2SO_4$	NH_4NO_3	$CO(NO_2)_2$	$Ca(NO_3)_2$	$NaNO_3$	Control
Mean	36.25	32.38	29.42	31.02	30.70	25.35

[a]Standard error = 0.86 kg/plot.

3.4 SUBSAMPLING

In some field trials it is inconvenient to measure yield on the entire plot. In a fertilizer trial on a forage crop, for example, the mass of forage produced might be too great to be readily handled. In this situation it is common to take one or more subsamples from within the plot and to measure yield on the subsamples. This can introduce a complication in the analysis of the results from the trial.

If a single subsample is taken from each plot, the analysis of variance is unchanged. Yield of the subsample is expressed on a unit area basis and the analysis proceeds in the usual way. If two or more subsamples are taken from each plot, there are two alternatives. One is to compute the average yield, per plot, of the subsamples. The plot averages, expressed on a unit area basis, are then analyzed in the usual way. The second alternative is to subject the subsample yields themselves to analysis.

3.4.1 Analysis of Variance with Subsamples

The analysis proceeds in much the same way as if subsampling had not been done. The ANOVA table, however, contains one additional source of variation: variation among samples within plots. This source is often called *sampling error.*

To illustrate the process, suppose that we have a randomized block design with r blocks and p treatments. Suppose also that we have n subsamples from each plot. Let y_{ijk} represent the yield of the k-th subsample on the j-th treatment in the i-th block.

1. Construct a data table and compute totals and means as shown in Table 3.13.
2. Next, construct an ANOVA table. This takes the same form as the ANOVA table for the RBD except that it contains an additional source of variation for sampling error. The format for this table is displayed as Table 3.14.

The sums of squares (SS) in Table 3.14 are computed as follows:

1. First compute a correction term, C: $C = G^2/rpn$. Then
2. $SSTot = \Sigma_i \Sigma_j \Sigma_k y_{ijk}^2 - C$.
3. $SSR = (1/pn) \Sigma_i R_i^2 - C$.
4. $SST = (1/rn) \Sigma_j T_j^2 - C$.
5. $SSE = (1/n) \Sigma_i \Sigma_j y_{ij.}^2 - C - SSR - SST$.
6. $SSS = SSTot - SSR - SST - SSE$.

Table 3.13 **Data Table for RBD with Subsampling**[a]

		Treatment				
Block	Sample	1	2	...	p	Sum
1	1	y_{111}	y_{121}	...	y_{1p1}	
	2	y_{112}	y_{122}	...	y_{1p2}	
	.	.	.		.	
	.	.	.		.	
	.	.	.		.	
	n	y_{11n}	y_{12n}	...	y_{1pn}	
Sum		$y_{11.}$	$y_{12.}$	...	$y_{1p.}$	R_1
2	1	y_{211}	y_{221}	...	y_{2p1}	
	2	y_{212}	y_{222}	...	y_{2p2}	
	.	.	.		.	
	.	.	.		.	
	.	.	.		.	
	n	y_{21n}	y_{22n}	...	y_{2pn}	
Sum		$y_{21.}$	$y_{22.}$	...	$y_{2p.}$	R_2
	.	.	.		.	
	.	.	.		.	
	.	.	.		.	
r	1	y_{r11}	y_{r12}	...	y_{rp1}	
	2	y_{r12}	y_{r22}	...	y_{rp2}	
	.	.	.		.	
	.	.	.		.	
	.	.	.		.	
	n	y	y	...	y	
Sum		$y_{r1.}$	$y_{r2.}$	...	$y_{rp.}$	R_r
Grand Sum		T_1	T_2	...	T_p	G
Mean		$\bar{y}_1$	$^{-}{}_{2}y$	...	$\bar{y}_p$	$\bar{y}$

[a] $y_{ij\cdot} = \Sigma_k\, y_{ijk}$ = total yield of samples on the j-th treatment in the i-th block
$R_i = \Sigma_j\, y_{ij\cdot}$ = total yield of the samples in the i-th block
$T_j = \Sigma_i\, y_{ij\cdot}$ = total yield of samples on the j-th treatment
$G = \Sigma_i\, R_i = \Sigma_j\, T_j$ = grand total of all samples
$\bar{y}_j = T_j/rn$ = mean, on a sample basis, of the j-th treatment
$\bar{y}= = G/rpn$ = grand mean of all samples

Table 3.14 **ANOVA Table for a Randomized Block Design with Subsampling**

ANOVA

Source	df	SS	MS	F
Total	rpn−1	SSTot		
Blocks	(r−1)	SSR	MSR	
Treatment	(p−1)	SST	MST	F_T
Exp't. Error	(r−1)(p−1)	SSE	MSE	F_E
Samp. Error	rp(n−1)	SSS	MSS	

Mean squares (MS) are computed by dividing the sums of squares by the degrees of freedom on the same line in the ANOVA table.

1. MSR = SSR/(r − 1).
2. MST = SST/(p − 1).
3. MSE = SSE/(r − 1)(p − 1).
4. MSS = SSS/rp(n − 1).

The F ratios are

1. F_T = MST/MSE; p − 1, (r − 1)(p − 1) df.
2. F_E = MSE/MSS; (r − 1)(p − 1), pr(n − 1) df.

3.4.2 Significance Tests

To see what hypotheses are tested by the F ratios, we must consider the composition of the mean squares in the analysis of variance:

1. MSS = s_s^2 estimates the variation among samples, σ_s^2.
2. MSE = $s_s^2 + ns_e^2$ estimates the variation among samples, σ_s^2, plus the additional variation among plots treated alike, σ_e^2.
3. MST = $s_s^2 + ns_e^2 + rn \sum_j \tau_j^2$ estimates the variation among samples, σ_s^2, plus the plot-to-plot variation, σ_e^2, plus the variation among treatment means, $rn \sum_j \tau_j^2$.

We see that

$$F_E = \frac{MSE}{MSS} = \frac{s_s^2 + ns_e^2}{s_s^2}$$

This ratio follows the F distribution only if $\sigma_e^2 = 0$. Hence F_E provides a test of the significance of the variation among plots treated alike.

Similarly,

$$F_T = \frac{MST}{MSE} = \frac{s_s^2 + ns_e^2 + rn \sum_j \tau_j^2}{s_s^2 + ns_e^2}$$

This ratio follows the F distribution only if $\sum_j \tau_j^2 = 0$. Hence F_T is a test statistic for testing the significance of differences among the treatment means.

3.4.3 Means and Standard Errors

When we subsample in the randomized block design the true treatment means, μ_j, on a subsample basis, are estimated by the subsample means. We have

$$\mu_j \doteq \bar{y}_j = \frac{T_j}{rn}$$

The standard error, $s_{\bar{y}}$, of a treatment mean is

$$s_{\bar{y}} = \sqrt{\frac{MSE}{rn}} = \sqrt{\frac{s_s^2 + ns_e^2}{rn}}$$

$$= \sqrt{\frac{s_s^2}{rn} + \frac{s_e^2}{r}}$$

Note that r is a divisor of both components of $s_{\bar{y}}$ and n is a divisor for only one component. Hence for a fixed total number, $m = rn$, of observations on a treatment the standard error is minimized by setting $n = 1$ and $r = m$. If costs must be considered, however, this will not necessarily produce the most economical sampling scheme.

The $(1 - \alpha)100\%$ confidence interval estimate, $L(\mu_j)$, of a treatment mean is

$$L(\mu_j) = \bar{y}_j \pm t_\alpha \sqrt{\frac{MSE}{rn}}$$

where t_α is the two-tailed, α-level t with $(r - 1)(p - 1)$ degrees of freedom.

If we are interested in the difference between two treatment means the standard error, $s_{\bar{d}}$, of the difference is

$$s\bar{d} = \sqrt{\frac{2MSE}{rn}}$$

The $(1 - \alpha)100\%$ confidence interval estimate, $L(\mu_j - \mu_{j'})$, of a difference is given by

$$L(\mu_j - \mu_{j'}) = \bar{y}_j - \bar{y}_{j'}, \pm t_\alpha \sqrt{\frac{2MSE}{rn}}$$

where, again, t_α is the two-tailed t with $(r - 1)(p - 1)$ degrees of freedom.

Subsampling adds only a little complexity to the computations involved in the data analysis. Similarly, the interpretation of the analysis is relatively unaffected by subsampling. Hence we will not include a numerical example to illustrate the procedure when subsampling is used.

3.5 LATIN SQUARE DESIGN

Another type of experimental design, which is more restrictive than the RBD, is the Latin square. In the Latin square the number of plots is the square of the number of treatments. In this design the p^2 plots are first grouped into p groups of p plots each on the basis of one blocking factor. This is commonly called *row* grouping. The plots are then grouped into p groups of p plots each based on a different blocking factor. This is commonly called *column* grouping. Treatments are then assigned to the plots in such a way that each treatment appears once, and only once, in each row and column.

The basic pattern can be seen by considering a design for $p = 4$ treatments, labeled A, B, C, and D, in a 4×4 square, as shown in Figure 3.6. We see that each row and each column form a complete block in the sense of a randomized block design. By blocking on two sources of variation, we can sometimes achieve significant reduction in experimental error compared to that of a randomized block design.

3.5.1 Advantages, Disadvantages, and Uses

There is one primary advantage of the Latin square design:

It allows the experimenter to control two sources of variation.

The design has several *disadvantages*:

1. The error df value is small if there are only a few treatments.

Row	Column 1	Column 2	Column 3	Column 4
1	A	B	C	D
2	B	C	D	A
3	C	D	A	B
4	D	A	B	C

Figure 3.6 Basic arrangement of a 4 × 4 Latin square.

2. The experiment becomes very large if the number of treatments is large, because the number of plots is the square of the number of treatments.
3. The statistical analysis is complicated by missing plots and misassigned treatments.

The Latin square design has a limited number of uses in field experiments:

1. It is useful when two sources of variation must be controlled.
2. For practical purposes its use is restricted to trials with more than four but fewer than 10 treatments.

In field experiments the Latin square could be used to remove the effect of two gradients, say, slope and fertility, at right angles to each other. Here the best field plan would be square, with square plots. Such a plan is diagrammed in Figure 3.7.

The Latin square design can also be used to impose a double control on a linear gradient. We can group the plots into p groups of p plots each along the gradient for the row classification. We can then use order within the row grouping for the column classification as illustrated in Figure 3.8.

3.5.2 Randomization

To randomize a Latin square design, we start with a basic square:

1. First row in alphabetic order.

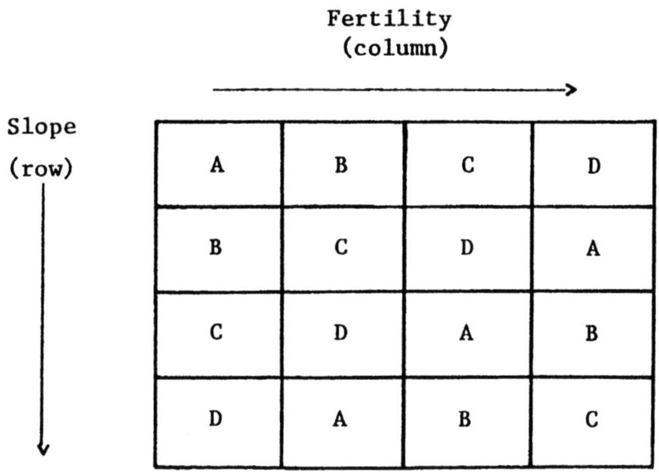

Figure 3.7 Plan for a Latin square to remove two perpendicular gradients.

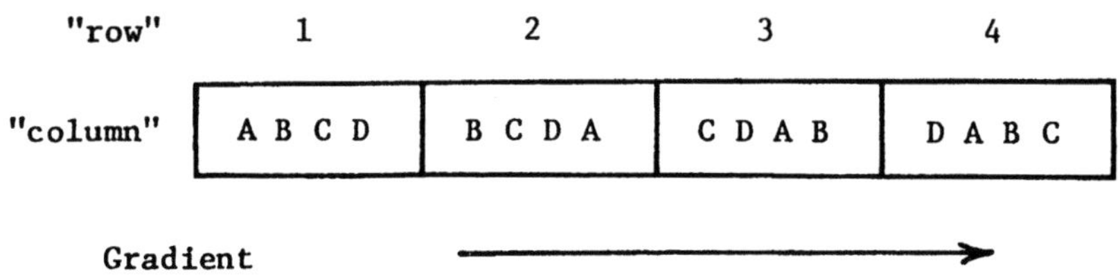

Figure 3.8 Using a Latin square to control a linear gradient.

2. Subsequent rows with letters shifted one position to the left of the one in the row immediately above:

   A B C D
   B C D A
   C D A B
   D A B C

   Then randomize the order of the rows, say, 2 1 3 4:

   B C D A
   A B C D
   C D A B
   D A B C

Finally, randomize the order of the columns, say, 4 2 3 1:

A C D B
D B C A
B D A C
C A B D

The last randomization gives the assignment of treatments to the plots on the field plan. Note that the Latin square properties have been preserved.

3.5.3 Analysis

The analysis of data from a Latin square follows much the same pattern as for other designs. Suppose that we have a p × p square. Let $y_{ij(k)}$ represent the observation in the i-th row and j-th column receiving the k-th treatment. The analysis proceeds as follows:

1. Set up a two-way row-by-column table and compute the row and column totals as shown in Table 3.15. Note that $R_i = \Sigma_j y_{ij}$, $C_j = \Sigma_i y_{ij}$, and $G = \Sigma_i R_i = \Sigma_j C_j$.
2. Complete Table 3.16, a table of treatment totals and means.
3. Set up an ANOVA table as shown in Table 3.17.

   Computations:

   a. *Sources of variation*: Total variation is partitioned into four components: (1) rows, (2) columns, (3) treatment, and (4) error.

Table 3.15 **Row-by-Column Table of Data from a Latin Square**

	Column				
Row	1	2	...	p	Sum
1	y_{11}	y_{12}	...	y_{1p}	R_1
2	y_{21}	y_{22}	...	y_{2p}	R_2
.	.	.		.	.
.	.	.		.	.
.	.	.		.	.
p	y_{p1}	y_{p2}	...	y_{pp}	R_p
Sum	C_1	C_2	...	C_r	G

Table 3.16 **Treatment Totals and Means from a Latin Square**[a]

Treatment	1	2	...	p	Sum
Sum	T_1	T_2	...	T_p	G
Mean	$\bar{y}_1$	$\bar{y}_2$	...	$\bar{y}_p$	$\bar{y}$

[a]T_k = sum of observations on the k-th treatment
$\bar{y}_k = T_k/p$ = mean of the k-th treatment
$G = \Sigma_k T_k$ = grand total
$\bar{y} = G/p$ = grand mean

Table 3.17 **ANOVA Table for Observations from a Latin Square Design**

ANOVA

Source	df	SS	MS	F
Total	p-1	SSTot		
Row	p-1	SSR	MSR	F_R
Column	p-1	SSC	MSC	F_C
Treatment	p-1	SST	MST	F_T
Error	(p-1)(p-2)	SSE	MSE	

b. *Degrees of freedom*: There are p^2 observations, for a total of $p^2 - 1$ df. There are p rows, p columns, and p treatments each with p − 1 df. Error df, (p − 1)(p − 2), may be obtained by difference.

c. *Sums of squares*:

(1) Correction term, $C = G^2/p^2$.

(2) $SSTot = \Sigma_i \Sigma_j y_{ij}^2 - C$.

(3) $SSR = (1/p) \Sigma_i R_i^2 - C$.

(4) $SSC = (1/p) \Sigma_j C_j^2 - C.$

(5) $SST = (1/p) \Sigma_k T_k^2 - C.$

(6) $SSE = SSTot - SSR - SSC - SST.$

d. *Mean squares*: Divide the sums of squares by their df.

(1) $MSR = SSR/(p - 1).$
(2) $MSC = SSC/(p - 1).$
(3) $MST = SST/(p - 1).$
(4) $MSE = SSE/(p - 1)(p - 2).$

e. *F values*: Divide the other mean squares by MSE.

$$F_R = \frac{MSR}{MSE}$$

$$F_C = \frac{MSC}{MSE}$$

$$F_T = \frac{MST}{MSE}$$

f. *Coefficient of variation*, CV:

$$CV = \left(\sqrt{\frac{MSE}{\bar{y}}}\right)100$$

4. *Significance tests*: F_T, with (p − 1) and (p − 1)(p − 2) df, is a test statistic for the significance of the differences among treatment means. It tests the hypothesis that all treatment means are the same against the alternative that at least one mean differs from the others. We compare F_T with the F in the table at p − 1; (p − 1)(p − 2) df, to determine whether the differences are highly significant (1%), significant (5%), or not significant (NS). F_R and F_C, with (p − 1) and (p − 1)(p − 2) df, are not exact tests but they provide a measure of whether or not the row or column groupings are effective in reducing experimental error.

3.5.4 Means and Standard Errors

a. As might be expected the true mean, μ_k, of the k-th treatment is estimated by the sample mean. We have

$$\mu_k \doteq \bar{y}_k = \frac{T_k}{p}$$

The standard error of a treatment mean, $s_{\bar{y}}$, is given by

$$s_{\bar{y}} = \sqrt{\frac{MSE}{p}}$$

This is the same for all means.

The $(1 - \alpha)100\%$ confidence interval estimate, $L(\mu_k)$, of the true mean of the k-th treatment is computed as

$$L(\mu_k) = \bar{y}_k \pm t_\alpha \sqrt{\frac{MSE}{p}}$$

where t_α is the two-tailed α-level t with $(p - 1)(p - 2)$ df.

b. The difference between two treatment means, $\mu_k - \mu_{k'}$, is estimated by the difference between the sample means. That is,

$$\mu_k - \mu_{k'} \doteq \bar{y}_k - \bar{y}_{k'}$$

As usual, $\bar{y}_k$ and $\bar{y}_{k'}$ are the sample means of treatments k and k′.

The standard error, $s_{\bar{d}}$, of a difference between two means is given by

$$s_{\bar{d}} = \sqrt{\frac{2MSE}{p}}$$

We can use this standard error, as before, for interval estimates and hypothesis tests. The $(1 - \alpha)100\%$ interval estimate of the true difference between the means of treatments k and k′ is

$$L(\mu_k - \mu_{k'}) = (\bar{y}_k - \bar{y}_{k'}) \pm t_\alpha \sqrt{\frac{2MSE}{p}}$$

To test the significance of the difference between means we can use as a test statistic

$$t = \frac{\bar{y}_k - \bar{y}_{k'}}{\sqrt{2MSE/p}}$$

This is a test statistic for

$$H_0: \mu_k = \mu_{k'}$$

against

$$H_a: \mu_k \neq \mu_{k'}$$

We compare t with the tabled t at $(p - 1)(p - 2)$ df. If $|t| > t_{0.01}$, the difference is highly significant; if $t_{0.01} > |t| > t_{0.05}$, the difference is significant; and if $|t| < t_{0.05}$, the difference is not significant.

3.5.5 Missing Plots

If data are missing from a Latin square, we can proceed in much the same way as with a randomized block design. We compute a substitute value, $\hat{y}_{ij(k)}$, to use in the analysis of variance. The formula is

$$\hat{y}_{ij(k)} = \frac{p(R_i + C_i + T_k) - 2G}{(p-1)(p-2)}$$

where R_i, C_j, and T_k are the totals for the remaining observations in the row, column, and treatment with missing data. G is the grand total of the available data.

With more than one missing observation the iterative procedure described for the randomized block design is used. One degree of freedom is subtracted from the total and from the error for each substitute value supplied. Again, the substitute values are not used in computing the means.

3.5.6 Relative Efficiency

The effectiveness of either the row or column groupings in controlling experimental error may be estimated by computing the relative efficiency of either rows or columns as blocks in a randomized block design. The relative efficiency, RE, with columns as blocks is

$$RE = \frac{MSR + (p-1)MSE}{pMSE}$$

With rows as blocks the formula is

$$RE = \frac{MSC + (p-1)MSE}{pMSE}$$

Finally, if we want to compare the efficiency of the Latin square to that of a completely randomized design, we use the equation

$$RE = \frac{MSR + MSC + (p-1)MSE}{(p+1)MSE}$$

3.5.7 Numerical Example

A bacteriologist wanted to determine the effect of four different sources of seed inoculum, A, B, C, and D, and a control, E, on the dry matter yield of irrigated alfalfa. The plots were to be furrow irrigated and the agronomist suspected that there might be a differential effect of irrigation from one end of the furrow to the other. The field was also bordered by a row of

A 33.8	B 33.7	D 30.4	C 32.7	E 24.4
D 37.0	E 28.8	B 33.5	A 34.6	C 33.4
C 35.8	D 35.6	A 36.9	E 26.7	B 35.1
E 33.2	A 37.1	C 37.4	B 38.1	D 34.1
B 34.8	C 39.1	D 32.7	E 37.4	A 36.4

Water Flow

Figure 3.9 Field plan and dry matter yields (kg/plot) for an inoculum source trial on irrigated alfalfa.

trees and he suspected a differential effect of shading as distance from the tree line increased. He decided to use square plots in a Latin square design with rows across the irrigation gradient and columns across the shade gradient. The field plan and total dry matter yields (kg/plot) are shown in Figure 3.9.

The analysis begins by constructing the row by column table shown in Table 3.18. Treatment totals and means are summarized in Table 3.19 and the analysis of variance is presented in Table 3.20.

Computations:

1. *Sums of squares*:

$$C = \frac{G^2}{p^2} = \frac{(852.7)^2}{25} = 29{,}083.89$$

$$SSTot = \sum_i \sum_j y_{ij}^2 - C = 29{,}380.55 - 29{,}083.89 = 296.66$$

$$SSR = \frac{1}{p} \sum_i R_{ij}^2 - C = \frac{1}{5}(155.0^2 + \cdots + 180.4^2) - C$$
$$= 87.40$$

$$SSC = \frac{1}{p} \sum_j C_{ij}^2 - C = \frac{1}{5}(174.6^2 + \cdots + 163.4^2) - C$$
$$= 16.56$$

Table 3.18 **Row-by-Column Summary Table of Data from a Latin Square Design**

	Column					
Row	1	2	3	4	5	Sum
1	33.8	33.7	30.4	32.7	24.4	155.0
2	37.0	28.8	33.5	34.6	33.4	167.3
3	35.8	35.6	36.9	26.7	35.1	170.1
4	33.2	37.1	37.4	38.1	34.1	179.9
5	34.8	39.1	32.7	37.4	36.4	180.4
Sum	174.6	174.3	170.9	169.5	163.4	852.7

Table 3.19 **Summary of Treatment Totals and Means of Dry Matter from a Latin Square Trial on Alfalfa**

Treat.	A	B	C	D	E	Sum
Total	178.8	175.2	178.4	174.5	145.8	852.7
Mean	35.76	35.04	35.68	34.90	29.16	34.11

$$SST = \frac{1}{p}\sum_k T_k^2 - C = \frac{1}{5}(178.8^2 + \cdots + 145.8^2) - C$$
$$= 155.90$$
$$SSE = SSTot - SSR - SSC - SST$$
$$= 296.66 - 87.40 - 16.56 - 155.90 = 36.80$$

Table 3.20 **Analysis of Variance of Dry Matter Yield**[a]

ANOVA

Source	df	SS	MS	F
Total	24	296.66		
Rows	4	87.40	21.85	7.12
Columns	4	16.56	4.14	1.35
Treatments	4	155.90	38.97	12.69**
Error	12	36.80	3.07	

[a] $CV = (\sqrt{3.07}/34.11)100 = 5.1\%$.
**Significant at the 1% level.

2. *Mean squares*:

$$MSR = \frac{SSR}{p-1} = \frac{87.40}{4} = 21.85$$
$$MSC = \frac{SSC}{p-1} = \frac{16.56}{4} = 4.14$$
$$MST = \frac{SST}{p-1} = \frac{155.90}{4} = 38.97$$
$$MSE = \frac{SSE}{(p-1)(p-2)} = \frac{36.80}{12} = 3.07$$

3. *F ratios*:

$$F_R = \frac{MSR}{MSE} = \frac{21.85}{3.07} = 7.12$$
$$F_C = \frac{MSC}{MSE} = \frac{4.14}{3.07} = 1.35$$
$$F_T = \frac{MST}{MSE} = \frac{38.97}{3.07} = 12.69$$

4. *Significance tests*:

$F_{0.05(4,12)} = 3.26 \qquad F_{0.01(4,12)} = 5.41$

Since $12.69 > 5.41$, treatment differences were highly significant.

Since 7.12 > 5.41, rows were effective in reducing error.
Since 1.35 < 3.26, columns were not effective in reducing error.

5. *Standard errors*:

   a. Treatment means:

$$s_{\bar{y}} = \sqrt{\frac{MSE}{p}} = \sqrt{\frac{3.07}{5}} = 0.78$$

   b. Difference between two treatment means:

$$s_{\bar{d}} = \sqrt{\frac{2MSE}{p}} = \sqrt{\frac{(2)(3.07)}{5}} = 1.11$$

Report of Statistical Analysis

Statistical analysis of data from an experiment designed to measure the effect of different sources of seed inocculum on the dry matter yield of irrigated alfalfa indicates that differences among treatment means were highly significant. On considering mean dry weight, shown in Table 3.21, it is seen that there is little difference among inocula. However, inoculation regardless of source produced more dry matter than did no inoculation.

There appeared to be no differential effect of irrigation from one end of the furrow to the other. Distance from shade, however, appeared to make a real contribution to experimental error.

3.6 FUNDAMENTAL RELATIONSHIP FOR INTERVAL ESTIMATION AND HYPOTHESIS TESTS

In the preceding discussions of basic experimental designs we have presented interval estimates of treatment means and of differences between treatment means. We have also looked at tests of the hypothesis that two treatment means are equal. These have been examples of more general

Table 3.21 **Mean Dry Matter Yield (kg/plot) of Irrigated Alfalfa Treated with Various Sources of Added Inoculum**[a]

P Source	A	B	C	D	None
Mean Yield	35.67	35.04	35.68	34.90	29.16

[a]Standard error = 0.78.

relationships for interval estimation and hypothesis tests. These general relationships are described below.

If $\hat{\theta}$ is a normally distributed estimate of the parameter θ, and if $s_{\hat{\theta}}^2$ is the sample variance of $\hat{\theta}$, then, provided that $\hat{\theta}$ and $s_{\hat{\theta}}^2$ are independent,

$$t = \frac{\hat{\theta} - \theta}{\sqrt{s_{\hat{\theta}}^2}}$$

follows the t distribution with the same df as $s_{\hat{\theta}}^2$. Hence

1. $L(\theta) = \hat{\theta} \pm t_\alpha \sqrt{s_{\hat{\theta}}^2}$ is the $(1 - \alpha)100\%$ confidence interval estimate of θ.
2. $t = (\hat{\theta} - \theta_0)/\sqrt{s_{\hat{\theta}}^2}$ is a test statistic for H_0: $\theta = \theta_0$, where θ is the true value and θ_0 is the hypothetical value of the parameter.

3.6.1 Examples

A. Suppose that $\bar{y}$ is the normally distributed (central limit theorem) estimate of θ, and $s_{\bar{y}}^2 = s^2/n$ is the sample variance of $\bar{y}$ with $n - 1$ df. Then

$$t = \frac{\bar{y} - \mu}{\sqrt{s^2/n}}$$

follows the t distribution and

1. $L(\mu) = \bar{y} \pm t_\alpha \sqrt{s^2/n}$ is the $(1 - \alpha)100\%$ confidence interval estimate of μ, in which t is the two-tailed t at the α probability level with $n - 1$ df.
2. $t = (\bar{y} - \mu_0)/\sqrt{s^2/n}$, with $n - 1$ df, is a test statistic for H_0: $\mu = \mu_0$.

B. Suppose that $\bar{y}_1$ and $\bar{y}_2$ are the means of independent samples drawn from two populations with common variance, σ^2, and suppose that there are n observations in each sample. Then $(\bar{y}_1 - \bar{y}_2)$ is a normally distributed estimate of $(\mu_1 - \mu_2)$ and the sample variance of $(\bar{y}_1 - \bar{y}_2)$ is given by $s^2 = 2s^2/n$. In this case,

$$t = \frac{(\bar{y}_1 - \bar{y}_2) - (\mu_1 - \mu_2)}{\sqrt{2s^2/n}}$$

follows the t distribution with $2n - 2$ df, and:

1. $L(\mu_1 - \mu_2) = (\bar{y}_1 - \bar{y}_2) \pm t_\alpha \sqrt{2s^2/n}$ is the $(1 - \alpha)100\%$ confidence interval estimate of $(\mu_1 - \mu_2)$, where t_α is the two-tailed α-level t with $2n - 2$ df.

2. $t = (\bar{y}_1 - \bar{y}_2)/\sqrt{2s^2/n}$, with $2n - 2$ df, is a test statistic for H_0: $\mu_1 = \mu_2$, the two population means are equal.

FURTHER READING

Complete citations for books for further reading are given in the bibliography at the end of the book.

1. Cochran and Cox, Chap. 4.
2. LeClerg, Leonard, and Clark, Chap. 9.
3. Little and Hills, Chaps. 4, 5, and 7.
4. Petersen, Chaps. 2–4.
5. Snedecor and Cochran, Chaps. 12 and 14.
6. Steel and Torrie, Chaps. 7 and 9.

APPENDIX

A3.1 Missing Plots

The analysis of variance of data from a randomized block design is correct only if all of the treatments are equally replicated and if all of the blocks have the same number of observations. This introduces complications if observations are missing from the data. If only one observation is missing, the procedure is to calculate a substitute value, $\hat{y}_{ij}$, by means of the following formula:

$$\hat{y}_{ij} = (rR_i + pT_j - G)/\,[(r - 1)(p - 1)]$$

where

$\hat{y}_{ij}$ = substitute value for the missing observation on the j-th treatment in the i-th block
R_i = sum of the remaining observations in the i-th block
T_j = sum of the remaining observations on the j-th treatment
G = grand total of the available observations
r, p = number of blocks, treatments, respectively

The computed value, $\hat{y}_{ij}$, replaces the missing value and the analysis is completed as usual. However, the df for Total and Error must be reduced by 1. Further, the substitute value, $\hat{y}_{ij}$, is used only to obtain a valid ANOVA.

It does not add any information to the data, and it should not be used in the computation of means.

With two or three missing values the formula is used iteratively. Values are guessed for all but one of the missing values and the formula is used to calculate this one. Then using this value and all but one of the remaining guessed values, the formula is used again. This process is repeated until the computed values stabilize. This may take two or three iterations. The df for Total and for Error are reduced by 1 for each substituted value.

A3.2 Relative Efficiency

The goal of blocking in an RBD is to remove the variation among blocks from experimental error. This will reduce the standard errors of means and differences, with a consequent reduction in the width of confidence intervals and an increase in the size of the test statistics used for making significance tests. To the extent that blocking is successful in reducing error the randomized block will be more efficient than a completely randomized design for the same set of treatments and experimental plots.

Appendix

One measure of the effectiveness of blocking is the F ratio for blocks. A more precise measure of the efficiency of the randomized block design relative to the completely randomized design for a given situation is obtained by computing the relative efficiency, RE, of the RBD. This is computed as

$$RE = [(r - 1)MSR + r(p - 1)MSE]/(rp - 1)MSE$$

in which

r, p = number of blocks, treatments in the RBD
MSR, MSE = block, error mean squares from the RBD ANOVA

If RE > 1, the randomized block design is more efficient than a completely randomized design would have been in the same situation. The quantity (RE − 1)100 is the percent increase in efficiency resulting from the use of an RBD. The quantity r(RE) is the number of replications of a completely randomized design that would be required to obtain the same precision as that obtained with the randomized block design using r blocks.

4

Agronomy Trials

4.1 INTRODUCTION

We have considered in some detail the basic experimental designs and the analysis of data from them. We have not yet, however, given much thought to the treatments to be included in the experiment. The two principal types of experiment for the field scientist are agronomy trials and variety trials. We consider these classes of experiments separately because different design considerations and different interpretation problems are involved.

If our experiment is to be an agronomy experiment, then, presumably, we are interested in the factors of production. We might illustrate the types of questions that lead to an agronomy trial in the following ways. Suppose that we are interested in introducing a new crop, triticale for example, into an area in which it has never before been grown. Suppose, further, that we want to conduct an experiment, or series of experiments, to see what factors are involved in its production. There are a number of questions that must be answered before we can go on:

1. When should we plant?
2. What seed rate should we use?
3. Do we need to use fertilizer?
4. If so, how much of each of the major elements is needed?
5. Do we need to supply minor elements?

6. Is irrigation necessary?
7. If so, when and how much?
8. When should we harvest?

We could think of other factors to consider, but these illustrate the problem.

4.2 MAIN EFFECTS AND INTERACTIONS

The traditional approach is to consider one factor at a time. For example, we could design a planting date trial in which the treatments would be different dates of planting, with the other factors held constant. There are a number of problems with this approach. To study planting dates it is necessary to choose a seed rate, fertilizer schedule, rate of planting, and so on. Further, a planting date that produces maximum yield for one variety might not produce maximum yield with another variety.

Sometimes the factors act independent of each other. By this we mean, for example, that the effect of different row spacings is the same for all varieties. In this case the results can be summarized as the average of the difference from one row spacing to another averaged over varieties. This is called the *main effect* of row spacing.

Often, however, the effects of two or more factors are not independent of each other. A tall wheat variety might outyield a short one in closely spaced rows, whereas because of lodging of the tall variety, the short variety might produce higher yield in widely spaced rows. When two factors do not act independently, they are said to *interact* with each other.

Interaction between two factors is the failure of the differences in effect between levels of one factor to be the same at all levels of the other factor. Figure 4.1 illustrates two possible outcomes when three varieties are tested at four levels of nitrogen. In Figure 4.1A the differences between varieties are the same at all levels of N fertilization. There is no interaction between the amount of N and the varieties. On the other hand, Figure 4.1B indicates that in this situation the differences between varieties increase as the rate of nitrogen is increased. Here we have an N rate $\times$ variety interaction. We cannot say anything about varietal differences without, at the same time, specifying the level of N fertilization.

4.3 FACTORIAL EXPERIMENTS

If interactions exist (and they are fairly common), we should plan our experiments in such a way that they can be measured and tested. It seems clear that we cannot do this if we vary only one factor at a time. We need

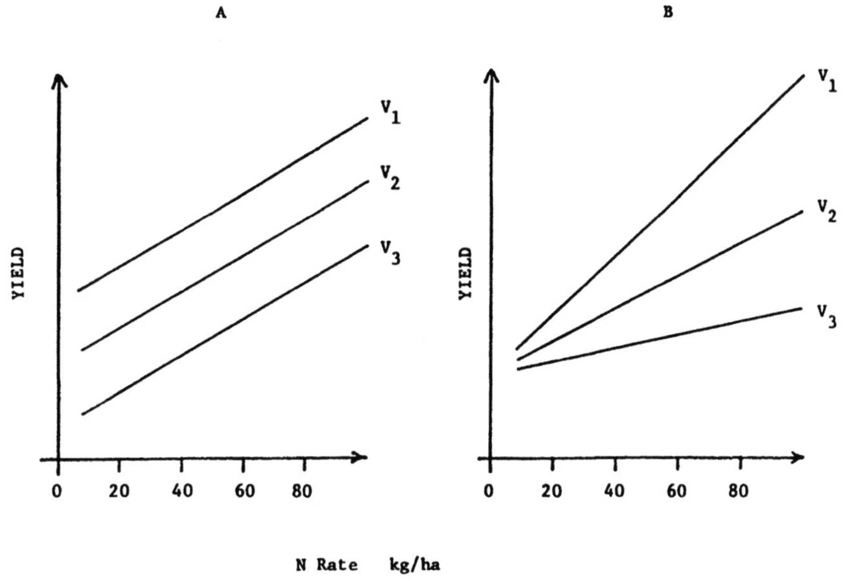

Figure 4.1 Possible yield patterns when three varieties are grown at four rates of nitrogen.

to include combinations of two or more factors each at two or more levels in the same experiment if we are concerned about interaction. One type of experiment that allows us to do this is the *factorial experiment*.

In factorial experiments the treatments consist of combinations involving two or more factors each at two or more levels. In particular, each level of every factor occurs together with each level of every other factor. The total number of treatments is the product of the number of levels of all factors. For example, we might want to conduct an agronomy trial on wheat with, as one factor, row spacing at four levels:

$S_1 = 15$ cm $\quad S_2 = 20$ cm $\quad S_3 = 25$ cm $\quad S_4 = 30$ cm

A second factor might be nitrogen at two levels:

N_1 = no nitrogen $\quad N_2 = 50$ kg/ha added N

(*Note*: "No nitrogen" is a legitimate level of the N factor for which "0 kg/ha" is the added level.) A full factorial set of treatments for this case would include $4 \times 2 = 8$ treatments. These would be (1) S_1N_1, (2) S_1N_2, (3) S_2N_1, (4) S_2N_2, (5) S_3N_1, (6) S_3N_2, (7) S_4N_1, and (8) S_4N_2.

It should be clear that the term *factorial experiment* refers to the nature of the treatments to be included in the experiment and not to the experimental design. (The term *factorial design* is incorrect.) The factorial set may be used in any design: CRD, RBD, Latin square, and so on.

4.3.1 Advantages, Disadvantages, and Uses

There are a number of advantages from using a factorial experiment. When the factors are independent, the factorial approach can result in a saving of time and material for two reasons:

1. Main effects are all that are required to describe the results.
2. Hidden replication: Each main effect is estimated as if the entire experiment had been devoted to that factor alone.

When factors are not independent, the factorial approach provides a systematic set of treatment combinations for estimating and testing interactions.

There are two primary disadvantages of factorial experiments:

1. As the number of factors increases, the total number of treatments rapidly becomes very large. For example, eight factors each at two levels requires $2^8 = 256$ plots for one replication of a full factorial set.
2. Large factorials may be difficult to interpret, particularly when interactions are present.

There are a number of uses for factorial experiments:

1. In exploratory work, where the aim is to discover which factors are important and which are not.
2. To study relationships among a number of factors.
3. For trials designed to lead to recommendations over a wide range of conditions. Some of the expected conditions may be included as factors even though they are not of primary interest.

4.3.2 Field Plan

To plan a factorial experiment we choose an experimental design that is appropriate for the site of the trial. Often, this is a randomized block design. We then use a randomization scheme that is specified for the design to allocate the factor combinations to the plots. The result is the field plan for the design. If the chosen design is a randomized block design, for example, we group the plots into blocks, then randomly allocate the treatments to the plots within the blocks using a separate randomization in each block.

4.3.3 Data Analysis

In the analysis of data from a factorial experiment we want to estimate and test the main effects of each factor and the interactions among the factors. The analysis, in general, follows the same pattern as for single-factor experiments:

1. Construct tables of totals.
2. Complete an ANOVA table.
3. Perform significance tests.
4. Compute appropriate means and standard errors.
5. Interpret the analysis and report the results.

4.4 TWO-FACTOR EXPERIMENTS

The simplest kind of factorial experiment is one in which there are only two factors. The agronomy trial on wheat that we used earlier to illustrate the idea of factorial combinations is an example. There were two factors: a spacing factor at four levels, and a nitrogen factor at two levels. There were a total of $4 \times 2 = 8$ treatments (combinations) as follows: (1) S_1N_1, (2) S_1N_2, (3) S_2N_1, (4) S_2N_2, (5) S_3N_1, (6) S_3N_2, (7) S_4N_1, and (8) S_4N_2. Suppose that the experiments were run using a randomized block design with three blocks of eight plots each. The field plan, after randomization, might appear as in Figure 4.2.

I		II		III	
S_4N_2	S_4N_1	S_3N_1	S_1N_1	S_3N_1	S_4N_2
S_3N_2	S_1N_1	S_1N_2	S_2N_1	S_4N_1	S_3N_2
S_2N_1	S_1N_2	S_2N_2	S_4N_2	S_2N_2	S_1N_1
S_2N_2	S_3N_1	S_3N_2	S_4N_1	S_1N_2	S_2N_1

Figure 4.2 Field plan for a 4×2 factorial experiment run as a randomized block design with three blocks.

Table 4.1 **Two-Way Table of Factor A and Factor B Totals**[a]

	Factor B				
Factor A	1	2	...	b	Sum
1	T_{11}	T_{12}	...	T_{1b}	A_1
2	T_{21}	T_{22}	...	T_{2b}	A_2
.	.	.		.	.
.	.	.		.	.
.	.	.		.	.
a	T_{a1}	T_{a2}	...	T_{ab}	A_a
Sum	B_1	B_2	...	B_b	G

[a]$A_j = \Sigma_k T_{jk}$ = sum of yields on the j-th level of A
$B_k = \Sigma_j T_{jk}$ = sum of yields on the k-th level of B
$G = \Sigma_j A_j = \Sigma_k B_k$ = grand total of all yields

4.4.1 Analysis

In the analysis of data from a two-factor factorial experiment we want to estimate and test two main effects and one two-factor (first-order) interaction. To illustrate the process, suppose that we have two factors: A at a levels and B at b levels, for a total of ab treatments. Assume that we use a randomized block design with r blocks. Let y_{ijk} represent the yield from the plot with the j-th level of factor A and the k-th level of factor B in the i-th block. Compute $T_{jk} = \Sigma_i y_{ijk}$ = sum of the yields on the j-th level of A and k-th level of B summed over blocks. Then construct the summary tables shown as Tables 4.1 and 4.2. We now construct the analysis of variance table given in Table 4.3.

Table 4.2 **One-Way Table of Block Totals**[a]

Block	1	2	...	r	Sum
Sum	R_1	R_2	...	R_r	G

[a]$R_i = \Sigma_j \Sigma_k y_{ijk}$ = sum of yields on the i-th block.

Table 4.3 **Format for the Analysis of Variance of a Two-Factor Factorial Experiment**

ANOVA

Source	df	SS	MS	F
Total	rab−1	SSTot		
Block	r−1	SSR	MSR	F_R
A	a−1	SSA	MSA	F_A
B	b−1	SSB	MSB	F_B
AB	(a−1)(b−1)	SSAB	MSAB	F_{AB}
Error	(r−1)(ab−1)	SSE	MSE	

1. The *sums of squares* in the table are computed as
   a. Correction term, C: $C = G^2/rab$. Then
   b. $SSTot = \sum_i \sum_j \sum_k y_{ijk}^2 - C$.
   c. $SSR = (1/ab) \sum_i R_i^2 - C$.
   d. $SSA = (1/rb) \sum_j A_j^2 - C$.
   e. $SSB = (1/ra) \sum_k B_k^2 - C$.
   f. $SSAB = (1/r) \sum_j \sum_k T_{jk}^2 - C - SSA - SSB$.
   g. $SSE = SSTot - SSR - SSA - SSB - SSAB$.
2. *Mean squares*: Computed by dividing the sums of squares by their associated degrees of freedom.
3. *F*: Computed by dividing the mean squares for the other sources of variation by the error mean square.
4. *CV*: The coefficient of variation, CV, is obtained as

$$CV = \left(\frac{\sqrt{MSE}}{\overline{\overline{y}}}\right)100$$

where $\overline{\overline{y}} = G/rab$, the grand mean of all yields.

4.4.2 Significance Tests

1. $F_A = MSA/MSE$, with $a - 1, (r - 1)(ab - 1)$ df, is a test statistic for the hypothesis that the main effect of factor A is zero. If $F_A >$

F_α, we conclude that there are significant differences among the means of the levels of factor A.
2. F_B = MSB/MSE, with b − 1, (r − 1)(ab − 1) df, is a test of the significance of differences among means for levels of factor B.
3. F_{AB} = MSAB/MSE, with (a − 1)(b − 1), (r − 1)(ab − 1) df, is a test statistic for testing the significance of the AB interaction. If $F_{AB} > F_\alpha$, we conclude that the interaction is significant. In this case the factors are not acting independently and the main effects have no meaning whether or not they test significant.
4. F_R = MSR/MSE, with r − 1, (r − 1)(ab − 1) df, is an approximate test of the significance of differences among block means. If $F_R > F_\alpha$, we conclude that blocking was effective in reducing experimental error.

4.4.3 Means and Standard Errors

There are three types of means in a two-factor factorial experiment:

1. A factor means
2. B factor means
3. Treatment (AB) means

Associated with each is a standard error, and the differences between means have their own standard errors as well.

1. Factor A means, $\bar{y}_{Aj} = A_j/rb$. The standard error, $s_{\bar{A}}$, is

$$s_{\bar{A}} = \sqrt{\frac{MSE}{rb}}$$

The (1 − α)100% confidence interval estimate of the mean at the j-th level of factor A, $L(\mu_{Aj})$, is given by $L(\mu_{Aj}) = \bar{y}_{Aj} \pm t_\alpha\sqrt{MSE/rb}$, where t_α is the two-tailed t at the α level with (r − 1)(ab − 1) df. The standard error of the difference between two factor A means, $s_{\bar{d}A}$, is given by

$$s_{\bar{d}A} = \sqrt{\frac{2MSE}{rb}}$$

To test the significance of the difference between two factor A means, we use

$$t = \frac{\bar{y}_{Aj} - \bar{y}_{Aj'}}{\sqrt{2MSE/rb}}$$

If $|t| > t_\alpha$, the α-level two-tailed t with (r − 1)(ab − 1) df, the difference is significant at the α level.

2. Factor B means, $\bar{y}_{Bk} = B_k/ra$. The standard error, $s_{\bar{B}}$, is

$$s_{\bar{B}} = \sqrt{\frac{MSE}{ra}}$$

The standard error of a difference between two factor B means, $s_{\bar{d}B}$, is computed as

$$s_{\bar{d}B} = \sqrt{\frac{2MSE}{ra}}$$

These standard errors may be used for interval estimates and significance tests in the same way as for the factor A means.

3. Treatment (AB) means, $\bar{y}_{jk} = T_{jk}/r$. The standard error of the mean at any combination of factor A and factor B, $s_{\bar{y}}$, is

$$s_{\bar{y}} = \sqrt{\frac{MSE}{r}}$$

The standard error of the difference between two treatment (AB) means, $s_{\bar{d}}$, is

$$s_{\bar{d}} = \sqrt{\frac{2MSE}{r}}$$

This may also be used for interval estimates and significance tests.

4.4.4 Interpretation

The interpretation of factorial experiments depends on the outcome of the significance tests. If the AB interaction is significant, the main effects have no meaning whether or not that test significant. This is because factor A and factor B are not acting independent of each other. The difference in yield between two levels of A changes from one level of B to another, and the mean for one level of A averaged over all levels of B has little meaning. In this case the results are best summarized in a two-way table of means for the various AB combinations.

If the interaction is not significant, the factors are acting independent of each other. The difference in yield between two levels of A is the same at all levels of B. In this case the results are summarized in one-way tables of means for the factors with significant main effects.

4.4.5 Example

A horticulturist wanted to study the effect of row spacing and phosphate fertilizer on the yield of bush beans. She decided to use a factorial set of treatments with spacing at three levels:

$S_1 = 45$ cm $\quad S_2 = 90$ cm $\quad S_3 = 135$ cm

and phosphate at two levels

$P_1 =$ no phosphate $\quad P_2 = 25$ kg/ha phosphate

She decided to use a randomized block experimental design with three blocks. The field plan and bean yields are presented in Figure 4.3. The

Block	I	II	III
	S_2P_1 60	S_1P_2 45	S_1P_1 55
	S_1P_1 65	S_3P_1 55	S_3P_1 51
	S_3P_2 66	S_3P_2 57	S_1P_2 43
	S_3P_1 59	S_1P_1 58	S_2P_1 54
	S_1P_2 56	S_2P_2 50	S_2P_2 45
	S_2N_2 62	S_2P_1 59	S_3P_2 50

Figure 4.3 Field plan and bean yields (kg/plot) for a spacing × phosphate trial on bush beans.

Table 4.4 **Phosphate × Spacing Totals**

Phosphate	Spacing S1	S2	S3	Sum
P1	178	173	165	516
P2	144	157	173	474
Sum	322	330	338	990

required summary tables are given as Tables 4.4 and 4.5. The analysis of variance of the bean yield data is given in Table 4.6.

Computations:

$$C = \frac{G^2}{rab} = \frac{(990)^2}{(3)(3)(2)} = 54{,}450$$

$$SSTot = \sum_i \sum_j \sum_k y_{ijk}^2 - C = 55{,}202 - 54{,}450 = 752.00$$

$$SSR = \frac{1}{ab} \sum_i R_i^2 - C = \frac{1}{(3)(2)}(368^2 + 324^2 + 298^2) - C$$
$$= 54{,}867.33 - 54{,}450.00 = 417.33$$

$$SSS = \frac{1}{rb} \sum_j A_j^2 - C = \frac{1}{(3)(2)}(322^2 + 330^3 + 338^2) - C$$
$$= 54{,}471.33 - 54{,}450.00 = 21.33$$

$$SSP = \frac{1}{ra} \sum_k B_k^2 - C = \frac{1}{(3)(3)}(516^2 + 474^2) - C$$
$$= 54{,}548.00 - 54{,}450.00 = 98.00$$

Table 4.5 **Block Totals**

Block	I	II	III	Sum
Sum	368	324	298	990

Table 4.6 **Analysis of Variance of Bean Yield (kg/plot) from a Spacing × Phosphate Trial on Bush Beans**

ANOVA

Source	df	SS	MS	F
Total	17	752.00		
Block	2	417.33	208.67	31.00
S	2	21.33	10.67	1.58 NS
P	1	98.00	98.00	14.56 **
SP	2	148.00	74.00	11.00 **
Error	10	67.33	6.73	

**Significant at the 1% level.

$$\text{SSSP} = \frac{1}{r}\sum_j \sum_k T_{jk}^2 - C - \text{SSS} - \text{SSP}$$

$$= \frac{1}{3}(178^2 + \cdots + 173^2) - 54{,}450.00 - 21.33 - 98.00$$

$$= 54{,}717.33 - 54{,}450.00 - 21.33 - 98.00 = 148.00$$

$$\text{SSE} = \text{SSTot} - \text{SSR} - \text{SSS} - \text{SSP} - \text{SSSP}$$

$$= 752.00 - 417.33 - 21.33 - 98.00 - 148.00 = 67.33$$

$$\text{MS} = \frac{\text{SS}}{\text{df}}$$

$$F_R = \frac{208.67}{6.73} = 31.00$$

$$F_A = \frac{10.67}{6.73} = 1.58$$

$$F_B = \frac{98.00}{6.73} = 14.56$$

$$F_{AB} = \frac{74.00}{6.73} = 11.00$$

Significance tests:

$F_{0.01(2,10)}$	$= 7.56 < 11.00$	SP is highly significant
$F_{0.05(2,10)}$	$= 4.10 > 1.58$	spacing is not significant
$F_{0.01(1,10)}$	$= 10.04 < 14.56$	P is highly significant
$F_{0.01(2,10)}$	$= 7.56 < 31.00$	blocking was highly effective

$$CV = \left(\sqrt{\frac{MSE}{\bar{y}}}\right)100 = \left(\sqrt{\frac{6.73}{55.00}}\right)100 = 4.7\%$$

Standard errors:

1. Phosphate mean, $s_{\bar{p}} = \sqrt{MSE/ra} = \sqrt{6.73/9} = 0.865$.
2. Treatment (SP) mean, $s_{\bar{y}} = \sqrt{MSE/r} = \sqrt{6.73/3} = 1.498$.

Report of Statistical Analysis

An experiment was conducted to determine the effect of row spacing and phosphate fertilizer on the seed yield of bush beans. Mean yields (kg/plot) at different spacings with and without phosphate are presented in Table 4.7. The table shows that the yield response to increased row spacing depends on whether or not additional phosphate has been supplied. If no phosphate is added, the yield decreases as spacing is increased. On the other hand, yields are increased if spacing is increased when 25 kg/ha of phosphate is applied. There was a substantial increase in precision in this trial, due to blocking. This has produced a satisfactorily low CV of 4.7%.

Table 4.7 **Yield of Bush Beans (kg/plot) at Various Levels of Phosphate and Row Spacing**[a]

	Spacing		
Phosphate	45 cm	90 cm	135 cm
None	59.33	57.67	55.00
25 kg/ha	48.00	52.33	57.67

[a]Standard error = 1.50.

4.5 THREE-FACTOR EXPERIMENTS

When three or more factors are included in a factorial experiment, the principles of design selection and randomization remain unchanged. The analysis of the resulting data, however, becomes somewhat more complicated. With a three-factor experiment we have to estimate and test three main effects, three two-factor (first-order) interactions, and one three-factor (second-order) interaction.

4.5.1 Analysis

The basic steps in the analysis follow the same pattern as before.

1. Construct tables of totals.
2. Complete an ANOVA table.
3. Perform significance tests.
4. Compute appropriate means and standard errors.
5. Interpret the results.

In general, suppose that we have three factors: A at a levels, B at b levels, and C at c levels. Suppose that the experiment is run using a randomized block design with r blocks. Let y_{ijkl} represent the yield of the j-th level of A, k-th level of B, and l-th level of C in the i-th block. Construct Table 4.8, the block totals. Then complete Table 4.9, a three-way table of treatment totals. From the three-way table we form three two-way tables of factor totals: Tables 4.10 to 4.12. (*Note*: It is easiest to begin with the three-way table and then use the entries in this table to construct the two-way tables.)

We now complete the analysis of variance presented in Table 4.13.

Computations:
First compute a correction term, $C = G^2/\text{rabc}$. Then

$$\text{SSTot} = \sum_i \sum_j \sum_k \sum_l y_{ijkl}^2 - C$$

$$\text{SSR} = \frac{1}{\text{abc}} \sum_i R_i^2 - C$$

Table 4.8 **Total Yields for Blocks**[a]

Block	1	2	. . .	r	Sum
Sum	R_1	R_2	. . .	R_r	G

[a]$R_i = \Sigma_j \Sigma_k \Sigma_l y_{ijkl}$ = sum of all yields in the i-th block
$G = \Sigma_i R_i$ = grand total of all yields

Table 4.9 **Three-Way Table of ABC Treatment Totals**[a]

		Factor C				
Factor A	Factor B	1	2	...	c	Sum
1	1	$T_{\cdot 111}$	$T_{\cdot 112}$	...	$T_{\cdot 11c}$	$T_{\cdot 11\cdot}$
	2	$T_{\cdot 121}$	$T_{\cdot 122}$	...	$T_{\cdot 12c}$	$T_{\cdot 12\cdot}$
	⋮	⋮	⋮		⋮	⋮
	b	$T_{\cdot 1b1}$	$T_{\cdot 1b2}$	...	$T_{\cdot 1bc}$	$T_{\cdot 1b\cdot}$
Sum		$T_{\cdot 1\cdot 1}$	$T_{\cdot 1\cdot 2}$	...	$T_{\cdot 1\cdot c}$	A_1
2	1	$T_{\cdot 211}$	$T_{\cdot 212}$	...	$T_{\cdot 21c}$	$T_{\cdot 21\cdot}$
	2	$T_{\cdot 221}$	$T_{\cdot 222}$	...	$T_{\cdot 22c}$	$T_{\cdot 22\cdot}$
	⋮	⋮	⋮		⋮	⋮
	b	$T_{\cdot 2b1}$	$T_{\cdot 2b2}$	...	$T_{\cdot 2bc}$	$T_{\cdot 2b\cdot}$
Sum		$T_{\cdot 2\cdot 1}$	$T_{\cdot 2\cdot 2}$	...	$T_{\cdot 2\cdot c}$	A_2
	⋮		⋮			⋮
a	1	$T_{\cdot a11}$	$T_{\cdot a12}$	...	$T_{\cdot a1c}$	$T_{\cdot a1\cdot}$
	2	$T_{\cdot a21}$	$T_{\cdot a22}$	...	$T_{\cdot a2c}$	$T_{\cdot a2\cdot}$
	⋮		⋮			⋮
	b	$T_{\cdot ab1}$	$T_{\cdot ab2}$	...	$T_{\cdot abc}$	$T_{\cdot ab\cdot}$
Sum		$T_{\cdot a\cdot 1}$	$T_{\cdot a\cdot 2}$	...	$T_{\cdot a\cdot c}$	A_a
Sum		C_1	C_2	...	C_c	G

[a] $T_{\cdot jkl} = \Sigma_i\, y_{ijkl}$ = sum for the j-th level of A, k-th level of B, l-th level of C summed over blocks

$T_{\cdot jk\cdot} = \Sigma_l\, T_{\cdot jkl}$ = sum for the j-th level of A, k-th level of B, summed over levels of C and blocks

$T_{\cdot j\cdot l} = \Sigma_k\, T_{\cdot jkl}$ = sum for the j-th level of A, l-th level of C, summed over levels of B and blocks

$A_j = \Sigma_k\, T_{\cdot jk\cdot} = \Sigma_l\, T_{\cdot j\cdot l}$ = sum for the j-th level of A

$C_l = \Sigma_j\, T_{\cdot j\cdot l}$ = sum for the l-th level of C

$G = \Sigma_j\, A_j = \Sigma_l\, C_l$ = grand total

Table 4.10 **Two-Way Table of AB Totals**[a]

Factor A	Factor B 1	2	...	b	Sum
1	$T_{\cdot 11 \cdot}$	$T_{\cdot 12 \cdot}$	...	$T_{\cdot 1b \cdot}$	A_1
2	$T_{\cdot 21 \cdot}$	$T_{\cdot 22 \cdot}$	...	$T_{\cdot 2b \cdot}$	A_2
.	.	.		.	.
.	.	.		.	.
.	.	.		.	.
a	$T_{\cdot a1 \cdot}$	$T_{\cdot a2 \cdot}$	...	$T_{\cdot ab \cdot}$	A_a
Sum	B_1	B_2	...	B_b	G

[a]$B_k = \Sigma_j T_{\cdot jk \cdot}$ = sum for the k-th level of B, and $T_{\cdot jk \cdot}$, A_j, and G are as in the three-way table.

Table 4.11 **Two-Way Table of AC Totals**[a]

Factor A	Factor C 1	2	...	c	Sum
1	$T_{\cdot 1 \cdot 1}$	$T_{\cdot 1 \cdot 2}$	...	$T_{\cdot 1 \cdot c}$	A_1
2	$T_{\cdot 2 \cdot 1}$	$T_{\cdot 2 \cdot 2}$	...	$T_{\cdot 2 \cdot c}$	A_2
.	.	.		.	.
.	.	.		.	.
.	.	.		.	.
a	$T_{\cdot a \cdot 1}$	$T_{\cdot a \cdot 2}$	...	$T_{\cdot a \cdot c}$	A_a
Sum	C_1	C_2	...	C_c	G

[a]$T_{\cdot j \cdot l}$, A_j, C_l, and G are as in the three-way table.

Table 4.12 **Two-Way Table of BC Totals**[a]

Factor B	Factor C 1	2	...	c	Sum
1	$T_{\cdot\cdot 11}$	$T_{\cdot\cdot 12}$	...	$T_{\cdot\cdot 1c}$	B_1
2	$T_{\cdot\cdot 21}$	$T_{\cdot\cdot 22}$	...	$T_{\cdot\cdot 2c}$	B_2
.	.	.		.	.
.	.	.		.	.
.	.	.		.	.
b	$T_{\cdot\cdot b1}$	$T_{\cdot\cdot b2}$	...	$T_{\cdot\cdot bc}$	B_b
Sum	C_1	C_2	...	C_c	G

[a]$T_{\cdot\cdot kl} = \sum_j T_{\cdot jkl}$ = sum for the k-th level of B, l-th level of C summed over A and blocks, and B_k, C_l, and G are as before.

Table 4.13 **Format for the Analysis of Variance of a Three-Factor Factorial Experiment**

ANOVA

Source	df	SS	MS	F
Total	rabc−1	SSTot		
Blocks	r−1	SSR	MSR	F_R
A	a−1	SSA	MSA	F_A
B	b−1	SSB	MSB	F_B
AB	(a−1)(b−1)	SSAB	MSAB	F_{AB}
C	c−1	SSC	MSC	F_C
AC	(a−1)(c−1)	SSAC	MSAC	F_{AC}
BC	(b−1)(c−1)	SSBC	MSBC	F_{BC}
ABC	(a−1)(b−1)(c−1)	SSABC	MSABC	F_{ABC}
Error	(r−1)(abc−1)	SSE	MSE	

$$\text{SSA} = \frac{1}{\text{rbc}} \sum_{j} \text{A}_j^2 - \text{C}$$

$$\text{SSB} = \frac{1}{\text{rac}} \sum_{k} \text{B}_k^2 - \text{C}$$

$$\text{SSAB} = \frac{1}{\text{rc}} \sum_{j} \sum_{k} \text{T}_{\cdot jk\cdot}^2 - \text{C} - \text{SSA} - \text{SSB}$$

$$\text{SSC} = \frac{1}{\text{rab}} \sum_{l} \text{C}_l^2 - \text{C}$$

$$\text{SSAC} = \frac{1}{\text{rb}} \sum_{j} \sum_{l} \text{T}_{\cdot j\cdot l}^2 - \text{C} - \text{SSA} - \text{SSC}$$

$$\text{SSBC} = \frac{1}{\text{ra}} \sum_{k} \sum_{l} \text{T}_{\cdot\cdot kl}^2 - \text{C} - \text{SSB} - \text{SSC}$$

$$\text{SSABC} = \frac{1}{\text{r}} \sum_{j} \sum_{k} \sum_{l} \text{T}_{\cdot jkl}^2 - \text{C} - \text{SSA} - \text{SSB} - \text{SSAB} - \text{SSC} - \text{SSAC} - \text{SSBC}$$

$$\text{SSE} = \text{SSTot} - \text{SSR} - \text{SSA} - \text{SSB} - \text{SSAB} - \text{SSC} - \text{SSAC} - \text{SSBC} - \text{SSABC}$$

Mean squares: MS = SS/df.

F: F = MS/MSE.

4.5.2 Significance Tests

F_A, with a − 1, (r − 1)(abc − 1) df, is used to test the significance of the main effect of A.

F_B, with b − 1, (r − 1)(abc − 1) df, is used to test the significance of the main effect of B.

F_{AB}, with (a − 1)(b − 1), (r − 1)(abc − 1) df, is a test statistic for the AB interaction.

F_C, with c − 1, (r − 1)(abc − 1) df, tests the C main effect.

F_{AC}, with (a − 1)(c − 1), (r − 1)(abc − 1) df, is used to test the AC interaction.

F_{BC}, with (b − 1)(c − 1), (r − 1)(abc − 1) df, is a statistic for testing the BC interaction.

F_{ABC}, with (a − 1)(b − 1)(c − 1), (r − 1)(abc − 1) df, is used to test the second-order (ABC) interaction.

F_R, with r − 1, (r − 1)(abc − 1) df, is an approximate test statistic for the effectiveness of blocking.

4.5.3 Means and Standard Errors

There are a number of different types of means that may be computed for data from a three-factor factorial experiment. These, together with their standard errors, are presented in Table 4.14.

4.5.4 Interpretation

As with the two-factor factorial, interpretation of the three-factor factorial depends on the outcome of the significance tests on the main effects and interactions. We begin by considering the three-factor interaction. If it is significant, none of the factors are acting independently. In this case we summarize the results with a three-way table of treatment combination means.

If the ABC interaction is not significant, we look at the significance tests on the first-order (two-factor) interactions. If any of these interactions is significant, neither of the main effects for the factors in the significant interaction has any meaning because the action of these factors is not independent of each other. We then use two-way tables of means to summarize the results for these two factors.

If the interaction between any two factors is not significant, we then look at the tests of the main effects for these factors. In this case, factors with significant main effects are summarized in one-way tables of means.

Table 4.14 **Means and Standard Errors for a Three-Factor Factorial Experiment**

Factor	Mean	Standard error: Mean	Standard error: Difference
A	$\bar{y}_{Aj} = A_j \; /rbc$	$s_{\bar{A}} = \sqrt{MSE/rbc}$	$s_{\bar{d}A} = \sqrt{2MSE/rbc}$
B	$\bar{y}_{Bk} = B_k \; /rac$	$s_{\bar{B}} = \sqrt{MSE/rac}$	$s_{\bar{d}B} = \sqrt{2MSE/rac}$
AB	$\bar{y}_{AB} = T_{\cdot jk \cdot}/rc$	$s_{\bar{AB}} = \sqrt{MSE/rc}$	$s_{\bar{d}AB} = \sqrt{2MSE/rc}$
C	$\bar{y}_{Cl} = C_l \; /rab$	$s_{\bar{C}} = \sqrt{MSE/rab}$	$s_{\bar{d}C} = \sqrt{2MSE/rab}$
AC	$\bar{y}_{AC} = T_{\cdot j \cdot l}/rb$	$s_{\bar{AC}} = \sqrt{MSE/rb}$	$s_{\bar{d}AC} = \sqrt{2MSE/rb}$
BC	$\bar{y}_{BC} = T_{\cdot\cdot kl}/ra$	$s_{\bar{BC}} = \sqrt{MSE/ra}$	$s_{\bar{d}BC} = \sqrt{2MSE/ra}$
ABC	$\bar{y} = T_{\cdot jkl} \; /r$	$s_{\bar{y}} = \sqrt{MSE/r}$	$s_{\bar{d}} = \sqrt{2MSE/r}$

4.5.5 Example

A food legume agronomist wanted to study the effect of three production factors—variety, phosphorus fertilization, and weed control—on the yield of chick-peas. He believed that these factors might not act independently, so he decided to use a factorial experiment for his study. He chose the following factor levels:

1. Variety at two levels: V_1 and V_2
2. Phosphorus at three levels: P_1 = none, P_2 = 25 kg/ha, and P_3 = 50 kg/ha
3. Weed control at two levels: W_1 = none, W_2 = herbicide

The resulting set of treatments were (1) $V_1P_1W_1$, (2) $V_1P_2W_1$, (3) $V_1P_3W_1$, (4) $V_1P_1W_2$, (5) $V_1P_2W_2$, (6) $V_1P_3W_2$, (7) $V_2P_1W_1$, (8) $V_2P_2W_1$, (9) $V_2P_3W_1$, (10) $V_2P_1W_2$, (11) $V_2P_2W_2$, and (12) $V_2P_3W_2$.

The area in which the trial was to be run included three soil types, so he decided to use a randomized block design with soil type as a blocking factor. He located 12 plots on each of the soil types and assigned the treatment combinations at random to plots within types. The field plan and yields (kg/plot) of dry chick-peas are illustrated in Figure 4.4. The tables of totals needed to complete the analysis are given as Tables 4.15 to 4.19. The analysis of variance of the seed yield data from this trial is presented in Table 4.20.

Computations:

$$C = \frac{(2073)^2}{36} = 119{,}370.25$$

$$\text{SSTot} = \sum_i \sum_j \sum_k \sum_l y_{ijkl}^2 - C = 121{,}307.00 - C = 1936.75$$

$$\text{SSR} = \frac{1}{12}(735^2 + 682^2 + 656^2) - C = 270.17$$

$$\text{SSV} = \frac{1}{18}(984^2 + 1089^2) - C = 306.25$$

$$\text{SSP} = \frac{1}{12}(675^2 + 699^2 + 699^2) - C = 32.00$$

$$\text{SSVP} = \frac{1}{6}(318^2 + \cdots + 361^2) - C - \text{SSV} - \text{SSP} = 18.67$$

$$\text{SSW} = \frac{1}{18}(1026^2 + 1047^2) - C = 12.25$$

I

$V_1P_2W_1$ 60	$V_2P_1W_1$ 65
$V_1P_3W_2$ 66	$V_1P_3W_1$ 59
$V_2P_2W_1$ 60	$V_1P_1W_2$ 56
$V_1P_2W_2$ 60	$V_2P_3W_2$ 73
$V_2P_2W_2$ 62	$V_2P_3W_1$ 53
$V_1P_1W_1$ 61	$V_2P_1W_2$ 60

II

$V_2P_3W_2$ 57	$V_2P_1W_1$ 61
$V_2P_2W_2$ 68	$V_1P_1W_2$ 45
$V_2P_3W_1$ 53	$V_1P_3W_1$ 55
$V_2P_1W_2$ 61	$V_1P_3W_2$ 57
$V_1P_1W_1$ 58	$V_1P_2W_2$ 50
$V_2P_2W_1$ 58	$V_1P_2W_1$ 59

III

$V_1P_1W_1$ 55	$V_1P_3W_1$ 51
$V_2P_1W_2$ 50	$V_1P_1W_2$ 43
$V_2P_3W_2$ 77	$V_2P_3W_1$ 48
$V_2P_2W_1$ 56	$V_1P_2W_1$ 54
$V_2P_1W_1$ 60	$V_1P_2W_2$ 45
$V_2P_2W_2$ 67	$V_1P_3W_2$ 50

Figure 4.4 Randomization and seed yields (kg/plot) from a variety × phosphorus × herbicide trial on chick-peas.

Table 4.15 **One-Way Table of Block Totals**

Block	R_1	R_2	R_3	Sum
Sum	735	682	656	2073

Table 4.16 **Three-Way Table of VPW Totals**

Variety	Phosphorus	Herbicide W_1	W_2	Sum
V_1	P_1	174	144	318
	P_2	173	155	328
	P_3	165	173	338
Sum		512	472	984
V_2	P_1	186	171	357
	P_2	174	197	371
	P_3	154	207	361
Sum		514	575	1089
Sum		1026	1047	2073

Table 4.17 **Two-Way Table of VP Totals**

Variety	Phosphorus P_1	P_2	P_3	Sum
V_1	318	328	338	984
V_2	357	371	361	1089
Sum	675	699	699	2073

Table 4.18 **Two-Way Table of VW Totals**

Variety	Herbicide W_1	W_2	Sum
V_1	512	472	984
V_2	514	575	1089
Sum	1026	1047	2073

Table 4.19 **Two-Way Table of PW Totals**

Herbicide	Phosphorus P_1	P_2	P_3	Sum
W_1	360	347	319	1026
W_2	315	352	380	1047
Sum	675	699	699	2073

$$\text{SSVW} = \frac{1}{9}(512^2 + \cdots + 575^2) - \text{C} - \text{SSV} - \text{SSW} = 283.36$$

$$\text{SSPW} = \frac{1}{6}(360^2 + \cdots + 380^2) - \text{C} - \text{SSP} - \text{SSW} = 468.67$$

$$\text{SSVPW} = \frac{1}{3}(174^2 + \cdots + 207^2) - \text{C} - \text{SSV} - \text{SSP} - \text{SSVP} - \text{SSW} - \text{SSVW} - \text{SSPW} = 44.22$$

$$\text{SSE} = 1936.75 - 270.17 - 306.26 - 32.00 - 18.67 - 12.25 - 283.36 - 468.67 - 44.22 = 501.16$$

Table 4.20 **Analysis of Variance of Seed Yield Data from a Variety × Phosphorus by Herbicide Trial on Chick-Peas**[a]

ANOVA

Source	df	SS	MS	F
Total	35	1,936.75		
Block	2	270.17	135.08	5.93
Variety(V)	1	306.25	306.25	13.44 **
Phosphorus(P)	2	32.00	16.00	.70 NS
VxP	2	18.67	9.33	.41 NS
Herbicide(W)	1	12.25	12.25	.54 NS
VxW	1	283.36	283.36	12.44 **
PxW	2	468.67	234.33	10.29 **
VxPxW	2	44.22	22.11	.97 NS
Error	22	501.16	22.78	

[a]CV = $(\sqrt{22.78}/57.58)100 = 8.3\%$.
**Significant at the 1% level; NS, not significant.

Significance tests:

$$F_{0.01\,(1,22)} = 7.94 \qquad F_{0.05(2,22)} = 5.72$$

Sample F values for variety, variety × herbicide, and phosphorus × herbicide exceed the 1% F values. These effects are highly significant. None of the F values for the other effects exceed the 5% F. These effects are not significant. (*Note*: Since the VW interaction is highly significant, the test of significance for variety has no meaning.)

Report of Statistical Analysis

A trial was conducted to determine the effect of three levels of phosphorus fertilization on the seed yield of two varieties of chick-peas in the presence and absence of chemical weed control. The meaningful results are summarized in Tables 4.21 and 4.22. In Table 4.21 it is seen that the effect of herbicide depended on the variety. The addition of herbicide reduced the

Table 4.21 **Mean Seed Yield (kg/plot) from Two Varieties of Chick-Peas With and Without Herbicide**[a]

	Herbicide	
Variety	none	some
V_1	56.89	52.44
V_2	57.11	63.89

[a]Standard error = 1.59.

Table 4.22 **Mean Seed Yield (kg/plot) of Chick-Peas at Three Levels of Phosphorus Fertilization With and Without Herbicide**[a]

	Phosphorus		
Herbicide	none	25 kg/ha	50 kg/ha
none	60.00	57.83	53.17
some	52.50	58.67	63.33

[a]Standard error = 1.95.

yield for variety 1, whereas the yield of variety 2 was increased by the use of herbicide.

The response of chick-peas to added phosphorus depended on whether or not herbicide was used. If chemical weed control was not practiced, seed yield was reduced as the amount of added phosphorus was increased. On the other hand, when herbicide was applied, the seed yield increased as the rate of P fertilization was increased.

4.5.6 More than Three Factors

With more than three factors the basic principles of analysis that we have developed to this point still apply. The only difference is that we have more main effects and more interactions to estimate and test. Because of this the interpretation often becomes more complicated as additional factors are included in the experiment. However, since the basic analytical procedure remains the same, we will not consider higher-order factorial experiments at this point.

4.6 CONTRASTS

In agronomy trials we usually select our treatments to answer certain questions posed by the objectives of the experiment. To study the effect of irrigation on yield, for example, we might choose to vary the quantity of water applied and the time of application. The questions we would like to answer involve the effect of quantity and time and the interaction between the two. Often, we can find answers to these questions by looking at differences between means of groups of treatments grouped in different ways.

For example, suppose that we had conducted an irrigation experiment with the following treatments:

1. One hectare centimeter (ha-cm) of water applied early, μ_1
2. One ha-cm of water applied late, μ_2
3. Two ha-cm of water applied early, μ_3
4. Two ha-cm of water applied late, μ_4

The treatments themselves suggest some natural questions to be asked of the data:

1. Is there a yield difference between 2 ha-cm and 1 ha-cm of applied water?
2. Does it make a difference whether the water is applied early or late?
3. Does the difference between 1 and 2 ha-cm depend on whether the water is applied early or late?

We can answer these questions by looking at differences between means of the treatments grouped in different ways. For the first question we would want to compare

$\frac{1}{2}(\mu_1 + \mu_2)$	with	$\frac{1}{2}(\mu_3 + \mu_4)$
1 ha-cm		2 ha-cm

for the second question we would compare

$\frac{1}{2}(\mu_1 + \mu_3)$	with	$\frac{1}{2}(\mu_2 + \mu_4)$
early		late

and for the third the comparison would be

$(\mu_1 - \mu_2)$	with	$(\mu_3 - \mu_4)$
early vs. late		early vs. late
1 ha-cm		2 ha-cm

In each case we want to test the hypothesis that the difference between the grouped means is zero or, alternatively, that the means of the two groups are equal. For 1 inch vs. 2 inches we want to test:

1. H_0: $\frac{1}{2}(\mu_1 + \mu_2) = \frac{1}{2}(\mu_3 + \mu_4)$ or H_0: $\frac{1}{2}(\mu_1 + \mu_2) - \frac{1}{2}(\mu_3 + \mu_4) = 0$. Similarly,
2. H_0: $\frac{1}{2}(\mu_1 + \mu_3) = \frac{1}{2}(\mu_2 + \mu_4)$ or H_0: $\frac{1}{2}(\mu_1 + \mu_3) - \frac{1}{2}(\mu_2 + \mu_4) = 0$.
3. H_0: $(\mu_1 - \mu_2) = (\mu_3 - \mu_4)$ or H_0: $(\mu_1 - \mu_2) - (\mu_3 - \mu_4) = 0$.

4.6.1 Contrasts of Means

A systematic way of forming and testing hypotheses involving the comparison of means of grouped treatments is provided by the concept of contrasts among means. Contrasts can be considered to be particular linear functions of the means. Suppose that

$$L = k_1\mu_1 + k_2\mu_2 + \cdots + k_p\mu_p$$

is a linear function of a set of p means. The function, L, is determined by the values assigned to the coefficients, k_j. L is said to be a *contrast* of means if

$$\sum_j k_j = 0$$

provided that at least one $k_j \neq 0$. For example,

$$\begin{aligned} L &= \tfrac{1}{2}(\mu_1 + \mu_2) - \tfrac{1}{2}(\mu_3 + \mu_4) \\ &= \tfrac{1}{2}\mu_1 + \tfrac{1}{2}\mu_2 - \tfrac{1}{2}\mu_3 - \tfrac{1}{2}\mu_4 \end{aligned}$$

is a contrast because

$$\tfrac{1}{2} + \tfrac{1}{2} + (-\tfrac{1}{2}) + (-\tfrac{1}{2}) = 0$$

To use the concept of contrasts in hypothesis testing, we frame the null hypothesis to be tested in the form of a contrast and then test the hypothesis that the contrast is zero against the alternative that it is not. We have

$$H_0\text{: } L = 0 \quad \text{or} \quad H_0\text{: } k_1\mu_1 + k_2\mu_2 + \cdots + k_p\mu_p = 0$$

against

$$H_a\text{: } L \neq 0 \quad \text{or} \quad H_a\text{: } k_1\mu_1 + k_2\mu_2 + \cdots + k_p\mu_p \neq 0$$

For the purpose of making the test we require the sample estimate, say l, of L. This is obtained by substituting the sample value $\bar{y}_j$ for μ_j in L. That is,

$$L \doteq l = k_1\bar{y}_1 + k_2\bar{y}_2 + \cdots + k_p\bar{y}_p$$

Then, under the condition that the samples from the various treatments are independent, a property of balanced experimental designs, the variance, σ^2_l, of l is given by

$$\sigma_l^2 = \left(\frac{k_1^2}{n_1} + \frac{k_2^2}{n_2} + \cdots + \frac{k_p^2}{n_p}\right)\sigma^2$$

where n_j is the number of observations on the j-th treatment. The sample variance of l, V(l), is obtained by replacing σ^2 by its sample estimate:

$$V(l) = \left(\frac{k_1^2}{n_1} + \frac{k_2^2}{n_2} + \cdots + \frac{k_p^2}{n_p}\right) \text{MSE}$$

Note that when $n_1 = n_2 = \cdots = n_p = n$, equal replication

$$V(l) = \frac{1}{n}(\sum_j k_j^2)\ \text{MSE}$$

Under the normality assumptions about the experimental errors we can use the t distribution to construct an interval estimate of the contrast. The $(1 - \alpha)$ 100 % confidence interval estimate, L(L), of a contrast, L, is given by

$$L(L) = l \pm t_\alpha \sqrt{V(l)}$$

where, again, t_α is the α-level two-tailed t with the df as for MSE.

It is possible to form more than one meaningful contrast with a given set of treatments. In fact, with p treatments it is possible to form $p - 1$ contrasts that are statistically independent of each other. Now, for the $p - 1$ contrasts to be statistically independent, they must be mutually orthogonal. Hence we require the following definition: Two contrasts,

$$L_1 = k_{11}\mu_1 + k_{12}\mu_2 + \cdots + k_{1p}\mu_p$$

and

$$L_2 = k_{21}\mu_1 + k_{22}\mu_2 + \cdots + k_{2p}\mu_p$$

are *orthogonal* if, and only if,

$$\sum_j k_{1j}k_{2j} = 0$$

With p treatments it is always possible, though not always easy, to find $p - 1$ mutually orthogonal contrasts even though they may not all have meaning within the context of the experimental objectives. Further, more than one set of mutually orthogonal contrasts may be found among the p treatments, but each set can contain no more than p − 1 contrasts.

Examples When constructing contrasts it is usually easiest to reduce the coefficients, k_j, to integers. This does not affect the contrasts but it does make them easier to work with. Consider the comparison

$$\tfrac{1}{2}(\mu_1 + \mu_2) = \tfrac{1}{2}(\mu_3 + \mu_4)$$

This comparison may be expressed as the contrast

$$L = \tfrac{1}{2}(\mu_1 + \mu_2) - \tfrac{1}{2}(\mu_3 + \mu_4)$$
$$L = \tfrac{1}{2}\mu_1 + \tfrac{1}{2}\mu_2 - \tfrac{1}{2}\mu_3 - \tfrac{1}{2}\mu_4$$
$$= \mu_1 + \mu_2 - \mu_3 - \mu_4$$

Now, consider a two-factor factorial experiment with factors P and K, each at two levels. We might have the following treatments:

(1) = no P, no K
P = 20 kg/ha P
K = 20 kg/ha K
PK = 20 kg/ha P + 20 kg/ha K

We can construct a couple of mutually orthogonal contrasts among these four treatments. It is convenient to display the contrasts in tabular form with treatments as column headings, contrast descriptions as row designators, and coefficients, k_{ij}, of the contrasts in the body of the table. One set of orthogonal contrasts for the 2 × 2 PK factorial experiment is shown in Table 4.23. Note that for the table:

1. Each linear function is a contrast because $\sum_j k_{ij} = 0$ for each.

Table 4.23 **Coefficients Defining a Set of Orthogonal Contrasts Among Four Treatments**

	Treatment			
Contrast	(1)	P	K	PK
P vs K	0	+1	−1	0
Alone vs together	0	−1	−1	+2
None vs some	−3	+1	+1	+1

2. They are mutually orthogonal because $\Sigma_j k_{ij}k_{i'j} = 0$ for every possible pair of contrasts.
3. Plus and minus signs may be interchanged without changing the contrast.

A different set of three contrasts for the same set of treatments would be those displayed in Table 4.24. Note that for this table:

1. P compares treatments with P against treatments without P. This is the main effect of P.
2. K compares treatments with K against those without K. This is the K main effect.
3. PK compares the difference between P and no P in the presence and absence of K. This is the PK interaction.

Because contrasts are so useful in examining data from agronomy experiments, it might be worthwhile to look at another example of orthogonal contrasts. Consider the following fertilizer treatments:

(1) = none
PB = banded phosphate
PS = broadcast phosphate
NPB = nitrogen + banded P
NPS = nitrogen + broadcast P

One meaningful set of mutually orthogonal contrasts among these treatments would include those which answered the following questions, among others:

1. Is there a difference between fertilizer and none?

Table 4.24 **Coefficients Defining a Different Set of Orthogonal Contrasts Among Four Treatments**

	Treatment			
Contrast	(1)	P	K	PK
P	−1	+1	−1	+1
K	−1	−1	+1	+1
PK	+1	−1	−1	+1

2. Does the method by which P is applied make a difference?
3. Does added N make a difference?
4. Does the effect of N depend on the method of P application?

These questions are addressed by the contrasts shown in Table 4.25.

4.6.2 Contrasts of Totals

We could illustrate the computation and testing of contrasts among means with numerical examples. We will not do this because often it is easier to construct and test contrasts within the context of the analysis of variance. To do this we partition the treatment sum of squares into a number of components, each associated with a contrast. For this purpose we work with contrasts of totals instead of means, and we present a set of rules for constructing and testing these contrasts.

For designed experiments we make the following assumptions:

1. Treatments are equally replicated.
2. p = number of treatments.
3. r = number of replications per treatment.
4. T_j = sum of yields on the j-th treatment.
5. $SST = (1/r) \sum_j T_j^2 - (\sum_j T_j)^2/rp$ = treatment SS.

Under these assumptions the following rules apply:

1. $L_1 = k_{11}T_1 + k_{12}T_2 + \cdots + k_{1p}T_p$ is a contrast of totals if $\sum_j k_{1j} = 0$.
2. The sample variance, V(L), of a contrast of totals is

$$V(L) = (r \sum_j k_{ij}^2)s^2 = (r \sum_j k_{ij}^2)MSE$$

Table 4.25 **Coefficients Defining a Meaningful Set of Orthogonal Contrasts Among Five Treatments**

Contrast	Treatment (1)	PB	PS	NPB	NPS
None vs some	−4	+1	+1	+1	+1
N vs no N	0	−1	−1	+1	+1
Band vs bdcst P	0	+1	−1	+1	−1
N x (B vs S)	0	−1	+1	+1	−1

3. If L_1 and L_2 are contrasts of totals and if $\sum_j k_{1j}k_{2j} = 0$, then L_1 and L_2 are orthogonal.
4. If L_1 is a contrast of totals, then

$$S_1^2 = \frac{(\sum_j k_{1j}T_j)^2}{r\sum_j k_{1j}^2} = \frac{L_1^2}{r\sum_j k_{1j}^2}$$

is a single-degree-of-freedom component of the treatment sum of squares, SST.
5. If L_1 and L_2 are orthogonal, then

$$S_2 = \frac{L_2^2}{r\sum_j k_{2j}^2}$$

is a component of SST $-$ S_1^2.
6. If $L_1, L_2, \ldots, L_{p-1}$ are mutually orthogonal, then

$$S_1^2 + S_2^2 + \cdots + S_{p-1}^2 = \text{SST}$$

7. If $T_1, T_2, \ldots, T_q$ are treatment totals for a set of $q < p$ treatments, then

$$S_{q-1}^2 = \frac{1}{r}\sum_{j=1}^{q} T_j^2 - \frac{(\sum_{j=1}^{q} T_j)^2}{rq}$$

is a component of SST with $q - 1$ df.

By the preceding rules it is possible, but not always easy, to partition SST into $p - 1$ orthogonal components each representing one of the $p - 1$ degrees of freedom for treatments. Further, these $p - 1$ component sums of squares will add to SST. Such a partition is not unique. There are a number of sets of mutually orthogonal partitions, none of which contain more than $p - 1$ components.

It is not always necessary to decompose SST completely into $p - 1$ mutually orthogonal components. Sometimes only a subset is of interest. In this case we can break out the contrasts of interest and as long as they are mutually orthogonal, their sums of squares will constitute a proper subset of SST.

If we construct orthogonal contrasts of totals, we can compute a single-degree-of-freedom sum of squares, S_i^2, for each contrast. Then $F = S_i^2/\text{MSE}$, with 1, and dfE degrees of freedom, is a test statistic for

$$H_o\colon L_i = k_{i1}\mu_1 + k_{i2}\mu_2 + \cdots + k_{ip}\mu_p = 0$$

against

Table 4.26 **Partition of the Treatment Source of Variation for a 2 × 3 Factorial Set of Treatments**

Source	df	SS
A	1	SSA
B	2	SSB
AB	2	SSAB

$$H_a: L_i = k_{i1}\mu_1 + k_{i2}\mu_2 + \cdots + k_{ip}\mu_p \neq 0$$

4.6.3 Factorial Contrasts

Consider a 2 × 3 factorial set of treatments with factor A at two levels, a_1 and a_2, and B at three levels, b_1, b_2, and b_3. The ANOVA for treatments would be as shown in Table 4.26.

Ordinarily, we would compute the sums of squares from a two-way table of totals. An alternative but equivalent way to compute the SS is by means of contrasts of totals. Consider the set of orthogonal contrasts among the six treatment combinations in the factorial set that is displayed in Table 4.27. The contrasts in this table are interpreted in the following way:

Table 4.27 **Orthogonal Contrasts Among the Treatments in a 2 × 3 Factorial Set**

	Treatment						
Contrast	a_1b_1	a_1b_2	a_1b_3	a_2b_1	a_2b_2	a_2b_3	s_i^2
A	−1	−1	−1	+1	+1	+1	s_1^2
B1	0	−1	+1	0	−1	+1	s_2^2
B2	−2	+1	+1	−2	+1	+1	s_3^2
AB1	0	+1	−1	0	−1	+1	s_4^2
AB2	+2	−1	−1	−2	+1	+1	s_5^2

1. A compares all treatments with a_1 against those with a_2. This is the main effect of factor A. $S_1^2 = SSA$.
2. B1 and B2 are two orthogonal contrasts among the levels of factor B. These two sums of squares, each with 1 df, add to the sum of squares for the main effect of B:

   $S_2^2 + S_3^2 = SSB$

   Note that B1 and B2 are both orthogonal to A.
3. AB1 and AB2 are two orthogonal contrasts which measure the A by B interaction:

   $S_4^2 + S_5^2 = SSAB$

   (To get the coefficients for AB1 and AB2, we use, in turn, the product of the coefficients of A with those of B1 and B2. That is, $A \times B1 = AB1$ and $A \times B2 = AB2$)
4. We have constructed five mutually orthogonal contrasts among the six treatment totals. The sum of the five individual sums of squares is the treatment sum of squares. Further, the individual sums of squares can be combined to obtain the sums of squares for the main effects and the interaction.
5. Sums of squares for these contrasts are computed in just the same way as for any other contrast:

$$S_i^2 = \frac{(\Sigma^j k_{ij}\ T_j)^2}{r\ \Sigma_j\ k_{ij}^2} = \frac{L_i^2}{r\ \Sigma_j\ k_{ij}^2}$$

If we compute orthogonal contrasts of totals, we can compute a single-degree-of-freedom sum of squares, S_i^2, for each contrast. Then $F = S_i^2/MSE$, with 1, and dfE degrees of freedom, is a test statistic for

$$H_0\text{: } L_i = k_{i1}\mu_1\ + K_{i2}u_2 + \cdots\ k_{ip}\mu_p\ 0$$

against

$$H_a\text{: } L_i = k_{i1}\mu_1 + k_{i1}\mu_2 + \cdots + k_{ip}\mu_p \neq 0$$

4.6.4 Illustrations

Suppose that we had conducted the P by K experiment using three replications per treatment and observed the treatment totals shown in Table 4.28. One set of the orthogonal contrasts of totals for these treatments would be

Table 4.28 **Treatment Totals from a P × K Factorial Experiment**[a]

Treatment	(1)	P	K	PK
Total (3 reps) kg/plot	36	48	42	51

[a]SST = 44.25, 3 df.

those shown in Table 4.29. Note that in this table

$$L_i = \sum_j k_{ij} T_j$$

$$S_i^2 = \frac{L_i^2}{r \sum_j k_{ij}^2} = \frac{L_i^2}{3 \sum_j k_{ij}^2}$$

for example,

$$L_2 = (0)(36) + (-1)(48) + (-1)(42) + (2)(51) = 12$$

$$S_2^2 = \frac{(12)^2}{3[0^2 + (-1)^2 + (-1)^2 + 2^2]} = 8.00$$

A different set of orthogonal contrasts for the set of treatments in Table 4.28 is given in Table 4.30.

Note: The individual sums of square in each partition are different even though both sets are mutually orthogonal. However, the individual sums of squares add up to the treatment sum of squares in each case.

Table 4.29 **Orthogonal Contrasts Among Treatments in a P × K Factorial Experiment**

Contrast \ Treatment (Sum)	(1) 36	P 48	K 42	PK 51	L	s^2
P vs K	0	+1	−1	0	6	6.00
Alone vs together	0	−1	−1	+2	12	8.00
None vs some	−3	+1	+1	+1	33	30.25
					Sum	44.25

Table 4.30 **Alternative Set of Orthogonal Contrasts for a P × K Factorial Set of Treatments**

Contrast \ Treatment	(1)	P	K	PK		
Sum	36	48	42	51	L	S
P	−1	+1	−1	+1	21	36.75
K	−1	−1	+1	+1	9	6.75
PxK	+1	−1	−1	+1	−3	.75
					Sum	44.25

Example A weed scientist wanted to study the effect of a new, all-purpose herbicide to control grassy weeds in lentils. He decided to try the herbicide both as a preemergent and as a postemergent spray. As a standard against which to compare the herbicide, he decided to use hand weeding. He was also concerned about the effect of the addition of phosphorus fertilizer on the action of the weed control methods. Finally, he wanted to include a control, with neither weed control nor fertilizer, as a base treatment against which to judge the others. He selected the following treatments:

1. Control (no weed control, no P) (1)
2. Hand weeding W1
3. Preemergent spray W2
4. Postemergent spray W3
5. Hand weeding + 25 kg/ha P PW1
6. Preemergent spray + P PW2
7. Postemergent spray + P PW3

In an effort to control some of the soil variability in the experimental area the trial was run using a randomized block design with three blocks of seven plots each, blocked on soil variability. At harvest the weight of dry lentils on each plot was determined. The data may be summarzied as in Table 4.31. A preliminary analysis of variance of the yield data is displayed in Table 4.32.

$F_{0.01(6,\ 12)} = 4.82 < 117.30$. Clearly, the differences among treatment means are highly significant. We should examine these differences in greater detail. One way to do this is through the concept of contrasts among the treatment totals. One set of meaningful mutually orthogonal contrasts is

Table 4.31 **Yield (kg/plot) of Dry Lentils from a Weed Control Experiment in the Presence and Absence of Phosphorus**

Treatment	Block I	Block II	Block III	Sum
(1)	218	180	192	590
W1	357	353	345	1055
W2	325	311	297	933
W3	321	297	291	909
PW1	462	458	399	1319
PW2	407	409	381	1197
PW3	410	392	362	1164
Sum	2500	2400	2267	7167

Table 4.32 **Analysis of Variance of Dry Seed Yields (kg/plot) from a Weed Control × Phosphorus Trial on Lentils**

ANOVA

Source	df	SS	MS	F
Total	20	121,670.29		
Block	2	3,903.72	1,951.86	11.86
Treatment	6	115,792.29	19,298.72	117.30**
Error	12	1,974.28	164.52	

**Significant at 1%.

Table 4.33 **Orthogonal Contrasts Among Weed Control and P Treatment Totals for Lentils**[a]

	Treatment								
	(1)	W1	W2	W3	PW1	PW2	PW3		
Contrast	590	1055	933	909	1319	1197	1164	L_i	S_i^2
1.	−6	+	+	+	+	+	+	3037	73201.34
2.	0	−	−	−	+	+	+	783	34060.50
3.	0	+2	−	−	+2	−	−	545	8250.70
4.	0	0	−	+	0	−	+	−57	270.75
5.	0	−2	+	+	+2	−	−	9	2.25
6.	0	0	+	−	0	−	+	−9	6.75
								Sum	115792.29

[a]1, Check vs. others; 2, P vs. no P; 3, Hand weed vs. herbicide; 4, pre- vs. postemergence; 5, P × (hand vs. herb); 6, P × (pre- vs. post-); $L_i = \sum_j k_{ij}T_j$; $S_i^2 = L_i^2/3 \sum_j k_{ij}^2$.

presented in Table 4.33. The final analysis of variance is displayed in Table 4.34.

Report of Statistical Analysis

An experiment was run to determine the effect of three methods of weed control on the dry seed yield of lentils in the presence and absence of phosphorus fertilizer. The mean yields are summarized in Table 4.35.

Detailed analysis of the results reveals that, as expected, the treated plots outyielded the untreated check plots. Further, plots with 25 kg/ha added P outyielded the unfertilized plots by an average of about 87 kg/plot. Also, as expected, hand weeding resulted in a higher yield than did herbicide spray, no matter when applied. The difference was about 45 kg/plot in favor of hand weeding. There was no real difference in effect between pre- and postemergence application, nor did the differences among weed control treatments depend on the presence or absence of phosphorus fertilizer.

4.7 REGRESSION COMPONENTS OF THE TREATMENT SUM OF SQUARES

In agronomic experiments the treatments frequently consist of different levels of one or more quantitative variables. Some examples are:

Table 4.34 **Final Analysis of Variance of Lentil Yields**

ANOVA

Source	df	SS	MS	F
Total	20	121,670.29		
Block	2	3,903.72	1,951.86	
Treatment	6	115,792.29	19,298.72	117.30**
ck vs others	1	73,201.34	73,201.34	444.94**
P vs no P	1	34,060.50	34,060.50	207.03**
Hand. vs herb.	1	8,250.70	8,250.70	50.15**
pre- vs post-	1	270.75	270.75	1.64NS
PW interaction	2	9.00	4.50	.03NS
Error	12	1,974.28	164.52	

**Significant at 1%; NS, not significant.

Table 4.35 **Mean Seed Yields (kg/plot) of Lentils Under Different Weed Control and P Fertilizer Treatments**[a]

		no P			25 kg/ha P		
Treatment	Check	Hand	Pre-	Post-	Hand	Pre-	Post-
Mean	196.67	351.67	311.00	303.00	439.67	399.00	388.00

[a]Standard error = 7.40.

1. Varying amounts of a new fertilizer
2. Several different row spacings
3. Two or more depths of seeding

In these experiments the primary interest is in the way the plants respond to changes in the treatment level. In this case it would seem worthwhile to try to develop an equation, or curve, to describe the relationship between plant response and treatment level.

To illustrate the ideas involved, suppose that we have an experiment in which the treatments consist of four different amounts of nitrogen fertilizer applied to wheat. Suppose, further, that we assign each level of fertilizer to each of five randomly chosen wheat plots (completely randomized design, CRD, with $p = 4$, $r = 5$). At the conclusion of the trial a graph of mean wheat yield against amount of applied nitrogen might appear as in Figure 4.5.

The usual analysis of variance of data from this trial would provide a treatment mean square and an F ratio for testing the significance of the differences among the treatment means. It would not, however, provide much information about the way wheat yield changes as the amount of applied nitrogen is changed.

We could get some insight into the nature of the response if we approached the analysis of the results of the trial as a regression problem.

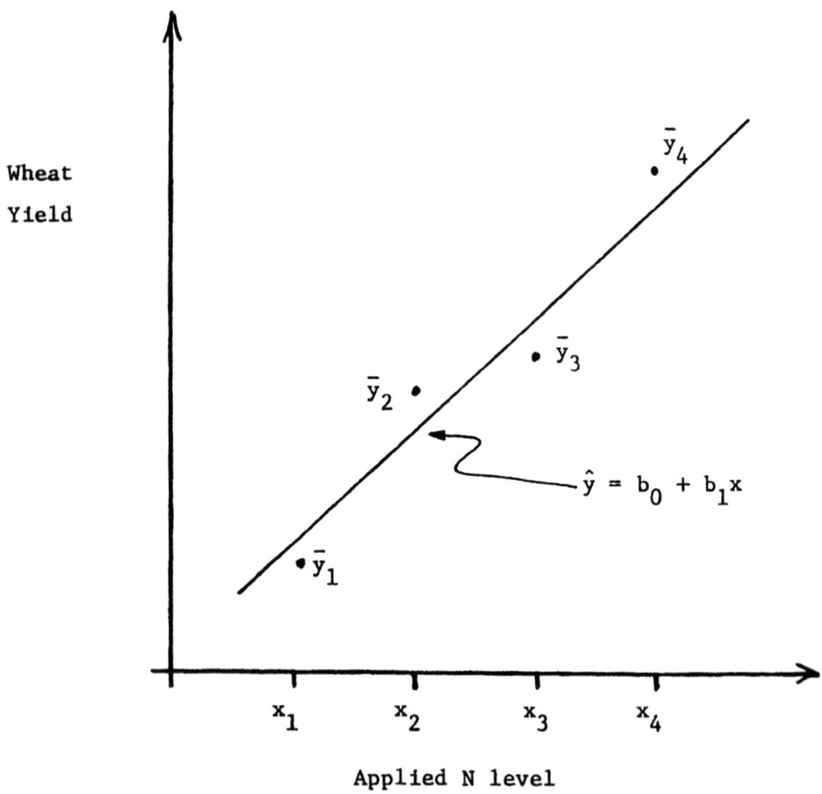

Figure 4.5 Relationship between mean wheat yield (y) and level of applied nitrogen (x).

With such an approach we could treat the level of applied nitrogen as an independent variable, x, and the yield of wheat as a dependent variable, y. We could then fit the model

$$y = \beta_0 + \beta_1 x + \epsilon$$

where

y = wheat yield
x = nitrogen level
β_0 = yield with no applied N
β_1 = change in yield per unit of applied N
ϵ = random error

With regression analysis we could test

$$H_0: \beta_1 = 0$$

to determine whether or not wheat yield is linearly related to the amount of applied N.

Ideally, we would like to perform this test within the framework of the analysis of variance. To do this we must partition the treatment sum of squares, SST, into two components:

1. SSR = Sum of squares associated with the linear regression of y on x.
2. SSL = sum of squares for the failure of the regression model to describe the relationship between y and x (lack of fit).

The analysis of variance with the treatment sum of squares partitioned in this way ($p = 4$, $r = 5$, CRD) takes the form shown in Table 4.36. The basic computations in the table are done in the usual way:

$$\text{SSTot} = \sum_i \sum_j y_{ij}^2 - \frac{(\Sigma_i \Sigma_j y_{ij})^2}{rp}$$

$$\text{SST} = \frac{1}{r} \sum_j T_j^2 - \frac{(\Sigma_i \Sigma_j y_{ij})^2}{rp}$$

$$\text{SSE} = \text{SSTot} - \text{SST}$$

4.7.1 Partitioning the Treatment Sum of Squares

The regression sum of squares, SSR, can be computed in two ways:

1. Treat the data as an ordinary regression problem and compute SSR in the usual way.
2. Find a set of coefficients that define a linear contrast among the treatment totals and compute S^2 for this contrast.

Table 4.36 **Analysis of Variance with Treatment Sum of Squares Partitioned for Regression Analysis**

ANOVA

Source	df		SS	MS	F
Total	19		SSTot		
Treatment	3		SST	MST	F_T
Linear Reg.		1	SSR	MSR	F_R
Lack of Fit		2	SSL	MSL	F_L
Error	16		SSE	MSE	

To form a linear contrast we can use the deviations of the treatment levels from the mean level of all treatments as the coefficients (k_j) of a contrast of totals. With this approach, if x_j is the level of the j-th treatment and $\bar{x}$ is the mean of all treatment levels, then

$$k_j = x_j - \bar{x}$$

and the linear contrast, L_R, is

$$L_R = \sum_j k_j T_j = \sum_j (x_j - \bar{x})T_j$$

[*Note*: $\Sigma_j (x_j - \bar{x}) = 0$, so that L_R is a contrast.] In the illustration we have four levels of N, and

$$L_R = (x_1 - \bar{x})T_1 + (x_2 - \bar{x})T_2 + (x_3 - \bar{x})T_3 + (x_4 - \bar{x})T_4$$

Further,

$$SSR = \frac{L_R^2}{r \Sigma_j (x_j - \bar{x})^2}$$

Having computed SSR, we have

$$SSL = SST - SSR$$

(*Note*: SSE corresponds to the sum of squares for pure error in the usual regression analysis.)

The mean squares in Table 4.36 are formed by dividing the sums of squares by their corresponding degrees of freedom. All of the F ratios have MSE as a denominator. These F ratios serve as test statistics for the following hypotheses:

1. F_T = MST/MSE tests:

   H_0: $\mu_1 = \mu_2 = \cdots = \mu_p$

   against

   H_a: $\mu_1 \neq \mu_2 \neq \cdots \neq \mu_p$

   This is the usual test of the significance of differences among the treatment means.
2. F_R = MSR/MSE is a test statistic for

   H_0: $\beta_1 = 0$

   against

   H_a: $\beta_1 \neq 0$

   This is a test of the hypothesis that there is no linear relationship between x and y against the alternative that a linear relation exists.
3. F_L = MSL/MSE is a test of

   H_0: $E(y) = \beta_0 + \beta_1 x$

   against

   H_a: $E(y) \neq \beta_0 + \beta_1 x$

   The hypothesis is that the simple linear regression model describes the data against the alternative that it does not.

SSL is the sum of squares of the deviations of the y means from the best-fitting regression line. The greater the deviations, the larger will be SSL and the stronger is the evidence that a simple linear regression equation does not describe the x, y relationship. If F_L is significant, there are two alternatives:

1. Fit a more complex equation.
2. Forget the regression approach.

If F_L is not significant but F_R is, the conclusion is that a linear relationship exists between y and x. This relation is described by the model

$E(y) = \beta_0 + \beta_1 x$ = expected value of y given x and the data are described by the sample equation

$$\hat{y}_j = b_0 + b_1 x_j = \bar{\bar{y}} + b_1(x_j - \bar{x})$$

where

$\bar{\bar{y}} = (\Sigma_j T_j)/rp$ = grand mean of y

x_j = value of x (treatment level) at which y is estimated

$b_1 = L_R/[r\Sigma_j (x_j - \bar{x})^2]$

In the fertilizer illustration we have four (p) levels of nitrogen replicated five (r) times. Then

$$b_1 = \frac{(x_1 - \bar{x})T_1 + (x_2 - \bar{x})T_2 + (x_3 - \bar{x})T_3 + (x_4 - \bar{x})T_4}{5[(x_1 - \bar{x})^2 + (x_2 - \bar{x})^2 + (x_3 - \bar{x})^2 + (x_4 - \bar{x})^2]}$$

The sample equation for the illustration is plotted on Figure 4.5.

4.7.2 Orthogonal Polynomials

When the relationship between plant response and treatment level is a simple linear one, the regression methods of Section 4.7.1 are fairly easy to apply. On the other hand, if the relationship is more complex than linear, the regression process is much more complicated. Under some conditions, however, curve fitting within the ANOVA framework can be simplified to a great extent by the use of orthogonal polynomial coefficients. Conditions necessary to use this approach are:

1. The number of observations at each treatment level must be the same at all levels (equal replication).
2. The levels of the treatment variable must be equally spaced. For example:
   a. 20, 40, 60, 80, 100 kg of fertilizer per hectare.
   b. 10, 20, 30, 40 cm spacing between rows.

When these conditions are met we can use a simple set of multipliers, called *orthogonal polynomial coefficients*, to fit a polynomial equation relating x and y. That is, we can fit models of the form

$$E(y) = \beta_0 + \beta_1 x + \beta_2 x^2 + \beta_3 x^3 + \cdots$$

where

$E(y)$ = expected value of y

x = level of the quantitative treatment variable

β_0 = yield with no treatment

$\beta_1, \beta_2, \beta_3, \ldots$ = change in y per unit change in $x, x^2, x^3, \ldots$

Orthogonal polynomial coefficients for up to eight treatment levels and for polynomial equations of up to the fourth order are provided in Table 4.37.

Table 4.37 **Orthogonal Polynomial Coefficients**[a]

No. of Levels	Order	Ordered Treatment Number 1	2	3	4	5	6	7	8	Divisor D_i	λ
3	1	−1	0	+1						2	1
	2	+1	−2	+1						6	3
4	1	−3	−1	+1	+3					20	2
	2	+1	−1	−1	+1					4	1
	3	−1	+3	−3	+1					20	10/3
5	1	−2	−1	0	+1	+2				10	1
	2	+2	−1	−2	−1	+2				14	1
	3	−1	+2	0	−2	+1				10	5/6
	4	+1	−4	+6	−4	+1				70	35/12
6	1	−5	−3	−1	+1	+3	+5			70	2
	2	+5	−1	−4	−4	−1	+5			84	3/2
	3	−5	+7	+4	−4	−7	+5			180	5/3
	4	+1	−3	+2	+2	−3	+1			28	7/12
7	1	−3	−2	−1	0	+1	+2	+3		28	1
	2	+5	0	−3	−4	−3	0	+5		84	1
	3	−1	+1	+1	0	−1	−1	+1		6	1/16
	4	+3	−7	+1	+6	+1	−7	+3		154	7/12
8	1	−7	−5	−3	−1	+1	+3	+5	+7	168	2
	2	+7	+1	−3	−5	−5	−3	+1	+7	168	1
	3	−7	+5	+7	+3	−3	−7	−5	+7	264	2/3
	4	+7	−13	−3	+9	+9	−3	−13	+7	616	7/12

[a]1. Order 1 = linear, 2 = quadratic, 3 = cubic, 4 = quartic.
2. $D_i = \sum_j k_{ij}^2$; $S_i^2 = (\sum_j k_{ij}T)^2/rD_i = L_i^2/rD_i$.

Other tables of orthogonal polynomial coefficients are available for as many as 100 treatment levels and for polynomials of order higher than 4.

In general, the orthogonal polynomial coefficients are used to define orthogonal contrasts among the treatment totals. In particular, with the coefficients in Table 4.37 we can construct contrasts associated with the linear (first-order), quadratic (second-order), cubic (third-order), and quartic (fourth-order) terms of the polynomial equation relating plant response, y, to treatment level, x. With these contrasts we can follow the usual procedures to extract single-degree-of-freedom sums of squares from the treatment SS, and use an F ratio to test the significance of each contrast. This is equivalent to a test of the hypothesis that the corresponding coefficient in the model is zero.

To illustrate the use of orthogonal polynomials in fitting and testing curves within the framework of the analysis of variance, let

y_{ij} = yield of the i-th replication on the j-th treatment level

p = number of equally spaced quantitative treatments

r = number of replications per treatment

x_j = level of the j-th treatment

$\bar{x}$ = mean level of all treatments

$= (\Sigma_j x_j)/p$

T_j = sum of the y's on the j-th treatment

$= \Sigma_i y_{ij}$

k_{ij} = (i-th)-order polynomial for the j-th treatment (Table 4.37)

Then

$$L_i = \sum_j k_{ij} T_j$$

is the contrast of totals associated with the i-th term of the polynomial, and

$$S_i^2 = \frac{(\Sigma_j k_{ij} T_j)^2}{(r\Sigma_j k_{ij}^2)} = \frac{L_i^2}{rD_i}$$

S_i^2 is the single-degree-of-freedom sum of squares associated with the (i-th)-order term. The ratio

$$F = \frac{S_i^2}{MSE}$$

is the test statistic for testing the hypothesis that the regression coefficient for the (i-th)-order term in the polynomial is zero.

Unless there is prior reason to believe that the equation is of particular order, it is customary to fit the terms of the polynomial sequentially starting with the linear term. If SST is the treatment sum of squares with p − 1 df, the procedure is as follows:

1. Compute S^2_{LIN} = linear sum of squares.
2. Compute "lack of fit," S^2_{LOF}: $S_{LOF} = SST - S^2_{LIN}$, with p − 2 df.
3. Test lack of fit against MSE.
4. Compute S^2_{QUAD} = quadratic sum of squares.
5. Compute lack of fit, S^2_{LOF}: $S^2 = SST - S^2_{LIN} - S^2_{QUAD}$, with p − 3 df.
6. Test lack of fit against MSE.

Continue in this way, adding one term at a time, until lack of fit is no longer significant. The final equation should contain all of the terms up to and including the term at which lack of fit first becomes nonsignificant.

To illustrate the computations involved, suppose that we have an experiment with five equally spaced levels of nitrogen fertilizer: 10, 30, 50, 70, and 90 kg/ha, each replicated four times in a randomized block design (p = 5, r = 4, RBD). Suppose, further, that we want to consider the quartic polynomial

$$E(y) = \beta_0 + \beta_1 x + \beta_2 x^2 + \beta_3 x^3 + \beta_4 x^4$$

The individual terms in the equation would be examined in the following way:

1. Linear

$$L_{LIN} = (-2)T_1 + (-1)T_2 + (0)T_3 + (1)T_4 + (2)T_5$$

$$S^2_{LIN} = \frac{L^2_{LIN}}{(4)(10)}$$

2. Quadratic

$$L_{QUAD} = (2)T_1 + (-1)T_2 + (-2)T_3 + (-1)T_4 + (2)T_5$$

$$S^2_{QUAD} = \frac{L^2_{QUAD}}{(4)(14)}$$

3. Cubic

$$L_{CUB} = (-1)T_1 + (2)T_2 + (0)T_3 + (-2)T_4 + (1)T_5$$

$$S^2_{CUB} = \frac{L^2_{CUB}}{(4)(10)}$$

4. Quartic

$$L_{QUAR} = (1)T_1 + (-4)T_2 + (6)T_3 + (-4)T_4 + (1)T_5$$

$$S^2_{QUAR} = \frac{L^2_{QUAR}}{(4)(70)}$$

With these sums of squares we could test the hypothesis that the individual terms belong in the model using an F test. We have

$$F = \frac{S_i^2}{MSE}$$

as a test statistic for

$$H_0: \beta_i = 0$$

The process of obtaining the polynomial equation that describes the sample data is somewhat complicated. For this purpose we need to define the following symbols in addition to those we have already defined:

$\hat{y}_j$ = estimated yield at the j-th treatment level

$\bar{y} = (\Sigma_j T_j)/rp$ = overall mean yield of y

$\Delta = x_j - x_{j-1}$ = difference between treatment levels

λ_i = (i-th)-order multiplier from the table of orthogonal polynomial coefficients

We first compute

$$b_i = \frac{\Sigma_j k_{ij} T_j}{r\Sigma_j k_{ij}^2} = \frac{L_i}{rD_i}$$

Then

$$a_i = \lambda_i b_i$$

We next compute

$$\xi_{1j} = \frac{x_j - \bar{x}}{\Delta}$$

$$\xi_{2j} = \left(\frac{x_j - \bar{x}}{\Delta}\right)^2 - \frac{p^2 - 1}{12}$$

$$\xi_{3j} = \left(\frac{x_j - \bar{x}}{\Delta}\right)^3 - \frac{(3p^2 - 7)[(x_j - \bar{x})/\Delta]^2}{20}$$

$$\xi_{4j} = \left(\frac{x_j - \bar{x}}{\Delta}\right)^4 - \frac{(3p^2 - 13)[(x_j - \bar{x})/\Delta]^2}{14} + \frac{3(p^2 - 1)(p^2 - 9)}{560}$$

Finally,

$$\hat{y}_j = \bar{y} + a_1\xi_{1j} + a_2\xi_{2j} + a_3\xi_{3j} + a_4\xi_{4j}$$

The process of estimating and testing the coefficients of a polynomial model within the framework of the analysis of variance is illustrated in the numerical example given in Section 4.7.3.

4.7.3. Numerical Example: Regression in the ANOVA

An experiment was conducted in Colorado to determine the effect of different rates of nitrogen (N) fertilizer on the yield of sugarbeet roots. The trial involved five levels of N: 0, 35, 70, 105, and 140 kg/ha. A randomized block experimental design with three blocks of five plots each was used. At harvest time the weight (megagrams/hectare, Mg/ha) was determined for each plot. The resulting yields are shown in Table 4.38. The preliminary analysis of variance of the sugarbeet root yield data is given in Table 4.39.

Since the differences in response to changes in the level of nitrogen fertilization are highly significant, we should examine the treatment source of variation in greater detail. We see that the yield of roots increases as the amount of added nitrogen is increased until the highest rate of fertilization is reached. With this in mind we should try to find, if possible, an equation describing root weight as a function of the amount of N fertilizer applied.

Note that rate of N fertilizer is a quantitative variable, that the levels are equally spaced, a difference of 35 kg/ha between levels, and that the five levels are equally replicated. Under these conditions we can use the orthogonal polynomial coefficients in Table 4.37 to partition SST into components associated with the terms of a polynomial equation relating

Table 4.38 **Yield (Mg/ha) of Sugarbeet Roots from a Nitrogen Fertilizer Trial**

	Fertilizer (kg/ha)					
Block	0	35	70	105	140	Sum
I	9.9	20.3	27.5	31.4	28.1	117.2
II	7.8	22.6	30.3	27.2	25.7	113.6
III	10.7	23.9	29.2	33.4	31.9	129.1
Sum	28.4	66.8	87.0	92.0	85.7	359.9
Mean	9.47	22.27	29.00	30.67	28.57	23.99

Table 4.39 **Preliminary Analysis of Variance of Beet Root Yields (Mg/ha)**

ANOVA

Source	df	SS	MS	F
Total	14	968.2493		
Block	2	26.3213	13.1607	
Nitrogen	4	913.5627	228.3907	64.41**
Error	8	28.3653	3.5457	

**Significant at the 1% level.

the weight of beet roots to the level of applied N. This partition is presented in Table 4.40.

We can now add one term at a time to the polynomial and test for lack of fit after each addition. As soon as lack of fit becomes nonsignificant, we stop. The equation we are looking for contains all of the terms added up to this point. This test is displayed in Table 4.41.

We see that after the quadratic term has been added, lack of fit is no longer significant. We can conclude that for these data the weight of sugarbeet roots

Table 4.40 **Orthogonal Partition of the Sum of Squares for Treatments**[a]

	N level (kg/ha)							
	0	35	70	105	140			
Order Sum	28.4	66.8	87.0	92.0	85.7	L_i	rD_i	S_i^2
Linear	−2	−1	0	+1	+2	139.8	30	651.4680
Quadratic	+2	−1	−2	−1	+2	−104.6	42	260.5038
Cubic	−1	+2	0	−2	+1	6.9	30	1.5870
Quartic	+1	−4	+6	−4	+1	0.9	210	.0039

[a] $L_i = \sum_j k_{ij}T_j$; $rD_i = r\sum_j k_{ij}^2$; $S_i^2 = L_i^2/rD_i$.
$S_1^2 + S_2^2 + S_3^2 + S_4^2 = 651.4680 + 260.5038 + 1.5870 + 0.0039$
$= 913.5627 = SST$

Table 4.41 **Sequential Test of Polynomial Nitrogen Effects**

Source	df	SS	MS	F
(1) Nitrogen	4	913.5627	228.3907	64.41**
(2) Linear	1	651.4680	651.4680	183.73**
Dev. from Lin.	3	262.0947	87.3649	24.64**
(3) Quadratic	1	260.5038	260.5038	73.47**
Dev. from Lin + Quad.	2	1.5909	.7955	.22NS

**Significant at the 1% level; NS, not significant.

is a quadratic function of the amount of applied nitrogen fertilizer. We can derive this quadratic polynomial equation for these data in the following way. Let y be the weight of sugarbeet roots and x the amount of added N. Then

$$\bar{y} = \frac{\sum_j T_j}{rp} = \frac{359.9}{(3)(5)} = 23.99$$

$$b_1 = \frac{L_1}{rD_1} = \frac{139.8}{30} = 4.66$$

$$b_2 = \frac{L_2}{rD_2} = \frac{-104.6}{42} = -2.49$$

$$\bar{x} = \frac{\sum_j x_j}{5} = \frac{350}{5} = 70$$

Note that $\Delta = 35$ and $\lambda_1 = \lambda_2 = 1$. From this information we can compute

$$\begin{aligned}\hat{y}_j &= \bar{y} + a_1\xi_{1j} + a_2\xi_{2j}\\ &= \bar{y} + \frac{\lambda_1 b_1(x_j - \bar{x})}{\Delta} + \lambda_2 b_2[(\frac{x_j - \bar{x}}{\Delta})^2 - \frac{p^2 - 1}{12}]\\ &= 23.99 + \frac{(1)(4.66)(x_j - 70)}{35} + (1)(-2.49)\,[(\frac{x_j - 70}{35})^2 - \frac{(5^2 - 1)}{12}]\\ &= 9.69 + 0.412x_j - 0.002x_j^2\end{aligned}$$

Table 4.42 **Observed and Expected Sugarbeet Root Weights (Mg/ha) from the Application of Nitrogen Fertilizer**

N Fertilizer (kg/ha)	0	35	70	105	140
Observed Mean Yield	9.46	22.27	29.00	30.67	28.57
Expected Mean Yield	9.69	21.66	28.73	30.90	28.17

where $\hat{y}_j$ is the expected root yield and x_j is the amount of applied nitrogen. With this equation we can determine the expected yield of beet roots for any rate of N application between 0 and 140 kg/ha. For the rates of nitrogen used in this study, the expected yields are compared with the actual mean yields in Table 4.42.

Report of Statistical Analysis

A trial was conducted to determine the effect of different rates of nitrogen fertilization on the yield of sugarbeet roots. As shown in Table 4.43, the yield of beet roots was increased by increasing the rate of N fertilization.

Further analysis revealed that the relationship between sugarbeet root yield and the amount of applied N over the range 0 to 140 kg/ha may be described by the equation

$$\hat{y} = 9.69 + 0.412x - 0.002x^2$$

where $\hat{y}$ is the expected yield (Mg/ha) of beet roots and x is the amount (kg/ha) of applied N.

4.8 SPLIT-PLOT DESIGNS

The most common procedure for experiments with factorial sets of treatments is to use a randomized block experimental design and assign the

Table 4.43 **Mean Yield of Sugarbeet Roots at Various Levels of Nitrogen Fertilizer**

N Fertilizer (kg/ha)	0	35	70	105	140	Std. Error
Root Yield (Mg/ha)	9.46	22.27	29.00	30.67	28.57	1.09

treatment combinations at random to plots within the blocks. With some factors, however, the requirement that all combinations be assigned at random can lead to difficulties:

1. An agronomist might be interested in tillage treatments as one factor and nitrogen levels as another. Because tillage treatments require machinery, they require large plots, while nitrogen, which can be applied by hand, can be applied to small plots.
2. Experiments involving planting date as a factor would be easier to conduct if all of the plots to be planted at one time could be grouped together.

These examples illustrate the types of treatments that require different sizes of plots for their efficient application. For this reason it would be convenient to have an experimental design in which the levels of one factor could be applied to large plots while the levels of another factor are applied to small plots. This is possible with split-plot designs or variations of them.

In the split-plot design the levels of one factor are assigned at random to large plots that are usually grouped into blocks in a randomized block design. The large plots are then divided into small plots and the levels of the second factor are applied at random to the small plots within the large plots. The larger units are usually called *whole plots* or *main plots*, while the smaller units are called *split plots* or *subplots*.

For example, the whole-plot factor might be three methods of tillage: T_1, T_2, and T_3, and the subplot factor might be four barley varieties: V_1, V_2, V_3, and V_4. If the experiment were run with whole plots in a randomized block design, the field plan for two replications might appear as in Figure 4.6.

Randomization The levels of the whole-plot factor are randomly assigned to the main plots within the blocks using a separate randomization for each block. The levels of the split-plot factor are randomly assigned to the subplots within the main plots using a separate randomization in each main plot.

Experimental Errors In the split-plot design the whole-plot factor effects are estimated from the large units, while the subplot factor effects and the interaction of the subplot factor with the whole-plot factor are estimated from the small units. Because there are two sizes of plot, there are two experimental errors, one for each size of plot. Generally, the subplot error is smaller than the whole-plot error. The result is that the whole-plot factor is estimated with less precision than is the subplot factor and its interaction with the whole-plot factor. This must be taken into account when deciding to use a split-plot design.

I

T_3	T_1	T_2
V_3	V_4	V_2
V_1	V_1	V_4
V_2	V_3	V_3
V_4	V_2	V_1

II

T_1	T_3	T_2
V_1	V_2	V_3
V_3	V_1	V_4
V_2	V_3	V_1
V_4	V_4	V_2

Figure 4.6 Field plan for two replications of a split-plot design.

4.8.1 Advantages, Disadvantages, and Uses

There are a number of advantages for the split-plot design:

1. It permits the efficient use of some factors that require different sizes of plot for their application.
2. It permits the introduction of new treatments into an experiment that is already in progress.
3. It provides increased precision in the estimation of some of the factorial effects.

There are a couple of disadvantages to this design:

1. The whole-plot factor is estimated with less precision than is the subplot. Hence large differences are often required for significance with the whole-plot factor, while the subplot factor might have significant effects that are too small to be of practical significance.
2. Statistical analysis is more complex than that required for the RBD because different standard errors are required for different comparisons.

There are a number of uses for the split-plot design:

1. In experiments in which one factor must be applied to larger plots than those required for another factor.
2. To introduce new factors into an experiment that is already in progress.

4.8.2 Data Analysis

The treatments in a split-plot design form factorial combinations of the whole plot and subplot factors. Hence the analysis of data from the design follows much the same pattern as that for the ordinary factorial experiment. We will want to estimate and test the appropriate main effects and interactions. The analysis proceeds as follows:

1. Construct tables of totals.
2. Complete an analysis of variance.
3. Perform significance tests.
4. Compute means and standard errors.
5. Interpret the analysis.

In general, suppose that we have factor A at a levels assigned at random to the whole plots and factor B at b levels assigned at random to the subplots. Assume that the experiment is conducted with the whole plots in a randomized block design with r blocks. Let y_{ijk} represent the yield of the j-th level of A, k-th level of B in the i-th block. We require two tables of totals, an A × block table, as shown in Table 4.44, and an A × B table, presented as Table 4.45. The analysis of variance of data from a split-plot design is shown in Table 4.46.

Table 4.44 **Factor A × Block Two-Way Table of Totals**[a]

	Factor A				
Block	1	2	...	a	Sum
1	$T_{11\cdot}$	$T_{12\cdot}$	...	$T_{1a\cdot}$	R_1
2	$T_{21\cdot}$	$T_{22\cdot}$	...	$T_{2a\cdot}$	R_2
.	.	.		.	.
.	.	.		.	.
.	.	.		.	.
r	$T_{r1\cdot}$	$T_{r2\cdot}$	...	$T_{ra\cdot}$	R_r
Sum	A_1	A_2	...	A_a	G

[a]$T_{ij\cdot} = \Sigma_k y_{ijk}$ = sum of the yields for the j-th level of A in the i-th block summed over levels of B
$R_i = \Sigma_j T_{ij\cdot}$ = sum for the i-th block
$A_j = \Sigma_i T_{ij\cdot}$ sum for the j-th level of factor A
$G = \Sigma_i R_i = \Sigma_j A_j$ = grand total of all yields

Table 4.45 **Factor A × Factor B Two-Way Table of Totals**[a]

Factor A	Factor B				Sum
	1	2	...	b	
1	$T_{\cdot 11}$	$T_{\cdot 12}$	...	$T_{\cdot 1b}$	A_1
2	$T_{\cdot 21}$	$T_{\cdot 22}$	...	$T_{\cdot 2b}$	A_2
.	.	.		.	.
.	.	.		.	.
.	.	.		.	.
a	$T_{\cdot a1}$	$T_{\cdot a2}$	...	$T_{\cdot ab}$	A_a
Sum	B_1	B_2	...	B_b	G

[a] $T_{\cdot jk} = \Sigma_i y_{ijk}$ = sum of yields for the j-th level of A, k-th level of B summed over blocks; $B_k = \Sigma_j T_{\cdot jk}$ = sum for the k-th level of B; and A_j and G are as for Table 4.44.

Table 4.46 **Analysis of Variance for a Split-Plot Design**

ANOVA

Source	df	SS	MS	F
Total	rab−1	SSTot		
Block	r−1	SSR	MSR	F_R
A	a−1	SSA	MSA	F_A
Error(a)	(r−1)(a−1)	SSEA	MSEA	
B	b−1	SSB	MSB	F_B
AB	(a−1)(b−1)	SSAB	MSAB	F_{AB}
Error(ab)	a(r−1)(b−1)	SSEAB	MSEAB	

Computations: The sources of variation and degrees of freedom are the same as those for a two-factor factorial experiment in a randomized block design except that the error is partitioned into two parts:

1. Error(a), with (r − 1)(a − 1) df, the whole-plot error.
2. Error(ab), with a(r − 1)(b − 1) df, the subplot error.

For the sums of squares we compute

$$C = \frac{G^2}{rab}$$

$$SSTot = \sum_i \sum_j \sum_k y^2_{ijk} - C$$

$$SSR = \frac{1}{ab} \sum_i R_i^2 - C$$

$$SSA = \frac{1}{rb} \sum_j A_j^2 - C$$

$$SSEA = \frac{1}{b} \sum_i \sum_j T^2_{ij\cdot} - C - SSR - SSA$$

$$SSB = \frac{1}{ra} \sum_k B_k^2 - C$$

$$SSAB = \frac{1}{r} \sum_j \sum_k T^2_{\cdot jk} - C - SSA - SSB$$

$$SSEAB = SSTot - SSR - SSA - SSEA - SSB - SSAB$$

Mean squares are computed as usual by dividing the sums of squares by their associated degrees of freedom.

The F ratio is computed somewhat differently than for other designs because there are two errors in the split-plot design:

$$F_R = \frac{MSR}{MSEA}$$

$$F_A = \frac{MSA}{MSEA}$$

and

$$F_B = \frac{MSB}{MSEAB}$$

$$F_{AB} = \frac{MSAB}{MSEAB}$$

Because there are two errors the CV is usually not computed for the split-plot design. If it were computed, there would be two CVs, one for each plot size.

4.8.3 Significance Tests

F_A, with a − 1, (r − 1)(a − 1) df, is used to test the significance of differences among the factor A means, the A main effect.
F_B, with b − 1, a(r − 1)(b − 1) df, tests the significance of the factor B main effect.
F_{AB}, with (a − 1)(b − 1), a(r − 1)(b − 1) df, is used to test the significance of the AB interaction.
F_R, with r − 1, (r − 1)(a − 1) df, is an approximate test for the effectiveness of blocking.

4.8.4 Means and Standard Errors

For the split-plot design the means are computed in the usual way. The standard errors, however, are somewhat more complicated, particularly when comparing one treatment with another.

Treatment means:

1. Factor A means, $\bar{y}_{Aj} = A_j/rb$
   Standard error, $s_{\bar{A}}$:

$$s_{\bar{A}} = \sqrt{\frac{MSEA}{rb}}$$

2. Factor B means, $\bar{y}_{Bk} = B_k/ra$
   Standard error, $s_{\bar{B}}$:

$$s_{\bar{B}} = \sqrt{\frac{MSEAB}{ra}}$$

3. Treatment AB means, $\bar{y}_{jk} = T_{\cdot jk}/r$
   Standard error, $s_{\bar{y}}$:

$$s_{\bar{y}} = \sqrt{\frac{MSEAB}{r}}$$

Differences between means:

1. Difference between two A means, $\bar{y}_{A2} - \bar{y}_{A1}$
   Standard error, $s_{\bar{d}A}$:

$$s_{\bar{d}A} = \sqrt{\frac{2MSEA}{rb}}$$

2. Difference between two B means, $\bar{y}_{B2} - \bar{y}_{B1}$
   Standard error, $s_{\bar{d}B}$:

$$s_{\bar{d}B} = \sqrt{\frac{2MSEAB}{ra}}$$

3. Difference between B means at same level of A, $\bar{y}_{A1B2} - \bar{y}_{A1B1}$
   Standard error, $s_{\bar{d}AB}$:

$$s_{\bar{d}AB} = \sqrt{\frac{2MSEAB}{r}}$$

4. Difference between A means at same or different level of B,
   $y_{A2B1} - y_{A1B1}$ or $y_{A2B2} - y_{A1B1}$
   Standard error, $s_{\bar{d}}$:

$$s_{\bar{d}} = \sqrt{\frac{2[(b - 1)MSEAB + MSEA]}{rb}}$$

Standard errors involving only MSEA have (r − 1)(a − 1) df, while those involving only MSEAB have a(r − 1)(b − 1) df. The standard error involving both MSEA and MSEAB is a special case. It has approximately

$$\frac{a(r - 1)(a - 1)[(b - 1)MSEAB + MSEA]^2}{(a - 1)(b - 1)MSEAB^2 + (a)MSEA^2}$$

degrees of freedom.

4.8.5 Interpretation

Interpretation proceeds in much the same way with the split-plot design as it does with the two-factor factorial experiment. First test the AB interaction. If it is significant, the main effects have no meaning whether or not they test significant. In this case the results are summarized in a two-way table of AB means. If the interaction is not significant, we then look at the significance of the main effects. In this case the results are summarized in one-way tables of means for the factors whose main effects are significant.

4.8.6 Numerical Example

A wheat breeder wanted to determine the effect of planting date on the yield of four varieties of winter wheat. She chose two factors as follows:

Planting date: D_1 = Oct. 15; D_2 = Nov. 1; D_3 = Nov 15
Variety: V_1; V_2; V_3; V_4

To facilitate the operations of planting and harvesting she decided to use a split-plot design with planting dates assigned at random to strips of plots within each of three blocks. Varieties were then assigned at random to plots within planting dates. The field plan and yields (kg/plot) are presented in Figure 4.7.

Data Analysis The preliminary steps in the data analysis involve the construction of two tables of totals. These are shown as Tables 4.47 and 4.48. The analysis of variance of the yield data is presented in Table 4.49.

Computations:

$$C = \frac{G^2}{rab} = \frac{(674)^2}{36} = 12{,}618.78$$

$$\text{SSTot} = \sum_i \sum_j \sum_k y_{ijk}^2 - C = 24^2 + \cdots + 19^2 - 12{,}618.78$$
$$= 1267.22$$

$$\text{SSR} = \frac{1}{ab} \sum_i R_i^2 - C = \frac{1}{12}(222^2 + 228^2 + 224^2) - C$$
$$= 1.55$$

Block	I			II			III		
Date	D_2	D_1	D_3	D_1	D_2	D_3	D_1	D_3	D_2
	V_2 24	V_1 25	V_1 17	V_4 14	V_2 20	V_4 13	V_1 28	V_4 8	V_3 16
	V_3 19	V_4 11	V_3 12	V_1 31	V_3 18	V_1 20	V_3 17	V_3 15	V_1 28
	V_1 30	V_2 19	V_4 8	V_3 20	V_1 32	V_2 16	V_2 16	V_1 19	V_2 24
	V_4 15	V_3 22	V_2 20	V_2 14	V_4 13	V_3 17	V_4 14	V_2 20	V_4 19

Figure 4.7 Field plan and yields (kg/plot) for a variety × planting date trial on wheat.

Table 4.47 **Two-Way Table of Block × Date Totals**

Date	Block I	II	III	Sum
D_1	77	79	75	231
D_2	88	83	87	258
D_3	57	66	62	185
Sum	222	228	224	674

Table 4.48 **Two-Way Table of Variety × Date Totals**

Date	Variety V_1	V_2	V_3	V_4	Sum
D_1	84	49	59	39	231
D_2	90	68	53	47	258
D_3	56	56	44	29	185
Sum	230	173	156	115	674

$$\begin{aligned} \text{SSA} &= \frac{1}{rb} \sum_j A_j^2 - C = \frac{1}{12}(231^2 + 258^2 + 185^2) - C \\ &= 227.05 \end{aligned}$$

$$\begin{aligned} \text{SSEA} &= \frac{1}{b} \sum_i \sum_j T_{ij\cdot}^2 - C - \text{SSR} - \text{SSA} \\ &= \frac{1}{4}(77^2 + \cdots + 62^2) - C - 1.55 - 227.05 \\ &= 14.12 \end{aligned}$$

Table 4.49 **Analysis of Variance of Yields (kg/plot) from a Variety × Planting Date Trial with Winter Wheat**

ANOVA

Source	df	SS	MS	F
Total	35	1267.22		
Block	2	1.55	.78	.22
Date	2	227.05	113.53	32.16**
Error(a)	4	14.12	3.53	
Variety	3	757.89	252.63	37.82**
Var by date	6	146.28	24.38	3.65*
Error(ab)	18	120.33	6.68	

*Significant at 5%; **significant at 1%.

$$\begin{aligned}
\text{SSB} &= \frac{1}{ra}\sum_k B_k^2 - C = \frac{1}{9}(230^2 + 173^2 + 156^2 + 115^2) - C \\
&= 757.89 \\
\text{SSAB} &= \frac{1}{r}\sum_j \sum_k T_{\cdot jk}^2 - C - \text{SSA} - \text{SSB} \\
&= \frac{1}{3}(84^2 + \cdots + 29^2) - C - 227.05 - 757.89 \\
&= 146.28 \\
\text{SSEAB} &= \text{SSTot} - \text{SSR} - \text{SSA} - \text{SSEA} - \text{SSB} - \text{SSAB} \\
&= 1267.22 - 1.55 - 227.05 - 14.12 - 757.89 - 146.28 \\
&= 120.33
\end{aligned}$$

Significance tests:

1. 1% F at 2,4 df = 18.00 < 32.16. Thus the main effect of the planting date is highly significant.
2. 1% F at 3,18 df = 5.09 < 37.82. Hence the main effect of the variety is highly significant.
3. 5% F at 6,18 df = 2.66 and the 1% F at 6,18 df = 4.01.

Since 2.66 < 3.65 < 4.01, the date × variety interaction is significant at the 5% level but not at the 1% level.

Standard errors:
Means:

1. Planting date means:

$$s_{\bar{A}} = \sqrt{\frac{MSEA}{rb}} = \sqrt{\frac{3.53}{12}} = 0.5424$$

2. Variety means:

$$s_{\bar{B}} = \sqrt{\frac{MSEAB}{ra}} = \sqrt{\frac{6.68}{9}} = 0.8615$$

3. Date × variety means:

$$s_{\bar{y}} = \sqrt{\frac{MSEAB}{r}} = \sqrt{\frac{6.68}{3}} = 1.4922$$

Differences:

1. Two date means:

$$s_{\bar{d}A} = \sqrt{\frac{2MSEA}{rb}} = \sqrt{\frac{(2)(3.53)}{12}} = 0.7670$$

2. Two variety means:

$$s_{\bar{d}B} = \sqrt{\frac{2MSEAB}{ra}} = \sqrt{\frac{(2)(6.68)}{9}} = 1.2184$$

3. Two variety means on same date:

$$s_{\bar{d}AB} = \sqrt{\frac{2MSEAB}{r}} = \sqrt{\frac{(2)(6.68)}{3}} = 2.1103$$

4. Two date means for same or different variety:

$$s_{\bar{d}} = \sqrt{\frac{2[(b-1)MSEAB + MSEA]}{rb}}$$

$$= \sqrt{\frac{2[(3)(6.68) + 3.53]}{12}} = 1.9812$$

Report of Statistical Analysis

The results of an experiment to examine the effect of planting date on the yield of four varieties of winter wheat are summarized in Table 4.50.

Table 4.50 **Mean Yields (kg/plot) of Four Winter Wheat Varieties Planted on Three Different Dates[a]**

Date	Variety 1	2	3	4	Mean
Oct. 15	28.00	16.33	19.67	13.00	19.25
Nov. 1	30.00	22.67	17.67	15.67	21.50
Nov. 15	18.67	18.67	14.67	9.67	15.42
Mean	25.55	19.22	17.33	12.78	18.72

[a]Standard errors: date = 0.542; variety = 0.862; variety × date = 1.492.

Although there was an indication that differences among varieties depended on planting date, the variety differences and date differences were highly significant. Except for variety 3, each variety produced its maximum yield when planted on November 1. On the average, the highest yield at every planting date was achieved by variety 1, while variety 4 produced the lowest yield for each planting date.

4.9 STRIP-PLOT DESIGNS

A number of variations of the basic split-plot design are possible. The subplots can be further subdivided into sub-subplots and a third factor introduced on these. Other variations, such as using a Latin square arrangement on the whole plots or on the subplots, are also possible. A particularly useful variation is called a *strip-plot design* (also called a *split-block design*).

Consider an experiment with methods of seed-bed preparation at three levels: S_1, S_2, and S_3, and a nitrogen fertilizer factor at four levels: N_1, N_2, N_3, and N_4. Suppose that we wanted to use machinery to apply the nitrogen as well as to do the seed-bed preparation. In this case it would be convenient to use large plots for both factors. This is possible with the strip-plot design.

In a strip-plot design the levels of one factor are assigned to strips of plots running through the block in one direction. A separate randomization is used in each block. The levels of the second factor are then applied to strips of plots that are oriented perpendicularly to the strips for the first factor. The field plan for two replications of a seed bed × nitrogen experiment run as a strip-plot design might appear as shown in Figure 4.8.

In the strip-plot design there are three sizes of plot:

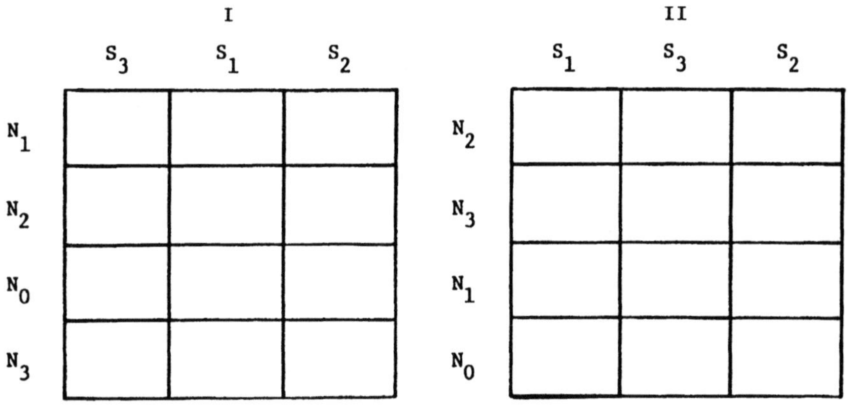

Figure 4.8 Field plan for a strip-plot design for a seed bed × nitrogen experiment.

1. The strip of four plots on which the seed-bed factor is applied.
2. The strip of three plots on which the levels of nitrogen are applied.
3. The small plots containing the NS combinations.

Because there are three sizes of plot, there are three experimental errors, one for each plot size. The interaction is measured with greater precision than are the main effects.

4.9.1 Advantages, Disadvantages, and Uses

There is one principal advantage of the strip-plot design:

1. It permits the efficient application of factors that would be difficult to apply to small plots.

The disadvantages of the strip-plot design are similar to those of the split-plot design:

1. Differential precision in the estimation of interaction and the main effects.
2. Complicated statistical analysis.

The uses of the strip-plot design are similar to the uses of the split-plot design:

1. For experiments involving factors that are difficult to apply to small plots.
2. To introduce new factors into an experiment already in progress.

4.9.2 Design Selection

The general rule for selecting an experimental design is to choose the simplest design that will estimate the effects with the desired precision or will permit the efficient conduct of the experiment. *In order of increasing complexity*, the designs we have looked at are:

1. Completely randomized design
2. Randomized block design
3. Split-plot design
4. Strip-plot design
5. Latin square

As the complexity of the design increases, so does the amount of restriction imposed on the randomization of the treatments to the plots. As randomization is restricted, degrees of freedom are lost from experimental error, and unless the restriction results in a reduction in the error sum of squares (reduced variation among plots treated alike), the error mean square will be increased. This results in a loss of precision for the experiment. For most purposes the randomized block design (RBD) provides the necessary control of experimental error. More complex designs should be used only when necessary.

4.9.3 Data Analysis

The analysis of data from a strip-plot design follows the same general pattern as for other designs:

1. Construct tables of totals.
2. Complete an analysis of variance.
3. Perform significance tests.
4. Compute means and standard errors.
5. Interpret the analysis.

To examine the pattern of analysis suppose that we have factor A at a levels assigned at random to a strips of b plots each, and factor B at b levels to b strips of a plots each crossing the strips used for the A factor. Assume the experiment is to be conducted using r blocks. Let y_{ijk} represent the yield of the j-th level of factor A, k-th level of factor B in the i-th block. For the strip-plot design we require the three summary tables shown as Tables 4.51 to 4.53. The format of the analysis of variance of data from a strip-plot experiment with factor A at a levels and factor B at b levels in r blocks is given in Table 4.54.

Table 4.51 **Two-Way Block × Factor A Table of Totals**[a]

Block	Factor A 1	2	...	a	Sum
1	$T_{11\cdot}$	$T_{12\cdot}$	...	$T_{1a\cdot}$	R_1
2	$T_{21\cdot}$	$T_{22\cdot}$	...	$T_{2a\cdot}$	R_2
⋮	⋮	⋮		⋮	⋮
r	$T_{r1\cdot}$	$T_{r2\cdot}$	...	$T_{ra\cdot}$	R_r
Sum	A_1	A_2	...	A_a	G

[a]$T_{ij\cdot} = \Sigma_k\, y_{ijk}$ = sum of the yields on the j-th level of A, in the i-th block summed over B

$R_i = \Sigma_j\, T_{ij\cdot}$ = sum of the yields for the i-th block

$A_j = \Sigma\, T_{ij\cdot}$ = sum of yields for the j-th level of factor A

$G = \Sigma_i\, R_i = \Sigma_j\, A_j$ = grand total of all yields

Table 4.52 **Two-Way Table of Factor B × Block Totals**[a]

Block	Factor B 1	2	...	b	Sum
1	$T_{1\cdot 1}$	$T_{1\cdot 2}$	...	$T_{1\cdot b}$	R_1
2	$T_{2\cdot 1}$	$T_{2\cdot 2}$	...	$T_{2\cdot b}$	R_2
⋮	⋮	⋮		⋮	⋮
r	$T_{r\cdot 1}$	$T_{r\cdot 2}$	...	$T_{r\cdot b}$	R_r
Sum	B_1	B_2	...	B_b	G

[a]$T_{i\cdot k} = \Sigma_j\, y_{ijk}$ = sum of yields on the k-th level of B in the i-th block summed over factor A, $B_k = \Sigma_i\, T_{i\cdot k}$ = sum of yields on the k-th level of factor B, and R_i and G are as in Table 4.51.

Table 4.53 **Two-Way Table of Factor A × Factor B Totals**[a]

	Factor B				
Factor A	1	2	...	b	Sum
1	$T_{\cdot 11}$	$T_{\cdot 12}$	...	$T_{\cdot 1b}$	A_1
2	$T_{\cdot 21}$	$T_{\cdot 22}$	...	$T_{\cdot 2b}$	A_2
.	.	.		.	.
.	.	.		.	.
.	.	.		.	.
a	$T_{\cdot a1}$	$T_{\cdot a2}$	...	$T_{\cdot ab}$	A_a
Sum	B_1	B_2	...	B_b	G

[a]$T_{\cdot jk} = \Sigma_i y_{ijk}$ = sum of yields on the j-th level of factor A, k-th level of factor B summed over blocks, and A_j, B_k, and G are as before.

Table 4.54 **ANOVA Table for a Strip-Plot Design**

ANOVA

Source	df	SS	MS	F
Total	rab−1	SSTot		
Block	r−1	SSR	MSR	
A	a−1	SSA	MSA	F_A
Error(a)	(r−1)(a−1)	SSEA	MSEA	
B	b−1	SSB	MSB	F_B
Error(b)	(r−1)(b−1)	SSEB	MSEB	
AB	(a−1)(b−1)	SSAB	MSAB	F_{AB}
Error(ab)	(r−1)(a−1)(b−1)	SSEAB	MSEAB	

Computations:

Sources of variation: Sources of variation are the same as for the two-factor factorial experiment except that the error has been partitioned into three parts:

1. Error(a), with (r − 1)(a − 1) df, the error for the factor A strips.
2. Error(b), with (r − 1)(b − 1) df, the error for the factor B strips.
3. Error(ab), with (r − 1)(a − 1)(b − 1) df, the error for the AB plots.

Sums of squares:

1. Correction term, $C = G^2/rab$.
2. $SSTot = \Sigma_i \Sigma_j \Sigma_k y_{ijk}^2 - C$.
3. $SSR = (1/ab) \Sigma_i R_i^2 - C$.
4. $SSA = (1/rb) \Sigma_j A_j^2 - C$.
5. $SSEA = (1/b) \Sigma_i \Sigma_j T_{ij\cdot}^2 - C - SSR - SSA$.
6. $SSB = (1/ra) \Sigma_k B_k^2 - C$.
7. $SSEB = (1/a) \Sigma_i \Sigma_k T_{i\cdot k}^2 - C - SSR - SSB$.
8. $SSAB = (1/r) \Sigma_j \Sigma_k T_{\cdot jk}^2 - C - SSA - SSB$.
9. $SSEAB = SSTot - SSR - SSA - SSEA - SSB - SSEB - SSAB$.

Mean squares: Divide the sums of squares by their associated df.
F ratio:

1. $F_A = MSA/MSEA$; a − 1, (r − 1)(a − 1) df.
2. $F_B = MSB/MSEB$; b − 1, (r − 1)(b − 1) df.
3. $F_{AB} = MSAB/MSEAB$; (a − 1)(b − 1), (r − 1)(a − 1)(b − 1) df.

4.8.4 Significance Tests

1. F_A tests the hypothesis that the factor A means are equal. This is the main effect of factor A.
2. F_B tests the significance of the main effect of factor B.
3. F_{AB} tests the significance of the AB interaction.

4.9.5 Means and Standard Errors

Treatment means:

1. Factor A means, $\bar{y}_{Aj} = A_j/rb$
   Standard error, $s_{\bar{A}}$:

$$s_{\bar{A}} = \sqrt{\frac{MSEA}{rb}}$$

2. Factor B means, $\bar{y}_{Bk} = B_k/ra$
   Standard error, $s_{\bar{B}}$:

$$s_{\bar{B}} = \sqrt{\frac{MSEB}{ra}}$$

3. Treatment (AB) means, $\bar{y} = T_{\cdot jk}/r$
Standard error, $s_{\bar{y}}$:

$$s_{\bar{y}} = \sqrt{\frac{MSEAB}{r}}$$

Differences between means:

1. Between two A means, $\bar{y}_{A2} - \bar{y}_{A1}$
Standard error, $s_{\bar{d}A}$:

$$s_{\bar{d}A} = \sqrt{\frac{2MSEA}{rb}}$$

2. Between two B means, $y_{B2} - y_{B1}$
Standard error, $s_{\bar{d}B}$:

$$s_{\bar{d}B} = \sqrt{\frac{2MSEB}{ra}}$$

3. Between two A means at the same level of B, $\bar{y}_{A2B1} - \bar{y}_{A1B1}$
Standard error, $s_{\bar{d}AB}$:

$$s_{\bar{d}AB} = \sqrt{\frac{2[(b - 1)MSEAB + MSEA]}{rb}}$$

4. Between two B means at the same level of A, $\bar{y}_{A1B2} - \bar{y}_{A1B1}$
Standard error, $s_{\bar{d}BA}$:

$$s_{\bar{d}BA} = \sqrt{\frac{2[(a - 1)MSEAB + MSEB]}{ra}}$$

5. Difference between A means at different levels of B, or between B means at different levels of A, $\bar{y}_{A2B2} - \bar{y}_{A1B1}$ or $\bar{y}_{A1B2} - \bar{y}_{A2B1}$
Standard error, $s_{\bar{d}}$:

$$s_{\bar{d}} = \sqrt{\frac{2[(ab - a - b)MSEAB + aMSEA + bMSE]}{rab}}$$

4.9.6 Interpretation

The interpretation of data from a strip-plot design depends on the significance of the AB interaction. If it is significant, the main effects of A and

B have no meaning whether or not they test significant. The results are then summarized in a two-way table of means. If the interaction is not significant, one-way tables of means are used to summarize the response to factors whose main effects are significant.

4.9.7 Numerical Example

A pasture specialist wanted to determine the effect of phosphorus and potash fertilizers on the dry matter production of barley to be used as a forage. He established three blocks in an experiment to study the following factors:

1. Potash: K_1 = none; K_2 = 25 kg/ha; K_3 = 50 kg/ha.
2. Phosphorus: P_1 = 25 kg/ha; P_2 = 50 kg/ha.

Because he wanted to use farm-scale fertilization equipment, he assigned rates of phosphorus to strips of three plots within each block. Potash rates were then assigned to strips of two plots that crossed the phosphorus plots at right angles. Separate randomizations were used in each block. The field plan and dry matter yields (kg/plot) were as shown in Figure 4.9.

Data Analysis The necessary summary tables are shown as Tables 4.55 to 4.57. The analysis of variance of this experiment is given in Table 4.58.

Computations:

$$C = \frac{G^2}{rab} = \frac{988^2}{18} = 54{,}230.22$$

$$SSTot = \sum_i \sum_j \sum_k y_{ijk}^2 - C = 56^2 + \cdots + 68^2 - C = 1833.78$$

$$SSR = \frac{1}{ab} \sum_i R_i^2 - C = \frac{1}{6}(316^2 + 338^2 + 334^2) - C$$
$$= 45.78$$

$$SSA = \frac{1}{rb} \sum_j A_j^2 - C = \frac{1}{6}(274^2 + 338^2 + 376^2) - C$$
$$= 885.78$$

$$SSEA = \frac{1}{b} \sum_i \sum_j T_{ij\cdot} - C - SSR - SSA$$
$$= \frac{1}{2}(86^2 + \cdots + 119^2) - C - 45.78 - 885.78$$
$$= 78.22$$

	K_3	K_1	K_2	Block
P_1	56	32	49	I
P_2	67	54	58	

	K_1	K_3	K_2	Block
P_2	38	62	50	II
P_1	52	72	64	

	K_2	K_1	K_3	Block
P_2	54	44	51	III
P_1	63	54	68	

Figure 4.9 Field plan and dry matter yields (kg/plot) from a strip-plot experiment designed to study rates of P and K fertilizer used on forage barley.

Table 4.55 **Potash × Block Two-Way Table of Totals**

	Block			
K	I	II	III	Sum
1	86	90	98	274
2	107	114	117	338
3	123	134	119	376
Sum	316	338	334	988

Table 4.56 **Phosphorus × Block Two-Way Table of Totals**

P	Block I	II	III	Sum
1	137	188	185	510
2	179	150	149	478
Sum	316	338	334	988

Table 4.57 **Potash × Phosphorus Two-Way Table of Totals**

P	Potash 1	2	3	Sum
1	138	176	196	510
2	136	162	180	478
Sum	274	338	376	988

$$\text{SSB} = \frac{1}{\text{ra}} \sum_k B_k^2 - C = \frac{1}{9}(510^2 + 478^2) - C = 56.89$$

$$\text{SSEB} = \frac{1}{a} \sum_i \sum_k T_{i\cdot k}^2 - C - \text{SSR} - \text{SSB}$$

$$= \frac{1}{3}(137^2 + \cdots + 180^2) - C - 885.78 - 56.89$$

$$= 693.78$$

$$\text{SSAB} = \frac{1}{r} \sum_j \sum_k T_{\cdot jk}^2 - C - \text{SSA} - \text{SSB}$$

$$= \frac{1}{3}(138^2 + \cdots + 180^2) - C - 885.78 - 56.89$$

$$= 19.11$$

Table 4.58 **Analysis of Variance of Dry Matter Yields of Forage Barley**

ANOVA

Source	df	SS	MS	F
Total	17	1,833.78		
Block	2	45.78	45.78	
Potash	2	885.78	442.89	22.64**
Error(a)	4	78.22	19.56	
Phosphorus	1	56.89	56.89	.16NS
Error(b)	2	693.78	346.89	
KP	2	19.11	9.56	.71NS
Error(ab)	4	54.22	13.55	

**Significant at 1%; NS, not significant.

$$\begin{aligned}\text{SSEAB} &= \text{SSTot} - \text{SSR} - \text{SSA} - \text{SSEA} - \text{SSB} - \text{SSEB} \\ &\quad - \text{SSAB} \\ &= 1833.78 - 45.78 - 885.78 - 78.22 - 56.89 - 693.78 \\ &\quad - 19.11 = 54.22\end{aligned}$$

Standard errors:

Potash means, $\bar{y}_{Aj}$:

$$s_{\bar{A}} = \sqrt{\frac{\text{MSEA}}{rb}} = \sqrt{\frac{19.56}{6}} = 1.80$$

Phosphorus means, $\bar{y}_{Bk}$:

$$s_{\bar{B}} = \sqrt{\frac{\text{MSEB}}{ra}} = \sqrt{\frac{346.89}{9}} = 6.21$$

KP means, $\bar{y}_{jk}$:

$$s_{\bar{y}} = \sqrt{\frac{\text{MSEAB}}{r}} = \sqrt{\frac{13.55}{3}} = 2.12$$

Difference between potash means:

$$s_{\bar{d}A} = \sqrt{\frac{2MSEA}{rb}} = \sqrt{\frac{(2)(19.56)}{6}} = 2.55$$

Difference between phosphorus means:

$$s_{\bar{d}B} = \sqrt{\frac{2MSEB}{ra}} = \sqrt{\frac{(2)(346.89)}{9}} = 8.78$$

Difference between potash means at same phosphorus level:

$$s_{\bar{d}AB} = \sqrt{\frac{2[(b - 1)MSEAB + MSEA]}{rb}}$$

$$= \sqrt{\frac{2[(1)(13.55) + 19.56]}{6}} = 3.32$$

Difference between phosphorus means at same potash level:

$$s_{\bar{d}BA} = \sqrt{\frac{2[(a - 1)MSEAB + MSEB]}{ra}}$$

$$= \sqrt{\frac{2[(2)(13.55) + 346.89]}{9}} = 9.12$$

Difference between potash means at different phosphorus levels:

$$s_{\bar{d}} = \sqrt{\frac{2[(ab - a - b)MSEAB + aMSEA + bMSEB]}{rab}}$$

$$= \sqrt{\frac{2[(6 - 3 - 1)(13.55 + (3)(19.56) + (2)(346.89)]}{18}}$$

$$= 9.22$$

Report of Statistical Analysis

Analysis of data from an experiment designed to determine the effect of potash and phosphorus fertilizers on the dry matter yield of forage barley indicates that only potash had a significant effect. The principal results are summarized in Table 4.59.

Each increment of added potash resulted in an increase in the yield of dry matter. This increase took place regardless of the level of phosphorus fertilizer. Increasing the amount of added phosphorus from 25 kg/ha to 50 kg/ha did not result in any real change in the yield of barley dry matter.

Table 4.59 **Effect of Different Amounts of Potash Fertilizer on the Dry Matter Yield (kg/plot) of Barley**

Potash	none	25 kg/ha	50 kg/ha	Std. error
Mean yield	45.67	56.33	62.67	1.80

FURTHER READING

Complete citations for books for further reading are given in the bibliography at the end of the book.

1. Cochran and Cox, Chaps. 5 and 7.
2. LeClerg, Leonard, and Clark, Chaps. 12 and 13.
3. Little and Hills, Chaps. 3, 9, and 10.
4. Petersen, Chaps. 5, 6, and 10.
5. Snedecor and Cochran, Chaps. 16 and 19.
6. Steel and Torrie, Chaps. 15 and 16.

5

Variety Trials

5.1 BREEDING PROGRAM

Plant breeding programs are characterized by the need to examine large numbers of selections for possible release as new varieties. The process takes several growing seasons. In the first season of a selection cycle there are a very large number of entries from which to choose. The numbers are reduced by selection in each successive season. The initial selections are made in a purely subjective way by the breeder, who has an idea of what the ultimate selection will be. At this stage the selection is based on the examination of single plants, and there is no possibility of making comparisons statistically.

As the number of possibilities is reduced by selection the amount of seed available from each of the remaining entries is increased so that single rows of each entry can be planted for evaluation. At this point rows of one or more standard, or check, varieties are included at intervals throughout the trial. The purpose of the check rows is to serve as a standard against which the new entries might be compared. At this stage yield may or may not be measured. No statistical analysis is possible, however, because the entries are not replicated.

When the number of entries has been reduced to 100 or 200, sufficient seed is usually available to permit the conduct of replicated yield trials.

When this point is reached, statistical analysis of the yield data becomes possible. In these trials the aim is to identify the highest-yielding selection, or group of selections, and to compare the yields of the new selections with the yield of the standards. For this purpose we require a statistical procedure that will permit pairwise comparisons of the selection means. There are a number of possibilities. The basic idea is developed in the following section.

5.2 COMPARISON OF MEANS

5.2.1 Least Significant Difference

Suppose that we want to compare the true means of two randomly chosen selections. That is, we want to test H_0: $\mu_1 = \mu_2$ against H_a: $\mu_1 \neq \mu_2$, where μ_1 and μ_2 are the true means of two random selections. We can use, as a test statistic,

$$t = \frac{\bar{y}_1 - \bar{y}_2}{\sqrt{V(\bar{y}_1 - \bar{y}_2)}}$$

where $\bar{y}_1$ and $\bar{y}_2$ are the sample means and $V(\bar{y}_1 - \bar{y}_2)$ is the sample variance of the difference. If $|t| > t_\alpha$, where $|t|$ is the sample value of t with sign ignored and t_α is the two-tailed t at the α significance level, we reject H_0 and conclude that the means are significantly different. Hence if

$$t_\alpha < |t| = \frac{|\bar{y}_1 - \bar{y}_2|}{\sqrt{V(\bar{y}_1 - \bar{y}_2)}}$$

or, equivalently, if

$$t_\alpha \sqrt{V(\bar{y}_1 - \bar{y}_2)} < |\bar{y}_1 - \bar{y}_2|$$

then μ_1 is significantly different from μ_2. Any difference between two sample means, $|\bar{y}_1 - \bar{y}_2|$, which is larger than $t_\alpha \sqrt{V(\bar{y}_1 - \bar{y}_2)}$ would indicate a significant difference between μ_1 and μ_2. Further, $t_\alpha \sqrt{V(\bar{y}_1 - \bar{y}_2)}$ is the smallest difference for which significance would be declared. Based on this line of reasoning we can define a *least significant difference* (LSD) as

$$LSD = t_\alpha \sqrt{V(\bar{y}_1 - \bar{y}_2)}$$

where t_α is the two-tailed α-level t.

In particular, with yield trials in which the entries are equally replicated, say n times, and MSE is the error mean square from the ANOVA table, we have

$$V(\bar{y}_1 - \bar{y}_2) = \frac{2MSE}{n}$$

and

$$LSD = t_\alpha \sqrt{\frac{2MSE}{n}}$$

where t_α is the two-tailed t at the α probability level with the df at which MSE is computed. (Use the t value from the column headed D in the appendix table, Table A2.)

In theory the LSD provides a valid test criterion in two situations:

1. Making comparisons planned in advance of seeing the data. This includes the comparison of each treatment with the control.
2. Comparing adjacent ranked means. This is, essentially, a procedure for picking winners.

Unless a previous F test has indicated that there are significant differences among treatment means, the LSD should not be used for:

1. Making all possible pairwise comparisons.
2. Making more comparisons than there are df for treatments.

5.2.2 Least Significant Increase

In the usual plant breeding program, standard varieties are included along with the new entries to provide a basis against which to measure progress in the breeding program. The goal is to find new entries that outyield the checks. The breeder is normally not interested in entries that yield less than the checks, whether or not the difference is significant. He would like to be able to reject H_0: $\mu_N = \mu_C$ in favor of H_a: $\mu_N > \mu_C$, where μ_N is the true mean of the new entry and μ_C is the true mean of the check. We could go through the same line of reasoning as we did in developing the LSD in order to find a suitable standard for comparison. This development would lead to a statistic we could call a *least significant increase* (LSI) against which we could compare yield increases for those entries that outyield the check. We have

$$LSI = t_\alpha \sqrt{\frac{2MSE}{n}}$$

where t_α is the one-tailed t (column head B on the t table, Table A2) at the α probability level for the df at which MSE is computed. (*Note*: The LSI provides, in fact, a one-sided significance test.)

In practice, if a new entry mean, $\bar{y}_N$, is greater than

$$\bar{y}_C + \text{LSI}$$

the increase in yield of the new entry over the check yield is said to be significant at the 100% level.

5.2.3 Numerical Examples

I. Pick the Winner

A plant breeder wanted to measure the resistance of six wheat varieties to a particular race of stem rust. He planted five seeds of each variety in each of four pots of well-mixed soil. He then assigned the 24 pots to random locations on a greenhouse bench. (Thus the experimental unit is a pot and the treatments are the six varieties. They are replicated four times in a completely randomized design.) After the plants had emerged he uniformly inoculated the experiment with a culture of the stem rust. At maturity he measured the seed yield (g/pot) in each pot. The results, arranged in order of decreasing mean yield, are shown in Table 5.1. An analysis of variance produced the results given in Table 5.2.

The breeder then used the error mean square to compute the 5% and

Table 5.1 **Ranked Mean Yields (g/pot) of Wheat from a Stem Rust Resistance Trial**

Variety	Rank(i)	Mean yield $\bar{y}_i$	Difference $\bar{y}_{i-1} - \bar{y}_i$
6	1	95.3	
4	2	94.0	1.3
5	3	75.0	19.0
2	4	69.0	6.0
1	5	50.3	18.7
3	6	24.0	26.3

Table 5.2 **Analysis of Variance of Wheat Yield Data (g/pot) from a Stem Rust Resistance Trial**

ANOVA

Source	df	MS	F
Variety	5	2,976.44	24.80
Error	18	120.00	

1% level LSD values:

$$\text{LSD} = t_\alpha \sqrt{\frac{2\text{MSE}}{n}} = 2.101 \sqrt{\frac{(2)(120.00)}{4}} = 16.27\ (5\%)$$

$$= 2.878 \sqrt{\frac{(2)(120.00)}{4}} = 22.29\ (1\%)$$

If the differences are compared to the LSD values, we see that the two highest-yielding varieties, 6 and 4, are not significantly different, but there is a significant ($\alpha = 0.05$) gap between varieties 4 and 5. Another gap ($\alpha = 0.05$) exists between varieties 2 and 1, while the difference between variety 3 and the others is highly significant ($\alpha = 0.01$).

II. Least Significant Increase

We can use the stem rust data from the "pick the winner" example to illustrate the LSI procedure. Suppose that variety 1 was a check variety and the others were new entries. We might want to see which of the new entries significantly ($\alpha = 0.05$) outyielded the check. The pertinent data are presented in Table 5.3.

To compute the LSI, we find that the one-sided, 5% t at 18 df is $t_{0.05} = 1.734$. Then

$$\text{LSI} = t_{0.05} \sqrt{\frac{2\text{MSE}}{r}} = 1.734 \sqrt{\frac{(2)(120.00)}{4}} = 13.43$$

Hence any new entry with a mean yield greater than

$$\bar{y}_C + \text{LSI} = 50.3 + 13.43 = 63.7$$

significantly outyields the check. Thus varieties 2, 4, 5, and 6 are significantly ($\alpha = 0.05$) better than the check.

Table 5.3 **Mean Yield (g/pot) of Five New Entries and a Check in a Stem Rust Resistance Trial**[a]

Variety	Mean yield, $\bar{y}$	Increase, $\bar{y}_N - \bar{y}_C$
1 (check)	50.3	
2	69.0	18.7
3	24.0	*
4	94.0	43.7
5	75.0	24.7
6	95.3	45.0

[a]MSE = 120.0, with 18 df.
*Not computed because $\bar{y}_3 < \bar{y}_C$.

5.2.4 Pairwise Comparison Procedures (Data Snooping)

There is probably no statistical technique that is more often misused than that of making multiple pairwise comparisons. The procedure is intended to be used only for data from experiments with unstructured treatments. Examples of appropriate experiments are variety trials and insecticide screening trials. Other experiments, such as factorial experiments, are more properly examined by forming groups of treatments suggested by the nature of the treatments or by the objectives of the experiment and comparing group means. These individual comparisons are the proper way to answer the specific questions posed by the objectives of the experiment (see, e.g., Chew, 1976; Petersen, 1977).

If a plant breeder conducts a variety trial with p varieties, he might want to look at all possible pairwise comparisons among the variety means. With p varieties there could be as many as p(p − 1)/2 comparisons, which is much greater than the p − 1 comparisons permitted using the ordinary LSD or the LSI. The breeder requires a procedure that will permit him to draw conclusions about each of a large number of differences and to assign a measure of confidence to those conclusions.

When a breeder compares two means, there are two possibilities for error in his conclusions. He can conclude that the two means are different when, in fact, they are not. This is called a *type I error*, and the rate at which this type of error is made is called the *significance level*, α, for the comparison. Alternatively, he can conclude that the means are the same when, in fact, they are different. This is called a *type II error*, and the probability of committing this type of error has been designated β. The probability that a comparison procedure will pick up a real difference when it exists is called the *power of the test* and is equal to $1 - \beta$. The breeder protects himself against a type I error when he chooses a significance level. He can do little about a type II error because it depends on the true, but unknown, magnitude of the difference. He must assume that he has chosen a test with sufficiently high power.

If he uses the ordinary LSD to judge all possible pairwise comparisons, the probability of a type I error is no longer the chosen significance level. Depending on the number of means being compared, the observed difference can exceed the 5% LSD as much as 90% of the time with 20 means among which there are no real differences. To overcome this difficulty a number of test procedures have been proposed for making multiple comparisons. Among these the more commonly used ones are:

1. Fisher's protected LSD (FPLSD) (Fisher, 1966)
2. Tukey's honestly significant difference (HSD) (Tukey, 1951)
3. Student–Newman–Keuls test (SNK) (Steel and Torrie, 1980)
4. Waller and Duncan's Bayes LSD (BLSD) (Waller and Duncan, 1969)

The first three procedures are more or less traditional. The breeder would choose a significance level, α, to protect himself against a type I error. He then would allocate his error rate, α, by his choice of comparison procedure. The FPLSD error rate is a comparisonwise error rate. It measures the proportion of all differences that are expected to be declared real when, in fact, they are not. The HSD uses an experimentwise error rate. It measures the proportion of experiments in which one or more differences are falsely declared to be significant.

With the SNK the error rate allocation varies with the degree of separation between the means being compared. For adjacent ranked means the error rate is comparisonwise and the difference required for significance is the LSD. For means with the greatest separation, highest versus lowest, the error rate is experimentwise and the difference required for significance is the HSD.

The criticism of comparisonwise error rate procedures is that the probability of falsely declaring at least one difference to be significant is high and increases as the number of means being compared increases. The difficulty with experimentwise error rate procedures is that their power to detect real differences decreases as the number of means increases.

The BLSD takes into consideration the homogeneity, as measured by the F ratio, of the means being compared. If the F is low, indicating a homogeneous set of means, the BLSD has the characteristics of an experimentwise test. However, with a larger F ratio ($F > 4$) the BLSD has the characteristics of a comparisonwise error rate procedure. Unlike the other procedures, the BLSD is not based on the probability of making a type I error. Rather, the significance level is based on a factor, k, that is the ratio of the seriousness of a type I error to that of a type II error. The advantage of the BLSD in comparison to the other procedures is that its power does not depend on the number of means being compared.

We can use the stem rust data, Table 5.1, and its analysis of variance, Table 5.2, to illustrate the multiple comparison procedures. The data are from a completely randomized design with six varieties replicated four times. We will use a 5% significance level, or its equivalent in the case of the BLSD.

Fisher's protected LSD (FPLSD)

1. Conduct the usual ANOVA.
2. Make the F test of treatments.
3. If $F < F_\alpha$, stop! Do not make any comparisons.
4. If $F > F_\alpha$, find t_α in the two-tailed t table and compute FPLSD = $t_\alpha \sqrt{\frac{2MSE}{r}}$.
5. For any pair of means, if $|\bar{y}_j - \bar{y}_{j'}| >$ FPLSD, the difference is significant.

For the example data we have

1. $F = MST/MSE = 2976.44/120.00 = 24.80$. $F_\alpha = 2.77 < 24.80$; F is significant.
2. $FPLSD = t_\alpha \sqrt{\frac{2MSE}{r}} = 2.101 \sqrt{\frac{(2)(120.00)}{4}} = 16.27.$
3. Any difference between means that is greater than 16.27 is declared to be significant.

Tukey's honestly significant difference (HSD)

1. From a table of percentage points of the studentized range (Appendix Table A6), select a value of Q_α that depends on n = number of means and γ = error df.
2. Compute HSD as $HSD = Q_\alpha \sqrt{\frac{MSE}{r}}$.
3. For any pair of means, if $|\bar{y}_j - \bar{y}_{j'}| >$ HSD, the difference is significant.

For the example we have:

1. $n = 6$, $\gamma = 18$, $Q_\alpha = 4.49$.
2. $\text{HSD} = Q_\alpha \sqrt{\dfrac{\text{MSE}}{r}} = 4.49 \sqrt{\dfrac{120.00}{4}} = 24.59.$
3. Any difference greater than 24.59 is significant.

Student–Newman–Keuls test (SNK)

1. Rank the p means from high, $\bar{y}_H$, to low, $\bar{y}_L$.
2. Compute p − 1 significant differences, SNK_j, using the HSD p − 1 times as

   $$\text{SNK}_j = Q_{\alpha j} \sqrt{\frac{\text{MSE}}{r}} \qquad j = 1,2, \ldots ,p - 1$$

   using $Q_{\alpha j}$ for γ df and, in turn, $n = 2,3, \ldots ,p$.
3. If $|\bar{y}_H - \bar{y}_L| < \text{SNK}$, no differences are significant.
4. If $|\bar{y}_H - \bar{y}_L| < \text{SNK}$, this difference is significant, and we proceed comparing $\bar{y}_H - \bar{y}_{L+1}$ and $\bar{y}_{H-1} - \bar{y}_L$ against SNK_{p-1}.
5. Continue in this way until all significant pairwise differences have been found.

For the example data we have

1. $n = 6,5, \ldots ,2$; $\gamma = 18$.
2. $\text{SNK} = Q_{\alpha j} \sqrt{\dfrac{\text{MSE}}{r}} = Q_{\alpha j} \sqrt{\dfrac{120.00}{4}}.$

n	2	3	4	5	6
Q	2.97	3.61	4.00	4.28	4.49
SNK	16.27	19.77	21.91	23.44	24.59

3. If the difference $|\bar{y}_H - \bar{y}_L| < 24.59$, none of the differences are significant. Any difference between adjacent ranked means smaller than 16.27 is not significant.
4. Range of differences required for significance: 16.27 − 24.59.

Waller–Duncan Bayes LSD (BLSD)

1. Do the ANOVA and compute F = MST/MSE with (to correspond to the table symbols) q and f degrees of freedom.
2. Choose an error weight ratio, k. Use k = 100 for an approximate 5% significance test or k = 500 for an approximate 1% test.

3. Obtain a minimum average risk t, t_B, from Table A7. This depends on k, F, q, and f.
4. Compute BLSD as $BLSD = t_B \sqrt{\frac{2MSE}{r}}$.
5. For any pair of means if $|\bar{y}_j - \bar{y}_{j'}| > BLSD$, the difference is a real difference.

For the example data we have:

1. Error weight ratio corresponding to $\alpha = 5\%$ is $k = 100$. $F = MST/MSE = 24.80$, with $q = 5$ and $f = 18$ df. $t_B = t_{(k,F,q,t)} = t_{(100,25,5,18)} = 1.93$
2. $BLSD = t_B \sqrt{\frac{2MSE}{r}} = 1.93 \sqrt{\frac{(2)(120.00)}{4}} = 14.95.$
3. Any difference greater than 14.95 is declared to be a real difference.

A few comments about multiple comparison procedures would seem to be appropriate here.

1. Carmer and Swanson (1973) conducted a computer simulation study involving 88,000 differences. In this study they compared LSD, FPLSD, HSD, SNK, and BLSD, among others. They concluded that there was little difference between FPLSD and BLSD in their ability to protect against a type I error and, more important, for their power to detect real differences if they exist. None of the other methods approached these two.
2. The preference of this author, and others (Carmer and Walker, 1982), is the BLSD. It is a single value that is larger when the F indicates that the means are homogeneous and small when the means appear to be heterogeneous. Further, its power to detect real differences does not depend on the number of means being compared.
3. If the necessary tables for using the BLSD are not available, the FPLSD, which uses the commonly available t table, is quite acceptable as a multiple comparison test.

5.3 EXPERIMENTAL DESIGNS FOR VARIETY TRIALS

In a variety trial the treatments consist of the new entries that are being evaluated and one or more checks, or standards, against which the new entries are compared. Any type of experimental design may be used for a variety trial. The most common, of course, is the randomized block design.

There are a couple of stages in the selection process, however, where a more specialized type of design might be useful. We shall take up this point shortly, but first we should comment on the type and number of checks to be included in the trial.

5.3.1 Type and Number of Checks

Checks, or standards, against which the new entries are compared are an integral part of any variety trial. Some general observations about the nature and number of checks in a trial are:

1. One or more checks may be used. One of these should be a long-term check that is unchanged from year to year. This is the standard that serves to monitor experimental conditions from one year to the next. It also serves as a baseline against which to measure the progress of the breeding program.
2. Other checks may be included for different purposes:
   a. For yield trials that are conducted at locations in several countries it is common to provide for the inclusion of a local variety to serve as a standard for local conditions.
   b. A newly released variety might be included to serve as the "best" current standard for comparison.
3. Because all of the new entries are compared to the same checks, the checks should be replicated at a higher rate than any of the new entries. Finney (1964) has suggested that the number of replications of a check should be the square root of the number of new entries in the trial. For example, with 100 new entries, $\sqrt{100} = 10$ replications of the check should be used for each replication of the new selections.

If the checks and new entries are not equally replicated, a modification must be made in the way the LSI is computed. Suppose that there are c replications of a check and r replications of each new entry. Then, to see whether the yield of a new selection is significantly greater than the check, we use

$$\text{LSI} = t_\alpha \sqrt{\left(\frac{1}{c} + \frac{1}{r}\right)\text{mse}} = t_\alpha \sqrt{\frac{(c + r)\text{MSE}}{cr}}$$

where t_α is the one-tailed t at the α level with the df for MSE.

5.4 AUGMENTED DESIGNS

In the early stages of the selection process there is insufficient seed of the new selections to permit the use of replicated yield trials. The usual procedure is to plant single rows of the new selections and, at regular intervals, plant a row of a check variety. The yields of the new selections are then compared subjectively with the yield of the nearest check. Because the new selections are not replicated, however, the usual statistical analysis is not possible. It would be useful if an objective comparison could be made at this point.

There are a couple of alternatives. One is to compute a measure of experimental error from the variation among observations of the check yields. Then, under the assumption that the variability of the new selections is similar to that of the checks, compute an LSI against which to compare the yield of the new selections with the mean of a check. For example, suppose that the check is replicated n times. Suppose that x_i represents the yield of the i-th observation on the check. Then

$$\bar{x}_i = \frac{\sum_i x_i}{n} = \text{check mean}$$

$$s^2 = \frac{\sum_i x^2_i - (\sum_i x_i)^2/n}{n - 1} = \text{variance of the check}$$

and

$$\text{LSI} = t_\alpha \sqrt{\left(1 + \frac{1}{n}\right)s^2} = t_\alpha \sqrt{\frac{(n + 1)s^2}{n}}$$

where t_α is the one-tailed α-level t at $n - 1$ df. Any new selection yield that is greater than $\bar{x}$ + LSI would be declared to be significantly greater than the check mean.

There are a couple of disadvantages to this procedure: (1) the checks are systematically placed, and (2) no provision is made to adjust yields for soil or other differences from one part of the experiment to another. A better method for the early part of the program is to use an experimental design called an *augmented design* which was developed by Federer (1956, 1961) and described by Federer and Ragavarao (1975).

5.4.1 Description and Construction

In the augmented design the area is divided into blocks in the usual way, the plots in the blocks being single rows. Two or more check varieties are then assigned at random to rows within the blocks, with the same check varieties appearing in every block. The remaining rows within the blocks are assigned to new selections, with a different set of new selections in each block. The number of new selections need not be the same in every block, but the design is most efficient if the block size is constant. Further, little is lost if one of the check varieties is placed systematically, say as the first entry in each block to serve as a block marker. The resulting design is one in which the check entries are replicated but the new entries are not.

The augmented design serves two purposes:

1. It provides an estimate of experimental error and an objective way of comparing new selections with the checks.
2. It provides a way of adjusting the yields of the new selections for differences from block to block.

The number of blocks is determined by the need to have at least 10 degrees of freedom for error in the analysis of variance of the checks. This restriction determines the minimum number of blocks. The final number of blocks is determined by the number of new entries and by the maximum acceptable block size for the experimental area. Suppose that c is the number of different check varieties to be used and that r is the number of blocks. In the ANOVA of the check varieties the degrees of freedom for error are (r − 1) (c − 1). As a result, the number of blocks must be such that the following inequality holds:

$$r \geq \frac{10}{c - 1} + 1$$

For example, with four checks

$$r \geq \frac{10}{4 - 1} + 1 = 4.33$$

In this case, the minimum number of blocks would be five. Each block would contain five or more plots, depending on the number of new selections to be tested.

5.4.2 Analysis

In the analysis of data from an augmented design the experimental error is estimated by treating the checks as if they were treatments in a randomized block design. An analysis of variance is conducted on the check yields and the error mean square from this analysis is used to construct standard errors for the various comparisons. Adjustments for block differences are based on the difference between the mean of the checks in a given block and the mean of the checks over the entire trial.

If block size is constant the following definitions and relationships hold:

c = number of different checks per block

v = total number of new selections
r = number of blocks
n = v/r = number of new selections per block
p = c + n = number of plots per block
N = rc + v = r(c + n) = total number of plots

We begin the analysis of data from an augmented design by first conducting an ANOVA on the check yields. Let x_{ij} represent the yield of the i-th check in the j-th block. The first step is to construct the two-way table of totals shown as Table 5.4. We can now compute the block adjustment, a_j, as

$$a_j = \bar{x}_j - \bar{\bar{x}}$$

(*Note*: $\Sigma_j\, a_j = 0$.) The yields of the new selections adjusted for the effect of the block in which it is grown may be tabulated as shown in Table 5.5.

Table 5.4 **Check Variety × Block Two-Way Table of Totals**[a]

Check Variety	Block 1	2	...	r	Sum	Mean
1	x_{11}	x_{12}	...	x_{1r}	C_1	$\bar{x}_{1\cdot}$
2	x_{21}	x_{22}	...	x_{2r}	C_2	$\bar{x}_{2\cdot}$
⋮	⋮	⋮		⋮	⋮	⋮
c	x_{c1}	x_{c2}	...	x_{cr}	C_c	$\bar{x}_{c\cdot}$
Sum	R_1	R_2	...	R_r	G	
Mean	$\bar{x}_{\cdot 1}$	$\bar{x}_{\cdot 2}$	...	$\bar{x}_{\cdot r}$		$\bar{\bar{x}}$
Adjustment	a_1	a_2	...	a_r		

[a] $C_i = \Sigma_j\, x_{ij}$ = sum of yields of the i-th check variety
$R_i = \Sigma_i\, x_{ij}$ = sum of check yields in the j-th block
$G = \Sigma_i\, C_i = \Sigma_j\, R_j$ = grand total of all checks
$\bar{x}_{i\cdot} = C_i/r$ = mean of the i-th check
$\bar{x}_{\cdot j} = R_j/c$ = mean of all checks in the j-th block
$\bar{\bar{x}} = G/rc$ = grand mean of all checks

Table 5.5 **Adjusted Yields of New Selections**[a]

Selection	Yield	
	Observed	Adjusted
1	y_{1j}	$\hat{y}_{1j}$
2	y_{2j}	$\hat{y}_{2j}$
.	.	.
.	.	.
.	.	.
v	y_{vj}	$\hat{y}_{vj}$

[a] y_{ij} = yield of new selection i grown in the j-th block
$\hat{y}_{ij} = y_{ij} - a_j$ = adjusted yield of the i-th selection (adjusted for block effect)

Table 5.6 **Analysis of Variance of Check Yields from an Augmented Design**

ANOVA

Source	df	SS	MS
Total	rc−1	SSTot	
Blocks	r−1	SSR	
Checks	c−1	SSC	
Error	(r−1)(c−1)	SSE	MSE

To obtain the estimate of experimental error we complete the standard analysis of variance for a randomized block design using the check yields for data. This is displayed in Table 5.6. The computations for the entries in this table are as follows:

$$\text{SSTot} = \sum_i \sum_j x_{ij}^2 - \frac{G^2}{rc}$$

$$\text{SSR} = \frac{1}{c} \sum_j R_j^2 - \frac{G^2}{rc}$$

$$SSC = \frac{1}{r} \sum_i C_i^2 - \frac{G^2}{rc}$$

$$SSE = SSTot - SSR - SSC$$

$$MSE = \frac{SSE}{(r - 1)(c - 1)}$$

An estimate, s^2, of experimental error is given by $s^2 = MSE$.

5.4.3 Standard Errors

There are a number of types of differences to be computed for the augmented design. These differences and their standard errors are as follows:

1. Difference between means of *two check varieties*:

   $$\text{standard error, } s_c = \sqrt{\frac{2MSE}{r}}$$

2. Difference between adjusted yields of *two new selections in the same block*:

   $$\text{standard error, } s_d = \sqrt{2MSE}$$

3. Difference between adjusted yields of *two new selections in different blocks*:

   $$\text{standard error, } s_v = \sqrt{\frac{2(c + 1)MSE}{c}}$$

4. Difference between adjusted yield of *a new selection and a check mean*:

   $$\text{standard error, } s_{vc} = \sqrt{\frac{(r + 1)(c + 1)MSE}{rc}}$$

These standard errors may be used for computing LSD or LSI values, as appropriate.

5.4.4 Numerical Example

A cereal breeder wanted to conduct a preliminary yield trial on 30 new selections of durum wheat. He wanted to compare the new selections against three standard varieties: (1) St = 'Stork', (2) Ci = 'Cimmaron', and (3) Wa = 'Waha'. He had only enough seed of the new varieties to plant a single 2.5-m row of each, so he decided to use an augmented design. Since he wanted to include c = 3 checks, he required a minimum of

$$r \geq \frac{10}{c - 1} + 1$$

$$r \geq \frac{10}{3 - 1} + 1$$

$$r \geq 6$$

blocks to have sufficient df for estimating experimental error. Using 6 blocks he had a design with the following characteristics:

$$c = 3$$

$$v = 30$$

$$r = 6$$

$$n = \frac{v}{r} = \frac{30}{6} = 5$$

$$p = c + n = 3 + 5 = 8$$

$$N = rc + v = 3 \times 6 + 30 = 48$$

Suppose that the field plan, after randomization, and the grain yields (kg/ha) were as given in Table 5.7. Note that the check variety 'Stork', ST, was used as the first entry in each block to serve as a marker.

To begin the analysis, Table 5.8 containing the summary data for the check varieties is constructed. The observed yields, y_{ij}, and adjusted yields, $\hat{y}_{ij}$, of the new selections are presented in Table 5.9. An analysis of variance, shown in Table 5.10, is now done on the yields of the check varieties. This provides a sample estimate of the experimental error for this trial: MSE = 91,103.

The standard errors for the various comparisons are:

1. Difference between check means:

$$s_c = \sqrt{\frac{2MSE}{r}} = \sqrt{\frac{(2)(91{,}103)}{6}} = 174$$

2. Difference between adjusted yields of two selection means in the same block:

$$s_d = \sqrt{2MSE} = \sqrt{(2)(91{,}103)} = 427$$

Table 5.7 **Field Plan and Yields (kg/ha) for an Augmented Design on Durum Wheat**

Block I		II		III	
Selection	Yield	Selection	Yield	Selection	Yield
St	2972	St	3122	St	2260
14	2405	Ci	3023	18	2603
26	2855	4	3018	27	2857
Ci	2592	15	2477	Ci	2918
17	2572	30	2955	25	2825
Wa	2608	3	3055	28	1903
22	2705	Wa	2477	5	2065
13	2391	24	2783	Wa	3107

Block IV		V		VI	
Selection	Yield	Selection	Yield	Selection	Yield
St	3348	St	1315	St	3538
9	2268	2	1055	29	2915
6	2148	21	1688	7	3265
Ci	2940	Wa	1625	Ci	3483
Wa	2850	Ci	1398	1	3013
20	2670	10	1293	Wa	3400
11	3380	8	1253	12	2385
23	2770	16	1495	19	3643

Table 5.8 **Yields, Totals, and Means of the Checks, and Adjustments for Block Differences (kg/ha)**

Variety	Block I	II	III	IV	V	VI	Sum	Mean
Stork	2972	3122	2260	3348	1315	3538	16555	2759
Cimmaron	2592	3023	2918	2940	1398	3483	16354	2726
Waha	2608	2477	3107	2850	1625	3400	16067	2678
Sum	8172	8622	8285	9138	4338	10421	48976	
Mean	2724	2874	2762	3046	1446	3474		2721
Adjust.*	3	153	41	325	-1275	753		

*$a_j = \bar{x}_{\cdot j} - \bar{\bar{x}}$; for example, $a_3 = 2762 - 2721 = 41$.

3. Difference between adjusted yields of two selection means in different blocks:

$$s_v = \sqrt{\frac{2(c + 1)MSE}{c}} = \sqrt{\frac{(2)(3 + 1)(91{,}103)}{3}} = 493$$

4. Difference between an adjusted selection yield and a check mean:

$$s_{vc} = \sqrt{\frac{(r + 1)(c + 1)MSE}{rc}} = \sqrt{\frac{(6 + 1)(3 + 1)\ (91{,}103)}{(6 \times 3)}} = 376$$

It is possible to compute LSD values for each of these types of comparison. For this purpose we would use a two-sided t_α with $(r - 1)(c - 1) = (6 - 1)(3 - 1) = 10$ df. At the 5% level this would be $t_{0.05} = 2.228$. For this trial, however, the most useful comparison would be between the adjusted yield of a new selection and a check mean. Here we would like to find those new selections that outyield the checks, and we would use the LSI for this purpose:

$$LSI = t_\alpha s_{vc}$$

Table 5.9 **Observed, y_{ij}, and Adjusted, $\hat{y}_{ij}$. Yields (kg/ha) of New Selections of Durum Wheat**[a]

Select.	Block	y_{ij}	$\hat{y}_{ij}$	Select.	Block	y_{ij}	$\hat{y}_{ij}$
1	6	3013	2260	16	5	1495	2770
2	5	1055	2330	17	1	2572	2569
3	2	3055	2902	18	3	2603	2562
4	2	3018	2865	19	6	3643	2890
5	3	2065	2024	20	4	2670	2345
6	4	2148	1823	21	5	1688	2963
7	6	3265	2512	22	1	2705	2702
8	5	1253	2528	23	4	2770	2445
9	4	2268	1943	24	2	2783	2630
10	5	1293	2568	25	3	2825	2784
11	4	3380	3055	26	1	2855	2852
12	6	2385	1632	27	3	2857	2816
13	1	2391	2388	28	3	1903	1862
14	1	2405	2402	29	6	2915	2162
15	2	2477	2324	30	2	2955	2802

[a] $\hat{y}_{ij} = y_{ij} - a_j$.

For the 5% LSI we have the 5% one-tailed t at 10 df:

$$t_{0.05} = 1.812$$

and

$$\text{LSI} = (1.812)(376) = 681$$

Then any adjusted selection yield greater than

1. 2759 + 681 = 3440 significantly outyields 'Stork'

Table 5.10 **Analysis of Variance of Check Yields in an Augmented Design to Estimate the Yields of 30 New Durum Wheat Selections**

ANOVA

Source	df	SS	MS	F
Total	17	7,899,564		
Blocks	5	6,986,486		
Checks	2	20,051		
Error	10	911,027	91,103	

2. 2726 + 681 = 3407 significantly outyields 'Cimmaron'
3. 2678 + 681 = 3359 significantly outyields 'Waha'

Report of Statistical Analysis

A preliminary yield trial was conducted to evaluate 30 new selections of durum wheat. The new selections were compared against three standard varieties using an augmented design with six blocks of eight plots each. The mean yields of the standard varieties and the yields, adjusted for block differences, of the new selections are given in Table 5.11. Although the adjusted yield of 10 of the new selections was greater than the yield of the highest check, 'Stork', none of the yields was significantly higher.

5.5 INCOMPLETE BLOCK DESIGNS

One of the basic principles of modern experimental design is to group the experimental units into blocks of homogeneous units to increase precision and to make comparisons under as nearly uniform conditions as possible. It is generally true, however, that precision increases as block size decreases, so that small blocks are preferable to large ones.

When a plant breeding program has reached the early yield trial stage there is usually sufficient seed to permit the use of replicated trials. However, the number of new selections to be tested is often quite large. In this situation it is often difficult to find sufficient uniform plots to form blocks large enough to contain a complete replication of all the new selections. Hence the effect of blocking is lost unless block size can be made smaller than the number of new selections to be tested. For these trials we require

Table 5.11 **Mean Yields of Checks and Adjusted Yields (kg/ha) of New Selections of Durum Wheat in Order of Decreasing Yields**[a]

Select.	yield	Select.	yield	Select.	yield
11	3055	Cimmaron	2726	13	2388
21	2963	22	2702	20	2345
3	2902	Waha	2678	2	2330
19	2890	24	2630	15	2324
4	2865	17	2569	1	2260
26	2852	10	2568	29	2162
27	2816	18	2562	5	2024
30	2802	8	2528	9	1943
25	2784	7	2512	28	1862
16	2770	23	2445	6	1823
Stork	2759	14	2402	12	1632

[a]5% LSI = 681 kg/ha.

a procedure for reducing the size of the blocks. This is possible with a class of designs called *incomplete block designs.* As the name implies, in incomplete block designs the plots are grouped into blocks that are not large enough to contain a full replication of all treatments.

Incomplete block designs fall into two broad classes: balanced designs and partially balanced designs. In *balanced designs* each treatment occurs together in the same block with every other treatment an equal number of times, usually once. In a balanced incomplete block design all pairs of treatments are compared with the same precision even though the differences between blocks may be large. It is possible to construct a balanced design for any number of treatments and any block size (plots per block). These two parameters fix the minimum number of replications required for balance. Often, this minimum is too large for the balanced design to be of practical significance.

Because balanced designs are not always practical, a class of designs called *partially balanced incomplete block designs* has been developed. In the most useful of these, some pairs of treatments occur together in the same block λ_1 times, and other pairs occur together λ_2 times, where λ_1 and λ_2 are integers (see Bose et al., 1954). Partially balanced designs, although more useful in a practical sense, are not as desirable as balanced designs. They are difficult to analyze statistically, and comparisons among some pairs of treatments are made with greater precision than for other pairs. A detailed examination of all types of incomplete block designs is beyond the scope of this book. For descriptions of designs not included here, the reader is referred to Cochran and Cox (1957).

5.5.1 Lattice Designs

Among the most useful of the incomplete block designs for the plant breeder is a class of designs called *lattices*. In the lattice designs the number of treatments must be a perfect square. The number of blocks and the number of plots in each block is the square root of the number of treatments. These incomplete blocks are combined into groups that form separate, complete replications. When an incomplete block design consists of two or more complete replications it has been shown that the design can be analyzed as if it were an ordinary randomized block design. Hence the blocking restriction can be ignored without destroying the validity of the analysis even though the precision may be reduced.

There are a number of types of lattice designs. These include, in general, balanced lattices and partially balanced lattices. In *balanced lattices* each treatment occurs together in the same block with every other treatment an equal number of times. However, if there are k blocks, there must be k + 1 replications to achieve balance. This is beyond the resources of most breeding programs. In *partially balanced lattices* the number of replications is reduced, but not all treatments occur together an the same block. This leads to a difference in the precision with which some comparisons are made relative to other comparisons. However, the reduced replication makes the trial manageable for most breeding programs while providing the possibility of increased precision associated with smaller blocks.

In all the lattice designs the number of treatments must be an exact square, and the number of plots per incomplete block is the square root of the number of treatments. The designs differ from each other primarily in the number of replications called for by the design. Plans for lattices in the range of treatment numbers that are useful to plant breeders are presented in Section 5.9. These plans are for $k^2 = 9$ to $k^2 = 100$ treatments with block sizes of $k = 3$ to $k = 10$. Plans are given for four replications where possible. However, designs with four replications that meet the lattice criteria are not possible for $k = 6$ and $k = 10$.

5.5.2 Field Arrangement and Randomization

The blocks in lattice designs should be composed of plots that are as homogeneous as possible. Further, blocks in the same replication should be as nearly alike as possible to maximize the variation among replications. This will result in maximum precision if the experiment is analyzed as a randomized block design. Randomization proceeds as follows:

1. Randomize the order of blocks within replications using a separate randomization in each replication.
2. Randomize the order of treatments within blocks. Be sure that the treatments which appear in a given block on the plan remain in the same block even though the order has been randomized within the block.

5.5.3 Types of Lattices

Simple Lattices These designs have two replications. Plans for the simple lattices may be obtained by taking the first two replications from the set of plans in Section 5.9. Although they exist, simple lattices for 9 and 16 treatments are not likely to be more precise than randomized block designs. This is because the blocking restriction reduces the error degrees of freedom below an acceptable level in 3×3 and 4×4 simple lattices.

Triple Lattices Triple lattices contain three replications. They are formed by using the first three replications from the plans in Section 5.9. Triple lattices are possible for all squares from 3×3 to 13×13 treatments.

Quadruple Lattices Quadruple lattices contain four replications. Where they exist quadruple lattices are obtained by using all four replications from the plans in Section 5.9. True quadruple lattices do not exist for 6×6 and 10×10 treatments. Four replications for these treatment numbers may be obtained by repeating the simple lattices. However, the data analysis in this case is different from the analysis to be presented here. Proper analysis for designs constructed by repeating the simple lattices may be found in most texts on experimental design (see, e.g., Petersen, 1985).

5.5.4 Analysis

The details for data analysis are the same for simple, triple, and quadruple lattices. Assume that we have a design for k^2 treatments replicated r times. There are k blocks of k plots each in each replication. Let $y_{ij(l)}$ represent the yield of the j-th treatment applied to a plot in the l-th block of the i-th replication. The data may be tabulated as shown in Table 5.12.

To continue the analysis it is also necessary to compute the unadjusted treatment totals. These may be tabulated, along with the adjusted treatment totals and means, as shown in Table 5.13. The format for the analysis of variance is given in Table 5.14. Except for SSB, the sums of squares are computed in the usual way, with SSE obtained by difference. We have

$$\text{SSTot} = \sum_i \sum_j y_{ij(l)}^2 - \frac{G^2}{rk^2}$$

$$\text{SSR} = \frac{1}{k^2} \sum_i R_j^2 - \frac{G^2}{rk^2}$$

$$\text{SSB} = \frac{1}{kr(r-1)} \sum_i \sum_l C_{il}^2 - \frac{1}{k^2 r(r-1)} \sum_i C_i^2$$

$$\text{SST} = \frac{1}{r} \sum_j T_j^2 - \frac{G^2}{rk^2}$$

$$\text{SSE} = \text{SSTot} - \text{SSR} - \text{SSB} - \text{SST}$$

and

$$E_b = \frac{\text{SSB}}{r(k-1)}$$

$$E_e = \frac{\text{SSE}}{(k-1)(rk-k-1)}$$

We now compare E_b with E_e. If $E_b \leq E_e$, the adjustment for blocks will have no effect. In this case we ignore the blocking restriction and analyze the data as if the design had been a randomized block design with replications as blocks.

If $E_b > E_e$, an adjustment factor, A, is computed for these designs,

$$A = \frac{E_b - E_e}{k(r-1)E_b}$$

This adjustment factor is used to compute AC_{il} in the data table, Table 5.12, and $\hat{T}_j$ in the table of totals, Table 5.13.

The effective error mean square, E_e', is now computed as

$$E\acute{e} = \left(\frac{1 + rkA}{k+1}\right)E_e$$

Except for small designs ($k = 3,4$), E_e' may be used in t tests and interval estimates.

Table 5.12 **Summary of Data from a Lattice Design**[a]

Rep.	Block	Yield				Sum B_{il}	Sum C_{il}	Adj. Sum C_{il}
1	1	$y_{11(1)}$	$y_{12(1)}$	...	$y_{1k(1)}$	B_{11}	C_{11}	AC_{11}
	2	$y_{1k+1(2)}$	$y_{1k+2(2)}$	...	$y_{1k+k(2)}$	B_{12}	C_{12}	AC_{12}
	.	.			.	.	.	.
	.	.			.	.	.	.
	.	.			.	.	.	.
	k	$y_{1k^2-k(k)}$		...	$y_{1k^2(k)}$	B_{1k}	C_{1k}	AC_{1k}
					Sum	R_1	C_1	
2	1	$y_{21(1)}$	$y_{22(1)}$	...	$y_{2k(1)}$	B_{21}	C_{21}	AC_{21}
	2	$y_{2k+1(2)}$	$y_{2k+2(2)}$	...	$y_{2k+k(2)}$	B_{22}	C_{22}	AC_{22}
	.	.			.	.	.	.
	.	.			.	.	.	.
	.	.			.	.	.	.
	k	$y_{2k^2-k(k)}$		...	$y_{2k^2(k)}$	B_{2k}	C_{2k}	AC_{2k}
					Sum	R_2	C_2	
r	1	$y_{r1(1)}$	$y_{r2(1)}$	...	$y_{rk(1)}$	B_{r1}	C_{r1}	AC_{r1}
	2	$y_{rk+1(2)}$	$y_{rk+2(2)}$	...	$y_{rk+k(2)}$	B_{r2}	C_{r2}	AC_{r2}
	.	.			.	.	.	.
	.	.			.	.	.	.
	.	.			.	.	.	.
	k	$y_{rk^2-k(k)}$		...	$y_{rk^2(k)}$	B_{rk}	C_{rk}	AC_{rk}
					Sum	R_r	C_r	
					Sum	G	0	0

[a]B_{il} = sum of yields of the k plots in the l-th block of the i-th replication
$R_i = \Sigma_l B_{il}$ = sum of yields in the i-th replication
C_{il} = sum (over all replications) of yields of all treatments in the l-th block of the i-th preplication minus rB_{il}
$C_i = \Sigma_l C_{il}$
AC_{il} = adjusted sum, where A is an adjustment factor computed after completing the ANOVA
$G = \Sigma_i R_i$ = grand total of all yields

Table 5.13 **Summary of Treatment Totals and Means from a Lattice Design**[a]

	Total		Mean
Treatment	Unadjusted	Adjusted	Adjusted
1	T_1	$\hat{T}_1$	$\bar{y}_1$
2	T_2	$\hat{T}_2$	$\bar{y}_2$
.	.	.	.
.	.	.	.
.	.	.	.
k	T_{k^2}	$\hat{T}_{k^2}$	$\bar{y}_{k^2}$

[a] $T_j = \Sigma_i\ y_{ij(l)}$ = sum of yields for the i-th treatment
$\hat{T}_j = T_j$ + (sum of the AC_{il} over all blocks in which the j-th treatment appears)
$\bar{y}_j = \hat{T}_j/r$

Table 5.14 **Analysis of Variance Format for a Lattice Design**

ANOVA

Source	df	SS	MS
Total	rk -1	SSTot	
Rep.	r-1	SSR	
Treat.	k -1	SST	
Block(adj.)	r(k-1)	SSB	E_b
Intrablock error	(k-1)(rk-k-1)	SSE	E_e

To test the significance of differences among adjusted treatment means, an adjusted treatment mean square is computed. To do this we first compute SSBu, the unadjusted blocks within reps sum of squares:

$$SSBu = \frac{1}{k} \sum_i \sum_l B^2_{il} - \frac{G^2}{rk^2} - SSR$$

Then the adjusted treatment sum of squares, SST(adj), is computed as

$$SST(adj) = SST - Ak(r - 1)\left[\frac{rSSBu}{(r - 1)(1 + kA)} - SSB\right]$$

Finally,

$$F = \frac{SST(adj)/(k - 1)}{E_e} \qquad \text{with} \quad k^2 - 1, (k - 1)(rk - k - 1) \text{ df}$$

is a test statistic for testing the differences among the adjusted treatment means.

5.5.5 Standard Errors

1. The standard error of an adjusted treatment mean, $s_{\bar{y}}$, is

$$s_{\bar{y}} = \sqrt{\frac{E_e'}{r}}$$

2. The standard error of the difference between adjusted means depends on whether or not the two treatments occur together in the same block. For treatments in the same block

$$s_{\bar{d}1} = \sqrt{\frac{2E_e}{r}[1 + (r - 1)A]}$$

For treatments in different blocks

$$s_{\bar{d}2} = \sqrt{\frac{2E_e}{r}(1 + rA)}$$

For k > 4 it is sufficient to use the effective error mean square for computing the standard error of a difference whether or not the treatments are in the same block. We have

$$s_{\bar{d}} = \sqrt{\frac{2E_e'}{r}}$$

5.5.6 Relative Precision

The precision of a lattice design relative to that of a randomized block design is found by first computing the error mean square, E_{RB}, of a randomized block design. This is

$$E_{RB} = \frac{SSB + SSE}{(k^2 - 1)(r - 1)}$$

Then the relative precision, RP, of the lattice is found to be

$$RP = \frac{E_{RB}}{E'_e}$$

5.5.7 Numerical Example

A cereal breeder wanted to develop a new, high-yielding variety of barley. As a part of the process she selected 22 promising lines for test against three check varieties in a replicated yield trial. Because of soil variability in the test area she decided to use a simple lattice design, a 5 × 5 lattice with two replications.

The field arrangement of the trial and the yields (kg/plot) are shown in Table 5.15. Also included are the C values and the AC_{il} adjustments (computed after completing the ANOVA) for each block. The unadjusted yield totals, T_j, together with the selection numbers in parentheses, are shown in Table 5.16.

The analysis of variance takes the form presented in Table 5.17. Computations for the entries in this table 5.17 are

$$\frac{G^2}{rk^2} = \frac{(691.5)^2}{(2)(25)} = 9563.44$$

$$\begin{aligned} SSTot &= \sum_i \sum_j y^2_{ij(l)} - \frac{G^2}{rk^2} = 18.2^2 + \cdots + 15.0^2 - 9563.44 \\ &= 805.42 \end{aligned}$$

$$\begin{aligned} SSR &= \frac{1}{k^2} \sum_j R_i^2 - \frac{G^2}{rk^2} = \frac{1}{25}(330.5^2 + 361.0^2) - 9563.44 \\ &= 18.60 \end{aligned}$$

$$\begin{aligned} SST &= \frac{1}{r} \sum_i T_j^2 - \frac{G^2}{rk^2} = \frac{1}{2}(30.0^2 + \cdots + 39.2^2) - 9563.44 \\ &= 621.82 \end{aligned}$$

$$\begin{aligned} SSB &= \frac{\sum_i \sum_l C_{il}^2}{kr(r - 1)} - \frac{\sum_i C_i^2}{k^2\, r(r - 1)} \\ &= \frac{1148.02}{(5)(2)(1)} - \frac{1860.50}{(25)(2)(1)} \\ &= 114.80 - 37.21 = 77.59 \end{aligned}$$

Table 5.15 **Yields (kg/plot) of Barley in a 5 × 5 Simple Lattice**[a]

Rep	Blk	Yield					Sum B_{11}	Sum C_{11}	adj. AC_{11}
I	1	(19)* 18.2	(16) 13.0	(18) 9.5	(17) 6.7	(20) 10.1	57.5	17.6	1.54
	2	(12) 13.3	(13) 11.4	(15) 14.2	(14) 11.9	(11) 13.4	64.2	4.2	.37
	3	(1) 15.0	(2) 12.4	(3) 17.3	(4) 20.5	(5) 13.0	78.2	5.3	.46
	4	(22) 7.0	(24) 5.9	(21) 14.0	(25) 19.2	(23) 7.8	54.0	-.5	-.04
	5	(9) 11.9	(7) 15.2	(10) 17.2	(8) 16.3	(6) 16.0	76.6	3.9	.34
						Sum	330.5	30.5	2.67
Rep	**Blk**						B_{21}	C_{21}	AC_{21}
II	1	(23) 7.7	(18) 15.2	(3) 19.1	(8) 15.5	(13) 14.7	72.2	-9.9	-.86
	2	(5) 15.8	(20) 18.0	(10) 18.8	(15) 14.4	(25) 20.0	87.0	-13.3	-1.17
	3	(22) 10.2	(12) 11.5	(2) 17.0	(17) 11.0	(7) 15.3	65.0	-10.4	-.91
	4	(14) 10.9	(24) 4.7	(9) 10.9	(4) 16.6	(19) 9.8	52.9	15.5	1.35
	5	(6) 20.0	(16) 21.1	(11) 16.9	(21) 10.9	(1) 15.0	83.9	-12.4	-1.08
						Sum	361.0	-30.5	-2.67
						Sum	691.5	0.0	0.00

*Selection number in parentheses.

[a] C_{il} = sum (over both reps) of all treatments in the block − $2B_{il}$; for example, $C_{11} = (18.2 + 9.8) + (13.0 + 21.1) + (9.5 + 15.2) + (6.7 + 11.0) + (10.1 + 18.0) - (2)(57.5) = 17.6$

Table 5.16 **Unadjusted Yield Totals (Selection Number in Parentheses)**

(1)	30.0	(2)	29.4	(3)	36.4	(4)	37.1	(5)	28.8
(6)	36.0	(7)	30.5	(8)	31.8	(9)	22.8	(10)	36.0
(11)	30.3	(12)	24.8	(13)	26.1	(14)	22.8	(15)	28.6
(16)	34.1	(17)	17.7	(18)	24.7	(19)	28.0	(20)	28.1
(21)	25.0	(22)	17.2	(23)	15.5	(24)	10.6	(25)	39.2
							Sum		691.5

Table 5.17 **Analysis of Variance of Barley Yields from a 5 × 5 Simple Lattice Trial**

ANOVA

Source	df	SS	MS
Total	49	805.42	
Replication	1	18.60	
Selection(unadj.)	24	621.82	
Block in rep(adj.)	8	77.59	$9.70 = E_b$
Intrablock error	16	87.41	$5.46 = E_e$

$$\begin{aligned} SSE &= SSTot - SSR - SSB - SST \\ &= 805.42 - 18.60 - 77.59 - 621.82 = 87.41 \end{aligned}$$

Since 9.70 > 5.46, blocking is effective. The adjustment factor, A, is computed as

$$\begin{aligned} A &= \frac{E_b - E_e}{k(r-1)E_b} = \frac{9.70 - 5.46}{(5)(1)(9.70)} \\ &= 0.0874 \end{aligned}$$

The block adjustments, AC_{il}, are now computed. They are given in Table

5.15. The adjusted treatment totals and means are now computed. These are shown in Table 5.18.

For this trial the effective error mean square, E'_e, is

$$E'_e = \left(\frac{1 + rkA}{k + 1}\right)E_e = \left(\frac{1 + (2)(5)(0.0874)}{6}\right)(5.46) = 6.26$$

To test the significance of the differences among adjusted treatment means, it is first necessary to compute the unadjusted blocks within replication sum of squares, SSBu:

$$\begin{aligned} SSBu &= \frac{1}{k}\sum_i \sum_l B_{il}^2 - \frac{G^2}{rk^2} - SSR \\ &= 9834.23 - 9563.44 - 18.60 = 252.18 \end{aligned}$$

Then

$$\begin{aligned} SST(adj) &= SST - Ak(r - 1)\left\{\left[\frac{rSSBu}{(r - 1)(1 + kA)}\right] - SSB\right\} \\ &= 621.82 - (0.0874)(5)(1)\left\{\left[\frac{(2)(252.18)}{(1)(1 + 5 \times 0.0874)}\right] - 77.59\right\} \\ &= 502.35 \end{aligned}$$

and

$$\begin{aligned} F &= \frac{SST(adj)/(k^2 - 1)}{E_e} = \frac{502.35/24}{5.46} \\ &= 3.83 \end{aligned}$$

Since $F_{0.01(24.16)} = 3.18 < 3.83$ differences among the adjusted means are significant at the 1% level.

The standard error of a selection mean, $s\bar{y}$, is

$$s\bar{y} = \sqrt{\frac{E'_e}{r}} = \sqrt{\frac{6.26}{2}} = 1.77$$

Since $k = 5$ the effective error mean square may be used for computing an LSI for determining which new selections outyield the checks. We have

$$LSI = t_\alpha \sqrt{\frac{2E'_e}{r}} = 1.746\sqrt{\frac{(2)(6.26)}{2}} = 4.37$$

(*Note*: t_α = one-tailed, 5% t with 16 df.)

Table 5.18 **Adjusted Treatment Totals and Means**

Selection	Total(adj.) *	Mean(adj.)
1	29.38	14.69
2	28.95	14.48
3	36.00	18.00
4	38.91	19.46
5	28.09	14.04
6	35.26	17.63
7	29.93	14.96
8	31.28	15.64
9	24.49	12.24
10	35.17	17.58
11	29.59	14.80
12	24.26	12.13
13	25.61	12.80
14	24.52	12.26
15	27.80	13.90
16	34.56	17.28
17	18.33	9.16
18	25.38	12.69
19	30.89	15.44
20	28.47	14.23
21	23.88	11.94
22	16.25	8.12
23	14.60	7.30
24	11.19	5.96
25	37.99	19.00

*$\text{Total(adj)}_j = T_j + (\Sigma\ AC_{il}$ for all blocks in which the j-th treatment appears); for example, $\hat{T}_1 = T_1 + (AC_{13} + AC_{25}) = 30.0 + 0.46 - 1.08 = 29.38$.

The precision of the simple lattice relative to that of a randomized block design is found by computing E_{RB}, the error mean square of a randomized block design:

$$E_{RB} = \frac{SSB + SSE}{(k^2 - 1)(r - 1)} = \frac{77.59 + 87.41}{(24)(1)}$$
$$= 6.88$$

Then

$$\% \text{ relative precision} = \left(\frac{E_{RB}}{E'_e}\right)100 = \left(\frac{6.88}{6.26}\right)100$$
$$RP = 110.0\%$$

Report of Statistical Analysis

A yield trial was run by a plant breeder to evaluate the yield potential of 22 new selections of barley relative to three check varieties. Because of variation in the experimental site, and because of economic considerations, the trial was run using a 5 × 5 simple lattice design. As was customary at this research center selections 1, 13, and 25 were assigned to the checks. Selection 13 was the long-time check, 1 was a variety widely grown in the area, and 25 was the latest variety released by the station. The resulting mean yields are shown in Table 5.19.

For this set of data the least significant increase (LSI) at the 5% level was found to be 4.37. Thus five new selections outyielded the long-time check, whose mean yield was 12.80 kg/plot, while one new selection (4) had a yield, 19.46 kg/plot, higher than the local check (1) with a mean yield of 14.69 kg/plot. None of the new selections outyielded the late

Table 5.19 **Adjusted Mean Yields (kg/plot) of 25 Selections of Barley**[a]

(1)	14.69	(2)	14.48	(3)	18.00	(4)	19.46	(5)	14.04
(6)	17.63	(7)	14.96	(8)	15.64	(9)	12.24	(10)	17.58
(11)	14.80	(12)	12.13	(13)	12.80	(14)	12.26	(15)	13.90
(16)	17.28	(17)	9.16	(18)	12.69	(19)	15.44	(20)	14.23
(21)	11.94	(22)	8.12	(23)	7.30	(24)	5.96	(25)	19.00

[a]Standard error = 1.77. Selection numbers in parentheses.

release, whose mean yield was 19.00 kg/plot. Use of the simple lattice resulted in a 10% increase in precision for this trial.

5.6 STABILITY ANALYSIS

5.6.1 Introduction

The goal of a plant breeding program is to produce genotypes that are, in some sense, optimum for the conditions under which they will be grown. There are at least two possible approaches to this goal. Ideally, the growing region could be divided into a number of agroclimatic zones which are uniform with respect to such characteristics as soil properties and weather. Genotypes could then be bred to meet the specific requirements of each zone. There are some drawbacks to this approach. While soil properties remain essentially constant at a given site, climate changes from year to year. This makes the delineation of uniform zones difficult even when the highly skilled people required to do the zoning are available. Further, particularly in the developing areas, even if it were possible to create the zones there are not enough trained plant breeders available to produce specific genotypes for each of the zones in a region.

An alternative and often more feasible approach is to attempt to develop genotypes that are widely adapted over a wide variety of environmental conditions. The breeder then selects those genotypes which, by some measure, show stability across environments. These stable genotypes provide a stock from which superior genotypes may be selected.

5.6.2 Stability Measures

A number of statistics have been proposed to measure genotypic stability. Several of these have been summarized and compared by Lin et al. (1986). All of these statistics can be computed using the observed means in a two-way genotype by environment table, as shown in Table 5.20.

Two of the stability statistics involve the variation within a genotype across environments. These statistics do not depend on the other genotypes that might be included in the trial. Hence they support broad-based inference. Because they provide no information about the response pattern over a range of environments, they are probably most useful over restricted geographical ranges. These statistics are:

1. Within-genotype mean square, s_i^2:

$$s_i^2 = \frac{\Sigma_i (\bar{y}_{ij} - \bar{y}_{i\cdot})^2}{q - 1}$$

Table 5.20 **Observed Means from a Trial Conducted Using p Genotypes at q Environments**[a]

Genotype	Environment 1	2	...	q	Mean
1	$\bar{y}_{11}$	$\bar{y}_{12}$	...	$\bar{y}_{1q}$	$\bar{y}_{1\cdot}$
2	$\bar{y}_{21}$	$\bar{y}_{22}$	...	$\bar{y}_{2q}$	$\bar{y}_{2\cdot}$
.	.	.		.	.
.	.	.		.	.
.	.	.		.	.
p	$\bar{y}_{p1}$	$\bar{y}_{p2}$	...	$\bar{y}_{\cdot q}$	$\bar{y}_{p\cdot}$
Mean	$\bar{y}_{\cdot 1}$	$\bar{y}_{\cdot 2}$		$\bar{y}_{\cdot q}$	$\bar{\bar{y}}$

[a]$\bar{y}_{ij}$ = observed mean of genotype i in environment j
$\bar{y}_{i\cdot}$ = mean, over environments, for genotype i
$= \Sigma_j \bar{y}_{ij}/q$
$\bar{y}_{\cdot j}$ = mean, over genotypes, for environment j
$= \Sigma_i \bar{y}_{is}/p$
$\bar{\bar{y}}$ = grand mean of all observations
$= \Sigma_i \Sigma_j \bar{y}_{ij}/pq$

2. Genotypic coefficient of variation, CV_i (used by Francis and Kannenberg, 1978):

$$CV_i = \left(\frac{\sqrt{S_i^2}}{\bar{y}_{i\cdot}}\right)100$$

For both of these statistics, the smaller the numerical value, the more stable is the genotype.

Two of the stability statistics measure the contribution of the i-th genotype in a trial to the genotype by environment interaction. Both of the statistics are dependent on the other genotypes that are included in the trial. Basically, they measure the deviation of the individual genotype from the location means of all genotypes on test. Genotypic stability should not be generalized using these statistics unless the genotypes used in the test are representative of those grown in the region. That is, a genotype that appears to be relatively stable when grown together with one set of other genotypes might be relatively unstable when grown with a different set. One of these statistics, W_i^2, is an index, and the other, σ_i^2, is an unbiased estimate of the variance of genotype i.

1. Wricke's (1962) ecovalence, W^2_i:

   $$W_i^2 = \Sigma_j (\bar{y}_{ij} - \bar{y}_{i\cdot} - \bar{y}_{\cdot j} + \bar{\bar{y}})^2$$

   [*Note*: $\Sigma_i W_i^2 = SS(GE)$ = genotype by environment SS.]
2. Shukla's (1972) stability variance, σ_i^2, an estimate of the variance of genotype i across environments based on residuals from the two-way (G × E) classification:

   $$\sigma_i^2 = \frac{p}{(p-2)(q-1)W_i^2} - \frac{SS(GE)}{(p-1)(p-2)(q-1)}$$

   where $SS(GE) = \Sigma_i W_i^2 = \Sigma_i \Sigma_j (\bar{y}_{ij} - \bar{y}_{i\cdot} - \bar{y}_{\cdot j} + \bar{\bar{y}})^2$.

With these statistics the genotype that is most stable is the one that minimizes the statistic.

One of the most frequently used stability measures is based on a regression model. This measure was first proposed by Yates and Cochran (1938). However, it was developed to describe the adaptation of individual varieties to changing environment by Finlay and Wilkinson (1963) and again proposed by Eberhardt and Russell (1966). The stability statistic in this instance is the slope, b_i, of the linear regression of the yield, $\bar{y}_{ij}$, of genotype i in environment j on the mean yield, $\bar{y}_{\cdot j}$, of all the genotypes in environment j. Thus the mean yield at environment j serves as an index of the overall yield potential at that environment. This statistic is also dependent on the other genotypes that are included in the trial and should not be taken as a general measure of stability. This statistic, b_i, is computed as

$$b_i = \frac{\Sigma_j (\bar{y}_{ij} - \bar{y}_{i\cdot})(\bar{y}_{\cdot j} - \bar{\bar{y}})}{\Sigma_j (\bar{y}_{\cdot j} - \bar{\bar{y}})^2}$$

With this statistic a genotype that has a slope of b = 1.00 is most stable over the environments in the trial as determined by the other genotypes included in the trial.

Finlay and Wilkinson (1963) point out that the slope provides an indication of regions of adaptability as well as stability. Genotypes that have a slope significantly greater than 1.0 are specifically adapted to high-yielding environments. On the other hand, genotypes with a slope less than 1.0 are insensitive to change in environment and are therefore better adapted to low-yielding environments.

The final two measures of stability are measures of the deviation from the regression of genotype means on the site index. With these measures a genotype is considered to be stable only if the residual mean square from regression is small. These measures are strictly dependent on the genotypes used in the trial and not on variables that can be measured and used to characterize the environment before the trial is run. These measures are difficult to justify because they only measure the goodness of fit of a linear regression model for the data in hand. One of these statistics, δ_i^2, corresponds to the residual mean square from a fitted regression model, while the other, R_i^2, measures the proportion of the variation in the mean yield of a genotype that is accounted for by the fitted model. We have

1. Eberhardt and Russell's (1966), δ_i^2:

$$\delta_i^2 = \frac{1}{q-2}\left[\Sigma_j (\bar{y}_{ij} - \bar{y}_{i\cdot})^2 - b^2 \Sigma_j (\bar{y}_{\cdot j} - \bar{\bar{y}})^2\right]$$

2. Coefficient of determination, R_i^2:

$$R_i^2 = \frac{b_i^2 \Sigma_j (\bar{y}_{\cdot j} - \bar{\bar{y}})^2}{\Sigma_j (\bar{y}_{ij} - \bar{y}_{i\cdot})^2}$$

A poor fit, as indicated by a large δ_i^2 or a small R_i^2, indicates only that the regression model does not describe the data and hence does not provide a real measure of stability.

In most breeding programs the breeder is interested in a particular set of genotypes and in how they perform over a more-or-less limited range of environments. From the selections under test he is interested in those that have a high yield and are relatively stable over the environments tested. For this purpose he should look for a high mean, $\bar{y}_{i\cdot}$, a relatively low ecovalence, W_i^2 (low contribution to the genotype by environment interaction) and a slope, b_i, of a linear regression on the environmental index that is close to 1.00.

5.6.3 Numerical Example

The International Center for Agricultural Research in the Dry Areas (ICARDA) has been assigned the principal responsibility for research on barley throughout the world. To meet this responsibility scientists at ICARDA are involved in developing and testing new barley genotypes. As part of the process the scientists prepare yield trials containing one or two control varieties and from 25 to 30 new selections. These are sent to cooperating scientists at various locations throughout the world. The trials are grown by the cooperating scientists, who report the yield data and other pertinent information back to the barley breeder at ICARDA. The yield trials serve two purposes: they provide the cooperating scientists with a source of new genetic material for their own breeding programs, and they provide information to aid the ICARDA scientists in the selection of genotypes to be continued in the breeding program as possible candidates for release as new varieties.

For the 1979–1980 growing season, ICARDA developed and sent to the cooperators the Tenth Regional Barley Yield Trial (RBYT). The trial, grown by each cooperator using a randomized block experimental design with three blocks, consisted of 23 new selections, as well as a long-time check, 'Beecher', and a local check supplied by the cooperator. The trial was sent to 23 cooperators, primarily in the Mediterranean region but including one in Afghanistan, two on the Arabian peninsula, and one in Oregon. From this set of trials data for 10 genotypes at 12 locations have been selected to illustrate the analysis of stability. These are summarized in Table 5.21.

A wide variety of environments are represented by these data. Mean yields ranged from 430 kg/ha under hot, dry, rain-fed conditions in Saudi Arabia to 6860 kg/ha under cool, damp, irrigated conditions in Oregon. Although some of the genotypes appeared to be widely adapted, others appeared to do better under more restricted environments. For example, "Bco.Mr. Godiva" seems to thrive under the high-rainfall, cool-temperature conditions in parts of Tunisia, Morocco, Greece, and Oregon. On the other hand, 'TJ 70' seems better adapted to the hot, dry environment represented by Oman and Saudi Arabia.

Stability statistics for these data are given in Table 5.22. These statistics are computed from the data in Table 5.21. The computations may be illustrated using genotype 1, 'Beecher', as an example. Because certain quantities enter several of the stability measures it is useful to compute

$$\begin{aligned}
\Sigma_j (\bar{y}_{1j} - \bar{y}_{1\cdot})^2 &= (2724 - 3569)^2 + (6681 - 3569)^2 + \cdots \\
&\quad + (7540 - 3569)^2 \\
&= 45{,}770{,}550 \\
\Sigma_j (\bar{y}_{\cdot j} - \bar{\bar{y}})^2 &= (2286 - 3528)^2 + (6134 - 3528)^2 + \cdots \\
&\quad + (6860 - 3528)^2 \\
&= 35{,}960{,}200 \\
\Sigma_j (\bar{y}_{1j} - \bar{y}_{\cdot j})(\bar{y}_{\cdot j} - \bar{\bar{y}}) &= (2724 - 3569)(2286 - 3528) + \cdots + (7540 \\
&\quad - 3569)(6860 - 3528) \\
&= 39{,}477{,}100
\end{aligned}$$

Table 5.21 **Observed Mean Yields, y (kg/ha), of 10 Barley Genotypes Grown at 12 Locations in the ICARDA 10th RBYT, 1979–1980**

	Location						
Genotype*	Afghan-istan	Cyprus	Egypt	Greece	Jordan	Lebanon	Morocco
1	2724	6681	4583	2966	3500	1583	3109
2	2240	6014	3817	4330	2725	2282	4576
3	3204	5929	4006	2817	3400	2394	3807
4	2103	5492	3381	2529	2808	1847	2312
5	2190	5389	4372	2476	3183	3028	2258
6	2112	7008	3470	2159	3133	2602	2704
7	2638	7685	4806	3644	3933	2042	3068
8	918	5441	2822	2697	2825	2208	2934
9	2425	6501	4639	3183	4150	1458	3555
10	2308	5201	3167	3841	2458	2296	3488
Mean, $\bar{y}_{\cdot j}$	2286.2	6134.1	3906.3	3064.2	3211.5	2174.0	3181.1

(*Note*: All computations are carried only to six places. Rounding error might be a problem when trying to reproduce these results.) Then the stability statistics are as follows:

1. Genotype mean, $\bar{y}_{1\cdot}$:

$$\bar{y}_{1\cdot} = \frac{\sum_j \bar{y}_{ij}}{q} = \frac{2724 + 6681 + \cdots + 7540}{12} = 3569$$

2. Within-genotype mean square, s_1^2:

$$s_1^2 = \frac{\sum_j (\bar{y}_{1j} - \bar{y}_{1\cdot})^2}{q-1} = \frac{45{,}770{,}500}{11} = 4{,}160{,}590$$

3. Coefficient of variation, CV_1:

$$CV_1 = \left(\frac{\sqrt{s_1^2}}{\bar{y}_{1\cdot}}\right)100 = \left(\frac{\sqrt{4{,}160{,}590}}{3569}\right)100 = 57.2\%$$

Table 5.21 **Continued**

Genotype	Oman	Saudi Arabia	Syria	Tunisia	Oregon U.S.A.	Mean $\bar{y}_{i\cdot}$
1	2037	332	3171	4602	7540	3569.00
2	2593	322	4704	5951	9390	4078.67
3	1698	302	4250	4979	7283	3672.42
4	2469	249	3800	4467	7233	3224.17
5	2994	824	3558	3623	5691	3298.83
6	2037	574	3363	4618	2667	3037.25
7	1975	507	5042	5736	7242	4026.50
8	1389	342	4713	4223	7178	3140.83
9	1975	542	3950	5750	5537	3638.75
10	1420	303	4075	5714	8838	3591.58
Mean, $\bar{y}_{\cdot j}$	2058.7	429.7	4062.6	4966.3	6859.9	3527.88 $\bar{\bar{y}}$

*Genotype: 1, Beecher (check); 2, Bco.Mr.Godiva; 3, Composite 29; 4, CI 13575; 5, TJ 70; 6, Cr 370/45/2; 7, ER-Apam; 8, Api-CM 67; 9, WI 2198; 10, Carina.

4. Ecovalence, W_1^2:

$$W_1^2 = \Sigma_j (\bar{y}_{1j} - \bar{y}_{1\cdot} - \bar{y}_{\cdot j} + \bar{\bar{y}})^2$$
$$= (2724 - 3569 - 2286 + 3528)^2 + \cdots + (7540 - 2569 - 6860 + 3528)^2$$
$$= 2{,}775{,}980$$

5. Stability variance, σ_1^2:

$$\sigma^2{}_1 = \frac{p}{(p-2)(q-1)W^2{}_1} - \frac{SS(GE)}{(p-1)(p-2)(q-1)}$$
$$SS(GE) = \Sigma_i W_i^2 = 2{,}775{,}980 + \cdots + 7{,}544{,}070 = 56{,}835{,}850$$
$$= \frac{10}{(8)(11)}(2{,}755{,}980) - \frac{56{,}835{,}850}{(9)(8)(11)}$$
$$= 243{,}589$$

Table 5.22 **Stability Statistics for 10 Genotypes of Barley Grown at 12 Locations in the ICARDA 10th RBYT, 1979–1980**

	Stability Measure*				
Genotype(i)	$\bar{y}_{i\cdot}$	s_i^2	CV_i	W_i^2	σ_i^2
1	3,569	4,160,951	57.2%	2,677,510	233,617
2	4,079	5,544,754	57.7	8,273,880	869,569
3	3,671	3,447,454	50.8	1,556,010	106,174
4	3,224	3,379,875	57.0	1,592,770	110,352
5	3,299	1,882,690	41.6	6,509,900	669,116
6	3,037	2,518,298	52.2	17,524,700	1,920,798
7	4,026	4,710,981	53.9	3,006,380	270,989
8	3,096	3,723,306	62.3	2,817,500	249,525
9	3,639	3,310,315	50.0	4,465,320	436,778
10	3,592	5,022,133	62.4	7,527,180	784,716

6. Regression slope, b_1:

$$b_1 = \frac{\sum_j (\bar{y}_{1j} - \bar{y}_{1\cdot})(\bar{y}_{\cdot j} - \bar{\bar{y}})}{\sum_j (\bar{y}_{\cdot j} - \bar{\bar{y}})^2}$$

$$= \frac{39{,}477{,}100}{35{,}960{,}200} = 1.0978$$

7. Deviation from regression mean square, δ_1^2:

$$\delta_1^2 = \frac{1}{q-2}[\sum_j (\bar{y}_{1j} - \bar{y}_{1\cdot})^2 - b^2 \sum_j (\bar{y}_{\cdot j} - \bar{\bar{y}})^2]$$

$$= \frac{1}{10}[45{,}770{,}500 - (1.0978)^2\,(35{,}960{,}200)]$$

$$= 243{,}158$$

Table 5.22 **Continued**

Genotype(i)	Stability Measure* b_i	δ_i^2	R_i^2
1	1.1003	231,817	.949
2	1.2341	630,663	.897
3	1.0110	155,170	.959
4	.9955	159,204	.957
5	.6977	322,879	.844
6	.6417	1,291,480	.534
7	1.1797	184,679	.964
8	1.0310	278,287	.932
9	.9448	435,611	.880
10	1.1644	655,653	.881

*Stability measure:

1. $\bar{y}_{i\cdot} = \Sigma_j \bar{y}_{ij}/q$ = genotype mean
2. $s_i^2 = \Sigma_j (\bar{y}_{ij} - \bar{y}_{i\cdot})^2/q - 1)$ = genotype variance
3. $CV_i = (s_i^2/\bar{y}_{i\cdot})100$ = coefficient of variation
4. $W_i^2 = \Sigma_j (\bar{y}_{ij} - \bar{y}_{i\cdot} - \bar{y}_{\cdot j} + \bar{\bar{y}})$ = ecovalence
5. $\sigma_i^2 = [p/(p - 2)(q - 1)]W_i^2 - SS(GE)/(p - 1)(p - 2)(q - 1)$ = interaction variance
6. $b_i = [\Sigma_j (\bar{y}_{ij} - \bar{y}_{i\cdot})(\bar{y}_{\cdot j} - \bar{\bar{y}})]/\Sigma_j (\bar{y}_{\cdot j} - \bar{\bar{y}})^2$ = regression slope
7. $\delta_i^2 = [1/(q - 2)][\Sigma_j (\bar{y}_{ij} - \bar{y}_{i\cdot})^2 - b^2_u \Sigma_j (\bar{y}_{\cdot j} - \bar{\bar{y}})^2]$ = deviation mean square
8. $R_i^2 = [b^2_i \Sigma_j (\bar{y}_{\cdot j} - \bar{\bar{y}})^2]/[\Sigma_j (\bar{y}_{ij} - \bar{y}_{i\cdot})^2]$ = coefficient of determination

8. Coefficient of determination, R_i^2:

$$R_i^2 = \frac{b_i^2 \sum_j (\bar{y}_{\cdot j} - \bar{\bar{y}})^2}{\sum_j (\bar{y}_{1j} - \bar{y}_{1\cdot})^2}$$

$$= \frac{(1.0978)^2(35{,}960{,}200)}{45{,}770{,}500}$$

$$= 0.947$$

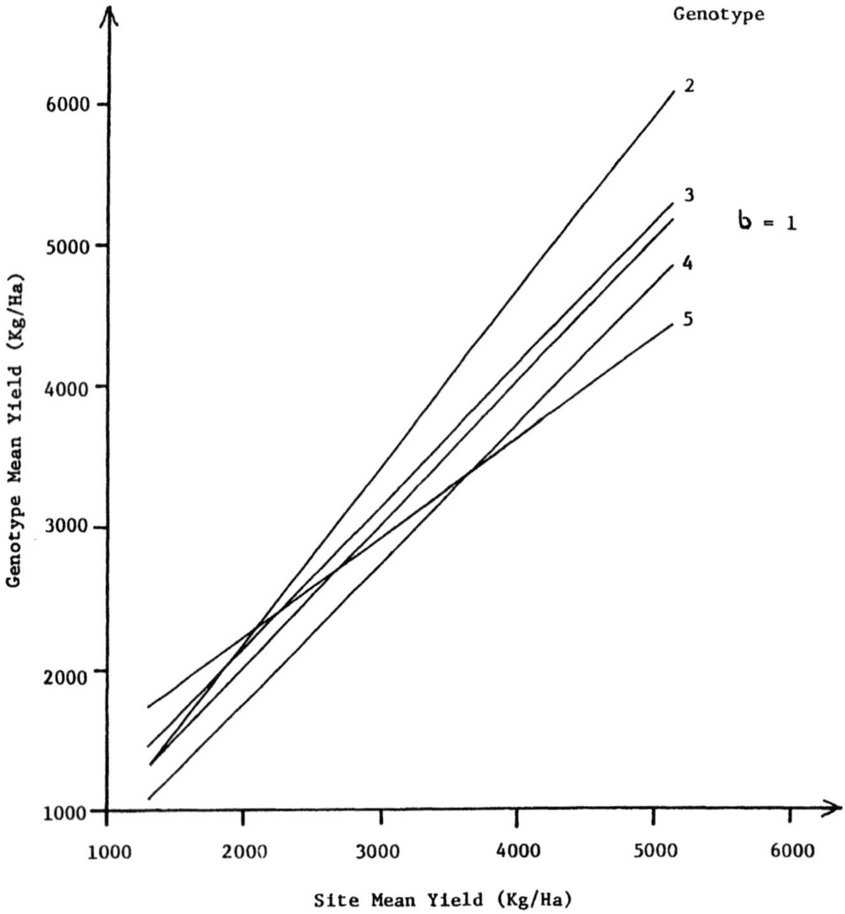

Figure 5.1 Linear relationship between mean yields of four barley genotypes and site mean yield. (Data from ICARDA 10th RBYT, 1979–1980.)

On consideration of Table 5.22, and using the slope of the regression of genotype mean on site mean yield, b_i, as a measure of stability, it is seen that genotype 3, $b_3 = 1.0099$, and genotype 4, $b_4 = 0.9949$, have slopes closest to 1.00 among the 10 genotypes in the data set. On this basis, 'Composite 29' (3) and 'CI13575' (4) would be judged the most stable of the 10 genotypes over the 12 environments represented in this trial. Further, these genotypes contributed the least to the genotype by environment interaction as measured by ecovalence, W_i^2, and the interaction variance, σ_i^2. In addition, these two genotypes have the smallest deviation from regression on site index, as measured by the deviation mean square, δ_i^2, of all genotypes. By these measures genotypes 3 and 4 represent the most stable of the 10 genotypes in the trial. Of the two, genotype 3, 'Composite 29', would be preferred because its mean yield is slightly higher than that of genotype 4. As mentioned previously, genotype 2. 'Bco.Mr.Godiva', appears to be adapted to better environments. This is supported by the slope for this genotype, $b_2 = 1.2332$, which is greater than 1.00. On the other hand, genotype 5, 'TJ 70', with a slope of $b_5 = 0.6971$, appears to be better adapted to poorer environments. Graphs of the regression of genotype mean on site mean for these four genotypes are shown in Figure 5.1.

5.7 PLANS FOR LATTICE DESIGNS

1. Block number is enclosed in parentheses.
2. Replication number is shown as a Roman numeral.
3. Treatment is indicated by number within block.
4. Use two reps for simple lattice, three reps for triple lattice, four reps for quadruple lattice.

5.7.1 9 Treatments, Block-Size = 3

	I		II		III		IV
(1)	1 2 3	(4)	1 4 7	(7)	1 5 9	(10)	1 8 6
(2)	4 5 6	(5)	2 5 8	(8)	7 2 6	(11)	4 2 9
(3)	7 8 9	(6)	3 6 9	(9)	4 8 3	(12)	7 5 3

5.7.2 16 Treatments, Block Size = 4

	I					II			
(1)	1	2	3	4	(5)	1	5	9	13
(2)	5	6	7	8	(6)	2	6	10	14
(3)	9	10	11	12	(7)	3	7	11	15
(4)	13	14	15	16	(8)	4	8	12	16

	III					IV			
(9)	1	6	11	16	(13)	1	14	7	12
(10)	5	2	15	12	(14)	13	2	11	8
(11)	9	14	3	8	(15)	5	10	3	16
(12)	13	10	7	4	(16)	9	6	15	4

5.7.3 25 Treatments, Block Size = 5

	I						II				
(1)	1	2	3	4	5	(6)	1	6	11	16	21
(2)	6	7	8	9	10	(7)	2	7	12	17	22
(3)	11	12	13	14	15	(8)	3	8	13	18	23
(4)	16	17	18	19	20	(9)	4	9	14	19	24
(5)	21	22	23	24	25	(10)	5	10	15	20	25

	III						IV				
(11)	1	7	13	19	25	(16)	1	12	23	9	20
(12)	21	2	8	14	20	(17)	16	2	13	24	10
(13)	16	22	3	9	15	(18)	6	17	3	14	25
(14)	11	17	23	4	10	(19)	21	7	18	4	15
(15)	6	12	18	24	5	(20)	11	22	8	19	5

5.7.4 36 Treatments, Block Size = 6

I						
(1)	1	2	3	4	5	6
(2)	7	8	9	10	11	12
(3)	13	14	15	16	17	18
(4)	19	20	21	22	23	24
(5)	25	26	27	28	29	30
(6)	31	32	33	34	35	36

II						
(7)	1	7	13	19	25	31
(8)	2	8	14	20	26	32
(9)	3	9	15	21	27	33
(10)	4	10	16	22	28	34
(11)	5	11	17	23	29	35
(12)	6	12	18	24	30	36

III						
(13)	1	8	15	22	29	36
(14)	31	2	9	16	23	30
(15)	25	32	3	10	17	24
(16)	19	26	33	4	11	18
(17)	13	20	27	34	5	12
(18)	7	12	21	28	35	6

5.7.5 49 Treatments, Block Size = 7

I							
(1)	1	2	3	4	5	6	7
(2)	8	9	10	11	12	13	14
(3)	15	16	17	18	19	20	21
(4)	22	23	24	25	26	27	28
(5)	29	30	31	32	33	34	35
(6)	36	37	38	39	40	41	42
(7)	43	44	45	46	47	48	49

II							
(8)	1	8	15	22	29	36	43
(9)	2	9	16	23	30	37	44
(10)	3	10	17	24	31	38	45
(11)	4	11	18	25	32	39	46
(12)	5	12	19	26	33	40	47
(13)	6	13	20	27	34	41	48
(14)	7	14	21	28	35	42	49

III							
(15)	1	9	17	25	33	41	49
(16)	43	2	10	18	26	34	42
(17)	36	44	3	11	19	27	35
(18)	29	37	45	4	12	20	28
(19)	22	30	38	46	5	13	21
(20)	15	23	31	39	47	6	14
(21)	8	16	24	32	40	48	7

IV							
(22)	1	37	24	11	47	34	21
(23)	15	2	38	25	12	48	35
(24)	29	16	3	39	26	13	49
(25)	43	30	17	4	40	27	14
(26)	8	44	31	18	5	41	28
(27)	22	9	45	32	19	6	42
(28)	36	23	10	46	33	20	7

5.7.6 64 Treatments, Block Size = 8

I

Block								
(1)	1	2	3	4	5	6	7	8
(2)	9	10	11	12	13	14	15	16
(3)	17	18	19	20	21	22	23	24
(4)	25	26	27	28	29	30	31	32
(5)	33	34	35	36	37	48	39	40
(6)	41	42	43	44	45	46	47	48
(7)	49	50	51	52	53	54	55	56
(8)	57	58	59	60	61	62	63	64

II

Block								
(9)	1	9	17	25	33	41	49	57
(10)	2	10	18	26	34	42	50	58
(11)	3	11	19	27	35	43	51	59
(12)	4	12	20	28	36	44	52	60
(13)	5	13	21	29	37	45	53	61
(14)	6	14	22	30	38	46	54	62
(15)	7	15	23	31	39	47	55	63
(16)	8	16	24	32	40	48	56	64

III

Block								
(17)	1	10	19	28	37	46	55	64
(18)	9	2	51	44	61	30	23	40
(19)	17	50	3	36	29	62	15	48
(20)	25	42	35	4	21	14	63	56
(21)	33	58	27	20	5	54	47	16
(22)	41	26	59	12	53	6	39	24
(23)	49	18	11	60	45	38	7	32
(24)	57	34	43	52	13	22	31	8

IV

Block								
(25)	1	18	27	44	13	62	39	56
(26)	17	2	35	60	53	46	31	16
(27)	25	34	3	12	45	54	23	64
(28)	41	58	11	4	29	22	55	40
(29)	9	50	43	28	5	38	63	24
(30)	57	42	51	20	37	6	15	32
(31)	33	26	19	52	61	14	7	48
(32)	49	10	59	36	21	30	47	8

5.7.7 81 Treatments, Block Size = 9

I									
(1)	1	2	3	4	5	6	7	8	9
(2)	10	11	12	13	14	15	16	17	18
(3)	19	20	21	22	23	24	25	26	27
(4)	28	29	30	31	32	33	34	35	36
(5)	37	38	39	40	41	42	43	44	45
(6)	46	47	48	49	50	51	52	53	54
(7)	55	56	57	58	59	60	61	62	63
(8)	64	65	66	67	68	69	70	71	72
(9)	73	74	75	76	77	78	79	80	81

II									
(10)	1	10	19	28	37	46	55	64	73
(11)	2	11	20	29	38	47	56	65	74
(12)	3	12	21	30	39	48	57	66	75
(13)	4	13	22	31	40	49	58	67	76
(14)	5	14	23	32	41	50	59	68	77
(15)	6	15	24	33	42	51	60	69	78
(16)	7	16	25	34	43	52	61	70	79
(17)	8	17	26	35	44	53	62	71	80
(18)	9	18	27	36	45	54	63	72	81

III									
(19)	1	20	12	58	77	69	34	53	45
(20)	10	2	21	67	59	78	43	35	54
(21)	19	11	3	76	68	60	52	44	36
(22)	28	47	39	4	23	15	61	80	72
(23)	37	29	48	13	5	24	70	62	81
(24)	46	38	30	22	14	6	79	71	63
(25)	55	74	66	31	50	42	7	26	18
(26)	64	56	75	40	32	51	16	8	27
(27)	73	65	57	49	41	33	25	17	9

IV									
(28)	1	11	21	31	41	51	61	71	81
(29)	19	2	12	49	32	42	79	62	72
(30)	10	20	3	40	50	33	70	80	63
(31)	55	65	75	4	14	24	34	44	54
(32)	73	56	66	22	5	15	52	35	45
(33)	64	74	57	13	23	6	43	53	36
(34)	28	38	48	58	68	78	7	17	27
(35)	46	29	39	76	59	69	25	8	18
(36)	37	47	30	67	77	60	16	26	9

5.7.8 100 Treatments, Block Size = 10

I

(1)	1	2	3	4	5	6	7	8	9	10
(2)	11	12	13	14	15	16	17	18	19	20
(3)	21	22	23	24	25	26	27	28	29	30
(4)	31	32	33	34	35	36	37	38	39	40
(5)	41	42	43	44	45	46	47	48	49	50
(6)	51	52	53	54	55	56	57	58	59	60
(7)	61	62	63	64	65	66	67	68	69	70
(8)	71	72	73	74	75	76	77	78	79	80
(9)	81	82	83	84	85	86	87	88	89	90
(10)	91	92	93	94	95	96	97	98	99	100

II

(11)	1	11	21	31	41	51	61	71	81	91
(12)	2	12	22	32	42	52	62	72	82	92
(13)	3	13	23	33	43	53	63	73	83	93
(14)	4	14	24	34	44	54	64	74	84	94
(15)	5	15	25	35	45	55	65	75	85	95
(16)	6	16	26	36	46	56	66	76	86	96
(17)	7	17	27	37	47	57	67	77	87	97
(18)	8	18	28	38	48	58	68	78	88	98
(19)	8	19	29	39	49	59	69	79	89	99
(20)	10	20	30	40	50	60	70	80	90	100

III

(21)	1	12	23	34	45	56	67	78	89	100
(22)	91	2	13	24	35	46	57	68	79	90
(23)	81	92	3	14	25	36	47	58	69	80
(24)	71	82	93	4	15	26	37	48	59	70
(25)	61	72	83	94	5	16	27	39	49	60
(26)	51	62	73	84	95	6	17	28	39	50
(27)	41	52	63	74	85	96	7	18	29	40
(28)	31	42	53	64	75	86	97	8	19	30
(29)	21	32	43	54	65	76	87	98	9	20
(30)	11	22	33	44	55	66	77	88	99	10

REFERENCES

Bose, R. C., W. H. Clatworthy, and S. S. Shirkhande, 1954. Tables of partially balanced designs with two associate classes. *N.C. Agric. Exp. Stn. Tech. Bull. 107.*

Carmer, S. G., and M. R. Swanson, 1973. An evaluation of ten pairwise multiple comparison procedures by Monte Carlo methods. *J. Am. Stat. Assoc.* 68:66–74.

Carmer, S. G., and W. M. Walker, 1982. Baby Bear's dilemma: a statistical tale, *Agron. J.* 74:122–124.

Chew, V., 1976. Comparing treatment means: a compendium. *HortScience* 11:348–357.

Cochran, W. G., and G. M. Cox, 1957. *Experimental Designs*, 2nd ed. Wiley, New York, Chaps. 9–13.

Duncan, D. B., 1955. Multiple range and multiple *F* tests. *Biometrics* 11:1–42.

Eberhardt, S. A., and W. A. Russell, 1966. Stability parameters for comparing varieties. *Crop Sci.* 6:36–40.

Federer, W. T., 1956. Augmented (or Hoonuiaku) designs. *Hawaii. Plant. Rec.* 55:191–208.

Federer, W. T., 1961. Augmented designs with one-way elimination of heterogeneity. *Biometrics* 17:447–473.

Federer, W. T., and D. Ragavarao, 1975. On augmented designs. *Biometrics* 31:29–35.

Finlay, K. W., and G. N. Wilkinson, 1963. The analysis of adaptation in a plant breeding programme. *J. Agric. Res.* 14:742–754.

Finney, D. J., 1964. Replication of variety trials. *Biometrics* 20:1–15.

Fisher, R. A., 1966. *The Design of Experiments*, 8th ed. Hafner, New York, Chap. 10.
Francis, T. R., and L. W. Kannenberg, 1978. Yield stability studies in short-season maize. 1. A descriptive method for grouping genotypes. *Can. J. Plant Sci.* 58:1029–1034.
Lin, C. S., M. R. Binns, and L. P. Lefkovitch, 1986. Stability analysis: where do we stand? *Crop Sci.* 26:894–900.
Petersen, R. G., 1977. Use and misuse of multiple comparison procedures. *Agron. J.* 60:205–208.
Petersen, R. G., 1985. *Design and Analysis of Experiments*. Marcel Dekker, New York.
Shukla, G. K., 1972. Some statistical aspects of partitioning genotype–environment components of variability. *Heredity* 29:237–245.
Steel, R. G. D., and J. H. Torrie, 1980. *Principles and Procedures of Statistics*, 2nd ed. McGraw-Hill, New York, pp. 186–187.
Tukey, J. W., 1951. Quick and dirty methods in statistics: Part II. Simple analyses for standard designs. *Proc. 5th Annu. Conv. Am. Soc. Quality Control*, pp. 189–197.
Waller, R. A., and D. B. Duncan, 1969. A Bayes rule for the symmetrical multiple comparison problem. *J. Am. Stat. Assoc.* 64:1484–1503.
Wricke, G., 1962. Über eine Methode zur Erfussung der okologischen Streubreite in Feldversuchen. *Z. Pflanzenzuecht.* 47:92–96.
Yates, F., and W. G. Cochran, 1938. The analysis of groups of experiments. *J. Agric. Sci.* 28:556–580.

FURTHER READING

Complete citations for books for further reading are given in the bibliography at the end of the book.

1. Cochran and Cox, Chap. 10.
2. LeClerg, Leonard, and Clark, Chap. 15.
3. Petersen, Chap. 13.

6

Combined Analysis of Several Experiments

6.1 INTRODUCTION

Agricultural research programs are conducted, essentially, for two broad purposes: (1) to obtain information to help explain the way plants and animals respond to changing conditions (basic research), and (2) to obtain information with which to solve problems or to make recommendations to producers (applied research).

For the most part, basic research involves a series of single experiments conducted in sequence. The researcher proposes a hypothesis about how a biological system might respond to changes. He conducts an experiment to obtain data to test the hypothesis. Based on the outcome of this experiment the researcher might modify the hypothesis and run another experiment to test the modified hypothesis. This process is continued sequentially, alternating between hypothesis and experiment, until a satisfactory endpoint is reached. Experiments are analyzed as they are completed, and because of the sequential nature of the process, combined analysis of the data from two or more experiments is seldom required.

Recommendations to producers, on the other hand, can seldom be effective if they are based on the analysis of data from a single isolated experiment. No matter how well a single experiment is designed and conducted, it can provide information only about a single location in a single season.

This is usually not enough information on which to base recommendations. For example, a plant breeder is often interested in new selections that produce well under a variety of conditions over a range of locations. The agronomist is usually interested in response to a new practice over a range of soil types. To meet these goals it is necessary that the experiment be conducted during several seasons (at least three) and at several locations within the domain over which the recommendations are to apply.

If maximum value is to be obtained from a program of applied research, combined analysis of the results from a number of experiments becomes an important, although often neglected part of the program. A combined analysis can provide information on how treatment effects, and differences among treatment effects, change in response to differences in soils and weather throughout a region. Moreover, a combined analysis is essential for detecting and measuring the interaction of treatment with location and the interaction of treatment with season in the recommendation domain.

In general, combined estimates will be valid only if the sites on which the experiments are conducted are a random sample of all possible sites in the area. If every possible site has an equal chance of being included in the set of experiments, the combined estimates will be unbiased over the area of recommendation. In this case the analytical methods presented here can be used without reservation. In the usual research program it is often impractical to adhere to the strict random selection requirement. Sites that are available, easy to reach, and so on, must be used because of cost and other considerations. In this case the methods discussed in this chapter may still be used, but the results should be interpreted in light of the restricted site selection.

6.2 PRELIMINARY ANALYSIS

Before beginning a combined analysis it is necessary to do some preliminary analyses:

1. An analysis of variance should be completed for each experiment.
2. The experimental errors of the several experiments should be examined for heterogeneity.
3. A preliminary estimate of the interaction of treatment with environment or season or both should be obtained.

Depending on the outcome of these preliminary steps and on the type of experiments involved, the combined analysis can be undertaken.

The purpose of the preliminary analysis is to determine whether or not all of the experiments can be included in the same combined analysis or whether the experiments should first be separated into more-or-less homogeneous groups. Further, the preliminary analysis should give some indication of whether generally applicable recommendations are possible or if different recommendations are required to meet different conditions.

6.3 SAME EXPERIMENT AT SEVERAL LOCATIONS: HOMOGENEOUS VARIANCE

When considering the combined analysis of data from several locations the first requirement is to assess the homogeneity of the error variance at the various locations. If the errors are homogeneous, the analysis can proceed. If they are not, the data can be transformed to produce homogeneous variance or the locations may be separated into groups within which the variance is homogeneous. We illustrate this, and other procedures, with numerical data.

6.3.1 Numerical Data

The data presented in Table 6.1 are from a number of on-farm trials on the factors affecting the production of lentils in Aleppo Province, Syria, during the crop year 1985–1986. The trials, conducted by the Farming Systems Program of ICARDA (International Center for Agricultural Research in the Dry Areas), consisted of a 2^3 factorial set of treatments conducted at eight locations using a randomized block design with two blocks at each location. The treatment factors were:

1. Sitona weevil control (S)

   S_0 = none S_1 = 1 kg/ha carbofuran

2. Fertilizer (F)

   F_0 = none F_1 = 50 kg/ha P_2O_5

3. Weed control (W)

   W_0 = none W_1 = herbicide spray

A number of variables were measured on each plot, including number of plants/plot, weight of 200 seeds, dry weight of weeds, grain yield, and total yield. At each location two trials were conducted using the same treatments in each trial. One trial was seeded early and the other was seeded late. We will use the data from the early seeded trials to illustrate the combined analysis of data from several locations in a single season. Mean grain yields (kg/ha) from the early seeded trials are given in Table 6.1.

Table 6.1 **Mean Grain Yield (kg/ha) of Early-Planted Lentils at Eight Locations in Aleppo Province, Syria, 1985–1986**

Treatment *	Location 1	2	3	4
(1)	976.5	1193.5	549.0	513.5
S	1166.5	1111.0	751.0	751.0
F	1074.5	1021.5	666.0	667.0
SF	1260.0	1058.0	843.5	1239.5
W	1091.0	1138.5	626.0	860.0
SW	1193.0	1176.5	1156.5	1445.0
FW	1107.0	1132.5	593.5	841.0
SFW	1310.0	1214.5	1077.0	1334.0

Treatment	Location 5	6	7	8
(1)	148.0	770.5	562.5	843.0
S	233.0	809.5	710.0	1133.5
F	305.0	920.0	579.5	1074.0
SF	656.5	878.0	649.0	1153.5
W	410.5	1005.5	744.5	963.0
SW	888.0	1120.0	814.5	1323.5
FW	791.5	880.0	929.0	935.5
SFW	1024.5	1093.5	919.5	1069.5

*(1), check; S, 1 kg/ha carbofuran; F, 50 kg/ha P_2O_5; SF, carb. + P_2O_5; W, herbicide; SW, carb. + herb.; FW, P_2O_5 + herb.; SFW, carb. + P_2O_5 + herb.

The analyses of variance of grain yields at each of the locations are shown in Table 6.2. The error mean squares in Table 6.2 are the sample estimates, s^2, of the error variance for these trials. These estimates provide the data for examining the homogeneity of variance.

6.3.2 Test of Homogeneity of Variance

A quick test of homogeneity of variance is provided by the ratio of the largest to the smallest s^2 in the set. With the table of critical values given in Table A4 of the Appendix it is often possible to draw a conclusion regarding homogeneity of variance without further testing. For the early planted lentil data we have

$$\frac{s^2(\max)}{s^2(\min)} = \frac{34{,}406}{3494} = 9.85$$

The upper 5% point for $a = k = 8$, $\gamma = 7$ from Table A4 is 12.7. Since $9.85 < 12.7$, we conclude that the variances are not heterogeneous.

Table 6.2 **Analysis of Variance of Grain Yields from Early-Planted Lentils at Eight Locations in Aleppo Province, Syria, 1985–1986**

		Mean Square at Location			
Source	df	1	2	3	4
Block	1	105	4970	2522	62750
Treatment	7	23113*	8779	102750	221985**
Error	7	5260	3494	34406	31031

		Mean Square at Location			
Source	df	5	6	7	8
Block	1	100172	176400	3220	155433
Treatment	7	214352**	32464	39984	44523
Error	7	27704	20190	21276	32759

*Significant at 5%; **significant at 1%.

An alternative procedure, which is more sensitive than the ratio test, is Bartlett's test of homogeneity of variance (Bartlett, 1937). This test, which is based on the natural logarithm of the sample variances, has been described by Snedecor and Cochran (1980). To perform this test, let

γ = error degrees of freedom for the individual trials

s_i^2 = error mean square at location i

a = number of locations

Then

$$M = \gamma[a(\ln \bar{s}^2) - \sum_i \ln s_i^2] \qquad \bar{s}^2 = \frac{\sum_i s_i^2}{a}$$

and

$$C = 1 + \frac{a + 1}{3a\gamma}$$

The ratio M/C is a test statistic for the null hypothesis that each s^2 is an estimate of σ^2. The ratio M/C is distributed as χ^2 with (a − 1) degrees of freedom. (Percentage points of the chi-square distribution are given in Table A5 of the Appendix.)

Computations for Bartlett's test are given in Table 6.3. From Table 6.3 we compute

$$M = \gamma\ [a(\ln \bar{s}^2) - \sum_i \ln s_i^2]$$
$$= (7)[(8)(9.994) - 77.90] = 13.734$$

and

$$C = 1 + \frac{a + 1}{3a\gamma} = 1 + \frac{8 + 1}{(3)(8)(7)} = 1.0536$$

Then

$$\chi^2 = \frac{M}{C} = \frac{13.734}{1.0536} = 13.064$$

Since $\chi^2_{0.05(7)} = 14.067 > 13.064$, we conclude that the variances are not heterogeneous.

6.3.3 Preliminary Analysis

Since our conclusion by both tests is that the variances are homogeneous, we can proceed with the combined analysis using the data from all eight locations. We first compute a preliminary combined analysis of variance.

Table 6.3 **Bartlett's Test of Homogeneity of Variance of Grain Yields from Early-Planted Lentils, 1985–1986**[a]

Location	s^2	$\ln(s^2)$
1	5,260	8.568
2	3,494	8.159
3	34,406	10.446
4	30,131	10.313
5	27,704	10.299
6	20,190	9.913
7	21,270	9.965
8	32,759	10.397
Sum	175,214	77.990

[a]With $\gamma = 7$ df for error and $a = k = 8$ locations.
$\bar{s}^2 = 175{,}214/8 = 21{,}902$; $\ln \bar{s}^2 = 9.994$.

With this analysis we can look at the magnitude of the among-locations variation, the variation among treatments, and, in particular, the location × treatment interaction. For the early planted lentil data the combined analysis of variance is shown in Table 6.4. (*Note*: The pooled error mean square for this analysis is the mean of the error mean squares from the individual analyses.)

The test of location × treatment interaction gives an indication of whether or not the treatments behave the same from one location to another. A significant interaction means that the effects of treatment vary from location to location. In this case the combined analysis of data from all locations has little meaning. However, subsets of locations may be found for which interaction is not a problem, and for which recommendations may be made.

A nonsignificant location × treatment interaction, on the other hand, does not necessarily mean that all of the meaningful comparisons among treatments are independent of location. To examine this point the location × treatment interaction should be partitioned into components correspond-

Table 6.4 **Preliminary Analysis of Variance of Grain Yields of Early-Planted Lentils Combined Over Eight Locations in Aleppo Province, 1985–1986**

Source	df	MS	F
Location	7	701.579	32.03**
Block in Locat.	8	63.197	
Treatment	7	386.166	17.63**
Loc. by Treat.	49	43.112	1.97 NS
Pooled Error	56	21.902	

**Significant at 1%; NS, not significant. Pooled error MS is used as a divisor for all F tests.

ing to the interaction of the meaningful treatment comparisons with location. These interaction components should then be tested for significance.

6.3.4 Detailed Analysis

To illustrate this type of partitioning, consider the nature of the treatments used in the early planted lentil experiments. The eight treatments in these experiments constitute a 2 × 2 × 2 factorial set of the three factors: sitona weevil control, S: fertilizer, F: and weed control, W. We can use the usual factorial analysis to examine the main effects and interactions among the treatments and the interaction of the factorial effects with location. The analysis of variance for this partition is given in Table 6.5. Note that the main effects and interactions among treatments each carry 1 degree of freedom. The interactions of these factorial effects with location each have 7 degrees of freedom.

The interaction components are tested against the pooled error mean square to determine which, if any, of the treatment comparisons differ from one location to another. For the early planted lentil data we see (Table 6.5) that the location × sitona and the location × weed interaction components are both significant. This is probably the result of differential infestations, at the eight sites, of both sitona weevils and weeds.

When an interaction component in the location × treatment variance source is significant, the implication is that the treatment component is subject not only to experimental error variation but also to variation due

Table 6.5 **Detailed Analysis of Variance of Grain Yield (kg/ha) of Early-Planted Lentils Combined Over Eight Locations in Aleppo Province, 1985–1986**

Source	df	MS	F	
Block in Location	8	63,197		
Location (L)	7	701,579		
Treatment	7			
Sitona (S)	1	1,545,280	11.93*	F1
Fertilizer (F)	1	94,395	4.31*	
S F	1	8,001	0.36	
Weed (W)	1	965,703	13.34**	F2
S W	1	44,179	2.02	
F W	1	40,970	1.87	
S F W	1	4,632	.21	
Location by Treatment	49			
L S	7	129,563	5.92*	
L F	7	37,948	1.73	
L S F	7	8,762	.40	
L W	7	72,395	3.30**	
L S W	7	17,715	.81	
L F W	7	24,882	1.14	
L S F W	7	10,519	.48	
Pooled Error	56	21,902		

*Significant at 5%; **significant at 1%. MS(LS) used as a divisor for F1; MS(LW) used as a divisor for F2; pooled error MS used as a divisor for all other F tests.

to location. In this case it is prudent to use the mean square for the interaction component in the denominator of the F ratio for testing the treatment comparison. This does two things: it uses a more nearly correct error variance in the test, and it reduces the degrees of freedom in the denominator. The effect is to make the test more conservative.

In the early planted lentil data the denominator mean square for the test of the sitona control effect is the L × S mean square giving an F with 1 and 7 df. Similarly, the L × W mean square is used to test the effect of weed control. Since none of the other interactions with location are significant, the pooled error mean square may be used for the test of the other factorial effects. These F values carry 1 and 56 df.

The results of the analysis of variance are used to determine which means should be used to summarize the combined data. The interpretation of a combined analysis follows the same pattern as for a factorial analysis. If treatment × location interactions exist, two-way tables should be used. If the interactions with location are not significant, one-way tables of means for the significant main effects should be used.

When dealing with data from several locations a problem arises as how best to combine the data to estimate the means. This problem was examined by Yates and Cochran (1938) and by Cochran (1954). Their conclusion was that except for cases in which there is extreme variation in precision among the experiments, the ordinary arithmetic mean across locations is to be preferred to any weighted mean. We use it in our presentation of the results of combined analyses. As is our usual practice, we present a *report of statistical analysis* to illustrate the interpretation of the early planted lentil data.

Report of Statistical Analysis

On-farm trials were conducted at eight locations in Aleppo Province, northern Syria, during the 1985–1986 growing season. The purpose of these trials was to evaluate three factors—sitona weevil control, phosphate fertilizer, and weed control—on the grain yield of early planted lentils. The principal results are presented in Tables 6.6 to 6.8. The addition of 50 kg/ha of P_2O_5 resulted in an average increase of about 54 kg/ha in grain yield. This small increase appeared to be independent of location.

The effect of sitona weevil control varied from one location to another. The largest difference, 567.5 kg/ha, between control and no control occurred at location 4. The smallest difference, 18.8 kg/ha, occurred at location 2. As with sitona control, the effect of weed control varied from one location to another. The largest difference between herbicide and none, 443.0 kg/ha, occurred at location 5; the smallest, 21.9 kg/ha, occurred at location 8.

Table 6.6 **Mean Grain Yield (kg/ha) from Early-Planted Lentils With and Without Phosphate Fertilizer, in Aleppo Province, 1985–1986**

Phosphate	None	50 kg/ha	Standard Error
Yield(kg/ha)	892.5	946.8	18.50

Table 6.7 **Mean Grain Yield (kg/ha) from Early-Planted Lentils With and Without Weevil Control at Eight Locations in Aleppo Province, 1985–1986[a]**

Sitona Control	Location 1	2	3	4
None	1062.2	1121.2	608.6	720.4
Carbo.	1232.4	1140.0	957.0	1287.9

Sitona Control	Location 5	6	7	8
None	413.8	894.0	703.9	953.9
Carbo.	700.5	975.2	773.2	1170.0

[a]Standard error = 127.26.

There is no indication that the effect of any of the three factors depended on the level of the other two.

6.4 SAME EXPERIMENT AT SEVERAL LOCATIONS: HETEROGENEOUS VARIANCE

When the error variance at several locations cannot be considered to be homogeneous, the combined analysis has some associated problems. Unless

Table 6.8 **Mean Grain Yield (kg/ha) from Early-Planted Lentils With and Without Weed Control at Eight Locations in Aleppo Province, 1985–1986**[a]

Weed Control	Location 1	2	3	4
None	1119.4	1096.0	702.4	888.2
Herbicide	1175.2	1165.5	863.2	1120.0

Weed Control	Location 5	6	7	8
None	335.6	844.5	625.2	1051.0
Herbicide	778.6	1024.8	851.9	1072.9

[a]Standard error = 95.13.

the variances are quite different, the simple means are valid for describing the response to treatments. However, significance tests using a pooled error are no longer valid. There are several alternatives in this case. The data may be transformed to produce homogeneous variances. The combined analysis is then run on the transformed data. This should be avoided if possible because of the resulting difficulty in interpreting the analysis. A second alternative is to partition the locations into groups that have homogeneous within-group variance. Combined analyses are then done on the data from each group. This method, too, poses problems because it makes formulation of recommendations over an area difficult.

A third alternative is to proceed with the analysis as if the variances were homogeneous. Cochran and Cox (1957) point out that the principal difficulty with this approach is that the F test of interaction against the pooled error is not valid if the variances are heterogeneous. The effect is to produce too many significant results. One solution is to determine the significance using reduced degrees of freedom for F. If there are p places, then F is distributed approximately with 1 and $p - 1$ degrees of freedom, which puts an upper limit on the level of F required for significance.

A valid and preferred alternative procedure is to partition the treatment sum of squares into a set of meaningful orthogonal components. The interaction sum of squares is then partitioned to obtain the component × location interaction for each component. We then test each treatment component against its own interaction with locations. Because the denominator degrees of freedom are reduced, this gives a conservative test of the individual treatment contrasts.

6.4.1 Numerical Example

We can illustrate the procedure involved by considering a set of data. A trial was conducted by the Oregon Agricultural Experiment Station to study the effect of various moisture regimes on the yield of eight selections of winter wheat. One of the selections, 'Stephens' (selection 5), was a well-established commercial variety. The others were new selections from the wheat breeding program of the station. One of the objects of the trial was to compare the stability of 'Stephens' with that of the other selections over a range of moisture regimes. The trial was planted in early November using a randomized block experimental design with eight selections in three

Table 6.9 **Mean Grain Yield (kg/ha) of Eight Winter Wheat Selections at Five Locations in Oregon**

	Location					
Selection	1	2	3	4	5	Mean
1	4902.7	5266.0	5143.0	6341.7	6248.0	5580.3
2	4810.0	5427.0	5732.3	6043.0	6146.3	5631.7
3	4827.0	5208.3	5363.0	6271.3	6288.7	5591.7
4	4893.0	5255.0	5392.7	6264.3	6225.7	5606.1
5*	5053.3	5390.7	5455.0	6513.0	6512.3	5784.9
6	4976.0	4927.3	5001.3	6162.0	6202.3	5453.8
7	4785.7	5015.7	5166.3	6300.3	6612.0	5576.0
8	4747.7	5302.3	5146.7	5824.7	6138.7	5486.0
Mean	4874.4	5224.0	5333.8	6215.0	6138.7	5588.7

*'Stephens'.

Table 6.10 **Analyses of Variance of Grain Yields of Winter Wheat Grown at Five Locations in Oregon**

Source	df	Mean Square at Location 1	2	3	4	5
Block	2	87.596	56.505	126.100	181.401	42.434
Selection	7	31.634*	89.789*	154.152	129.639	89.950**
Error	14	7.578	29.507	86.529	97.672	18.962

*Significant at 5%; **significant at 1%.

blocks at each of five locations. One of the locations (1) was in east central Oregon, two (2 and 3) were in the Columbia Basin, and two (4 and 5) were in the Willamette valley. Annual rainfall ranged from lowest at location 1 to highest at location 5. Mean grain yields are shown in Table 6.9.

The analyses of variance of grain yields at each of the five locations are given in Table 6.10, and Bartlett's test of homogeneity is presented in Table 6.11. Bartlett's test produced a chi-square value of 26.786. Since the

Table 6.11 **Bartlett's Test of Homogeneity of Variance of Yields of Winter Wheat Grown in Oregon**[a]

Location	s^2	$\ln(s^2)$
1	7.587	8.933
2	29.507	10.292
3	86.529	11.368
4	97.672	11.489
5	18.962	9.850
Sum	240.648	51.932

[a] $\bar{s}^2 = 48{,}050$; $\ln \bar{s}^2 = 10.780$; $M = (14)[(5)(10.780) - 51.932] = 27.551$; $C = (6)/[(3)(5)(14)] + 1 = 1.0286$; $\chi^2 = 27.551/1.0286 = 26.786$.

1% chi-square at 4 df is 13.277 < 26.786, the variances cannot be considered to be homogeneous.

6.4.2 Combined Analysis

The preliminary combined analysis of variance for the grain yield data at the five sites is given in Table 6.12. Because the error variances at the several sites are not homogeneous, we do not test the interaction against the pooled error. Rather, we proceed directly to partitioning the entry and the entry × location sums of squares.

The partition of greatest interest in this trial is the comparison of 'Stephens', entry 5, with the mean of the other entries. We can find the sum of squares for this comparison by means of a contrast of totals, as shown in Table 6.13. We can now partition the entry sum of squares as follows:

Source	df	SS
Entry	7	1,044,354
Stephens vs. others	1	658,952
Others	6	385,402

The interaction sum of squares may be partitioned in the same way by computing the contrast of 'Stephens' vs. others at each location. The sum of squares of these contrasts is the [location × ('Stephens' vs. others)] component of the location × entry interaction sum of squares. This computation is illustrated in Table 6.14.

We can now examine the partitioned combined analysis, as shown in Table 6.15. For this analysis the mean square for the interaction component serves as the denominator for the F ratio to test the significance of the

Table 6.12 **Preliminary Analysis of Variance of Yields of Winter Wheat at Five Locations**

Source	df	SS	MS
Location	4	38,443,038	9,610,760
Entry	7	1,044,354	149,193
L x E	28	2,421,792	86,492
Pooled Error	115		48,049

Table 6.13 **Comparison of 'Stephens' with the Other Entries**[a]

Entry	1	2	3	4	5	6	7	8
Total	83704	84476	83875	84092	86773	81087	83640	82290
Contrast	−1	−1	−1	−1	+7	−1	−1	−1

[a]$L = \Sigma_j T_j k_j = 24{,}247$
$S^2 = L^2/(1)(r)(\Sigma_j k_j^2) = (24{,}247)^2/(5)(3)(56) = 699{,}901$

Table 6.14 **Contrast of 'Stephens' vs. Others at Each of Five Locations**[a]

	Entry								
	1	2	3	4	5	6	7	8	
k_j	−1	−1	−1	−1	+7	−1	−1	−1	L_i
Site(i)									
1	14708*	14430	14481	14679	15160	14928	14357	14243	4294
2	15798	16281	15625	15765	16172	14782	15047	15907	3999
3	15429	17197	16089	16178	16365	15004	15499	16250	2909
4	19025	18129	18814	18793	19539	18486	18901	17474	7151
4	18744	18439	18866	18677	19537	18607	19836	18416	5174

[a]k_j = coefficient to define contrast

$L_i = \Sigma_j k_j T_{ij}$

$$\text{SS[location} \times \text{(Stephens vs. rest)]} = \frac{(\Sigma L_i^2)}{r\Sigma k_j^2} - \frac{(\Sigma L_i)^2}{r\Sigma k_j^2} - \frac{1}{(1)(r)(\Sigma_j k_j^2)(\sum_i L_j^2)}$$

$$= \frac{120{,}799{,}795}{(3)(56)} - \frac{553{,}519{,}729}{(5)(3)(56)}$$

$$= 719{,}046 - 658{,}952 = 60{,}094$$

*Site total.

Table 6.15 **Combined Analysis of Variance of Winter Wheat Yields at Five Oregon Locations**

Source	df	SS	MS	F
Location	4	38,443,038	9,610,760	
Entry	7	1,044,354		
Stephens vs				
Rest	1	699,901	699,901	46.58**
Remainder	6	344,453	57,409	.58
Loc by Entry	28	2,421,792		
L by(S vs Rest)	4	60,094	15,024	
Loc by Rem	24	2,361,698	98,404	

**Significant at 1%.

entry component. We have

$$F_{\text{S vs. rest}} = 46.58 = \frac{699{,}901}{15{,}024}$$

and

$$F_{\text{rest}} = 0.58 = \frac{57{,}409}{98{,}404}$$

Note that 'Stephens' vs. rest is highly significant even with the reduced df in the denominator of the F ratio. Variation among the other entries is not significant. The magnitude of the mean square for [location × ('Stephens' vs. rest)] relative to the pooled error mean square leads to the conclusion that there is no interaction between this contrast and location.

6.4.3 Alternative Partition

There is an alternative partition of the sums of squares that might shed some light on the stability of 'Stephens' across locations versus that of the other entries. The procedure used to obtain this partition was suggested by Yates and Cochran (1938). The approach serves as the basis of the stability analysis of Eberhardt and Russell (1966) (see Section 5.6). Basically, it involves computing the regression of the yields of the separate entries on the mean yield of all entries at a location. One way to do this is by means of a contrast using the deviations of the site mean from the grand mean as coefficients for the contrast (see Section 4.7).

In general, suppose that

$$
\begin{aligned}
l &= \text{number of locations} \\
p &= \text{number of entries} \\
r &= \text{number of replications per entry at each location} \\
T_{ij} &= \text{total yield of entry i at location j} \\
\bar{x}_j &= \text{mean of all entries at location j} \\
\bar{\bar{x}} &= \text{grand mean of all entries}
\end{aligned}
$$

Then

$$L_i = \sum_j (\bar{x}_j - \bar{\bar{x}})T_{ij}$$

is the contrast for the linear regression of entry i on the location means, and

$$b_i = \frac{L_i}{r \sum_j (\bar{x}_j - \bar{\bar{x}})^2}$$

is the coefficient of the regression of entry i on the location means.

The computation of the regression contrast, L_i, based on totals, for the linear relationship between the entry means and the location mean for the eight entries at the five locations is shown in Table 6.16. The corrected sum of squares of the contrasts, L_j, in Table 6.16 gives the component of the location × entry sum of squares (Table 6.12), which is accounted for by the differences in regressions. This is computed as

$$
\begin{aligned}
\text{SS(loc.} \times \text{reg.)} &= \frac{1}{(r) \sum_j (\bar{x}_j - \bar{\bar{x}})^2} \left[\sum_i L_i^2 - \frac{(\sum_i L_i)^2}{1} \right] \\
&= \frac{1}{(3)(1{,}601{,}154)} \left[\sum_i L_i^2 - \frac{(\sum_i L_i)^2}{8} \right] \\
&= 1{,}282{,}643
\end{aligned}
$$

This leads to the following partition of the location × entry interaction.

Source	df	SS
Location by entry	28	2,421,792
Differences of regressions	7	1,282,643
Deviations from regression	21	1,138,451

Table 6.16 **Contrasts of Totals for Regression of Entry Mean Yields on Location Mean Yield**[a]

Location	1	2	3	4	5	
Loc. Mean	4874.4	5224.0	5333.8	6215.0	6296.3	
Deviation	-714.3	-364.7	-254.9	626.3	707.6	L_i
Entry Total						
1	14708	15798	15429	19025	18744	4,978,305
2	14430	16281	17197	18129	18439	3,773,084
3	14481	15625	16089	18814	18866	4,989,488
4	14679	15765	16187	18793	18677	4,627,724
5	15160	16172	16365	19539	19537	5,163,502
6	14928	14782	15004	18486	18607	4,865,109
7	14357	15047	15499	18901	19836	6,810,109
8	14243	15907	16250	17474	18416	3,857,945

[a]Deviation = $(\bar{x}_j - \bar{\bar{x}})$ = deviation of site mean from grand mean
$\Sigma_j (\bar{x}_j - \bar{\bar{x}})^2 = 1{,}601{,}154$
$L_i = \Sigma_j (\bar{x}_j - \bar{\bar{x}})T_{ij}$; T_{ij} = total for entry i at site j
Example:
$L_1 = 4{,}978{,}305 = (-714.3)(14{,}708) + (-364.7)(15{,}798) + (-254.9)(15{,}429) + (626.3)(19{,}025) + (707.6)(18{,}744)$

Finally, the regression components can be partitioned to examine the comparison of 'Stephens' with the other entries. For this purpose we compute the linear regression of the 'Stephens' vs. rest contrast on the site index. This computation is shown in Table 6.17.

We can now further partition the location × entry sum of squares to obtain the final analysis of variance shown in Table 6.18. On considering Table 6.18 it is seen that there is a highly significant difference between the yield of 'Stephens' and the mean yield of the other entries. Consideration of Table 6.9 indicates that 'Stephens' outyields the other varieties by about 224 kg/ha. Whereas there are significant differences in the regression of variety means on location means, there is no real difference, in this respect, between 'Stephens' and the others.

Table 6.17 **Computation of Regression of 'Stephens' vs. Rest on Location Mean Yield**

Location	1	2	3	4	5	L_R
L_j = S vs Rest*	4294	3999	2909	7151	5174	2,872,650
k_j = Deviation	-714.3	-364.7	-254.9	626.3	707.6	

*L_j = contrast for comparing 'Stephens' with others (Table 6.14)
$k_j = (\bar{x}_j - \bar{\bar{x}})$ = deviation of site mean from grand mean
$L_R = \Sigma_j k_j L_j$

$$S_R^2 = \frac{L_R^2}{(3)(56)(1{,}601{,}154)} = \frac{(2{,}872{,}655)^2}{(168)(1{,}601{,}154)} = 30{,}678$$

One of the purposes of running these trials was to measure the stability of 'Stephens' and the other entries under different moisture regimes. We can use the computations leading to the analysis of variance in Table 6.18, particularly the data in Table 6.16, to obtain a number of the statistics that may be used to measure yield stability (see Section 5.6). These statistics are presented in Table 6.19.

Under the Eberhardt and Russell (1966) criterion, entries with a slope, b_i, near 1.0 are stable over environments; those for which $b_i > 1.0$ are adapted to better environments; and those with $b_i < 1.0$ are adapted to poorer environments. We see that for 'Stephens', $b_5 = 1.075$, indicating that 'Stephens' is relatively stable over the five stress environments used in this study. Further, its mean yield is higher than that of any other entry. Hence 'Stephens' would appear to be the entry of choice from among the eight winter wheat selections used in this study.

6.5 EXPERIMENTS AT SAME LOCATION OVER SEVERAL SEASONS

When considering the combined analysis of data from an experiment that has been conducted at a given location over several seasons the same principles apply as when combining data from an experiment that has been conducted at several locations in a single season. The only difference is that "season" replaces "location" in the analysis. The preliminary steps are the same in both cases. They include the analysis of variance of the data for each season and an assessment of the homogeneity of variance for the several seasons. Finally, a combined analysis is completed.

Table 6.18 **Final Combined Analysis of Variance of Winter Wheat Yields at Five Oregon Locations**

Source	df	SS	MS	F
Location	4	38,443,038	9,610,760	
Entry	7	1,044,354		
Stephens vs Rest	1	658,952	658,952	7.62 [b]
Remainder	6	385,402	64,243	
Location by Entry	28	2,421,094	86,492	
Regression on Location	7	1,282,643	183,235	3.38 [c]
Reg(Ste. vs Rest)	1	30,678	30,678	.34 [d]
Remainder	6	1,252,065	208,678	
Deviation from Regression	21	1,138,451	54,212	1.13 [e]
Dev(Ste. vs Rest)	7	628,274 [a]	89,753	
Remainder	14	510,177	36,441	
Pooled Error	115		48,044	

[a]628,274 = 658,952 − 30,678.
[b]F = 7.62 = 658,952/86,492: 1,28 df, significant at 1%.
[c]F = 3.38 = 183,235/54,212; 7,21 df, significant at 5%.
[d]F = .34 = 30,678/89,753; 1, 7 df, not significant.
[e]F = 1.13 = 54,212/48,049; 21,115 df, not significant.

6.5.1 Numerical Data

To illustrate the combined analysis of data from several seasons at the same location we will use more of the results from the lentil production trials conducted in Aleppo Province, Syria, in 1985–1986. These trials are described in Section 6.3.1. For this example we use data from a single location and treat the early planted trial as representing one season and the late planted trial as representing another. The data are given in Table 6.20.

The analysis of variance for each of the two planting dates is shown in Table 6.21. Note that there are significant treatment effects in each season.

Table 6.19 **Stability Measures for Eight Winter Wheat Entries Grown at Five Oregon Locations**

Entry	Stability Measure**					
	$\bar{y}_i$	s_i^2	CV_i	b_i	δ_i^2	R_i^2
1	1580.3	1,331,029	73%	1.036	54,868	.969
2	1631.7	870,926	57%	.777	173,325	.851
3	1591.7	1,298,778	72%	1.039	4,131	.998
4	1606.1	1,200,853	66%	.963	8,525	.994
5*	1784.9	1,394,050	66%	1.075	8,557	.995
6	1453.8	1,329,026	79%	1.013	129,249	.927
7	1576.0	2,028,286	90%	1.418	53,950	.980
8	1486.0	843,259	62%	.803	91,498	.919

**See Section 5.6 for definitions of the stability measures.
*'Stephens'.

The error mean squares in Table 6.21 are the estimates, s^2, of the error variance for these trials. These estimates are fairly close to being equal. As a quick test of the equality of variance we can use the ratio

$$\frac{s^2(\max)}{s^2(\min)} = \frac{6673.43}{5260.35} = 1.27$$

The upper 5% point for k = 2, $\tau = 7$ from Table A4 is 4.99. Since 1.27 < 4.99, there is clearly no evidence of unequal variance.

6.5.2 Preliminary Analysis

Since we have concluded that the variances are not different, we can run a preliminary analysis of the data combined over the two planting dates (seasons). This preliminary analysis is shown in Table 6.22. Again, the pooled error mean square is the simple average of the two error mean squares in the individual analyses.

We see that there is a significant difference in grain production from one season to the other. Further, differences among treatment means are highly significant. On the other hand, there is no evidence that the treatment

Table 6.20 **Mean Grain Yield (kg/ha) of Lentils from Two Seasons at a Single Location in Aleppo Province, Syria, 1985–1986**

Treatment[a]	Planting Date Early	Late
(1)	976.5	944.5
S	1166.5	1125.5
F	1074.5	1079.0
SF	1260.0	1236.0
W	1091.0	959.5
SW	1193.0	1076.0
FW	1107.0	898.5
SFW	1310.0	1241.0

[a]Treatment: (1), $S_0F_0W_0$; S, $S_1F_0W_0$; F, $S_0F_1W_0$; SF, $S_1F_1W_0$; W, $S_0F_0W_1$; SW, $S_1F_0W_1$; FW, $S_0F_1W_1$; SFW, $S_1F_1W_1$.

Table 6.21 **Analyses of Variance of Grain Yield from Lentils Planted in Two Seasons at One Location in Aleppo Province, Syria, 1985–1986**

Source	df	Mean Square for Early	Late
Block	1	576.00	105.06
Treatment	7	33,533.40*	23,113.10*
Error	7	6,673.43	5,260.35

*Significant at the 5% level.

Table 6.22 **Preliminary Analysis of Variance of Grain Yields of Lentils Grown at One Location Combined Over Two Seasons in Aleppo Province, Syria, 1985–1986**

Source	df	MS	F
Block in Season	2	340	
Season	1	47,818	8.01*
Treatment	7	51,690	8.66**
Season by Treat.	7	4,957	.83NS
Pooled Error	14	5,976	

*Significant at 5%; **significant at 1%; NS, not significant. Pooled error MS used as a divisor in all F tests.

differences vary from one season to the next (season × treatment interaction is not significant). We should decompose the treatment and the treatment × season interaction sources of variation to shed further light on the treatment differences.

6.5.3 Detailed Analysis

As noted in Section 6.3.4, the treatments used in the lentil study constitute a 2^3 factorial set of the three factors: (1) sitona weevil control (Si), (2) fertilizer (F), and (3) weed control (W). With this set of treatments we can decompose the treatment variance source into the factorial main effects and interactions. Similarly, we can examine the interaction of season with the factorial treatment effects. This partition is presented in the detailed analysis of variance in Table 6.23.

In examining this detailed analysis the first step is to test the season × treatment interaction components against the pooled error. When this is done we see that none of the interactions with season is significant. Since this is the case, we use the pooled error mean square as the denominator of the F ratio to test each treatment component.

The results of the tests of the main effects and interactions for the factorial treatment combinations indicate that only the main effects of sitona control and fertilizer were significant. The main effect of weed control was not significant, nor were any of the interactions significant.

Table 6.23 **Detailed Analysis of Variance of Grain Yields of Lentils Grown at One Location Combined Over Two Seasons in Aleppo Province, Syria, 1985–1986**

Source	df		MS	F
Block in Season	2		340	
Season (Se)	1		47,818	
Treatment	7			
Sitona (Si)		1	272,876	45.73**
Fertilizer (F)		1	56,700	9.50*
Si F		1	11,138	1.87
Weed (W)		1	23	.00
Si W		1	319	.05
F W		1	4,975	.83
Si F W		1	15,798	2.65
Season by Treat.	7			
Se Si		1	1,696	.28
Se F		1	75	.01
Se Si F		1	1,391	.23
Se W		1	23,490	3.94
Se Si W		1	4,584	.77
Se F W		1	850	.14
Se Si F W		1	2,610	.44
Pooled Error	7		5,967	

**Significant at 1%; *significant at 5%.

Table 6.24 **Mean Grain Yield (kg/ha) of Early- and Late-Planted Lentils Grown With and Without Sitona Weevil Control at a Single Location, Aleppo Province, Syria, 1985–1986**

Sitona Control	None	Some	Standard Error
Mean Yield	1016.3	1201.0	19.31

Report of Statistical Analysis

An experiment was planned to examine the effect of three factors: sitona weevil control, fertilizer, and weed control, on the grain yield of lentils. The trial was conducted at a single location in Aleppo Province, Syria, under both early and late planting conditions. The results of the trials, combined over both planting times, are presented in Tables 6.24 and 6.25.

Sitona weevil control resulted in an increased grain yield, when averaged over planting date, of about 185 kg/ha, while the addition of 50 kg/ha of phosphate resulted in an average increase of about 84 kg/ha of lentil grain. Weed control was apparently ineffective at this location. Although the mean grain yield for early planted lentils was about 75 kg/ha greater than for late-planted lentils, there was no indication that the differential response to the treatment factors was affected by the time of planting.

6.6. EXPERIMENTS REPEATED AT SEVERAL LOCATIONS OVER SEVERAL SEASONS

In the usual agricultural research program an experiment is conducted at a number of locations over several seasons before recommendations are made. In the simplest situation, and the only one considered here, the same

Table 6.25 **Mean Grain Yield (kg/ha) of Early- and Late-Planted Lentils Grown With and Without Fertilizer at a Single Location, Aleppo Province, Syria, 1985–1986**

Fertilizer	None	Some	Standard Error
Mean Yield	1066.6	1150.8	19.31

experiment is conducted at each location and repeated each season. At each location a new site is chosen each season and a new randomization is used for each repetition. Under these conditions the data from each trial may be considered to be statistically independent.

The locations chosen for the study and the seasons included are taken to be representative of all possible locations and seasons in the domain for which recommendations are to be made. The assumption is that the locations and seasons are a random sample of locations and seasons from the recommendation domain. If they are not, the results will apply only to the domain that is actually represented by the data. If this assumption is unrealistic, the analysis cannot lead to recommendations that have any practical value.

The basic steps in the combined analysis of data from experiments repeated in both time and space are the same as for those combined analyses discussed previously. Individual analyses of variance are computed for each location in each season. The error variances are then checked for heterogeneity. Finally, an appropriate combined analysis is completed and interpreted.

6.6.1 Numerical Data

To illustrate the combined analysis of data from experiments conducted at several locations over more than one season, we will use additional data from the lentil production trials conducted by ICARDA in Aleppo Province, Syria. These trials are described in detail in Section 6.3.1. We use data from all eight locations and treat the early-planted trials as representing one season and the late-planted trials as representative of a second season. The data for the early-planted trials are given in Table 6.1. The data from the late-planted trials are shown in Table 6.26.

The individual analyses of variance for both seasons at the eight locations are presented in Table 6.27. Consideration of the error mean squares from the 16 analyses reveals a range from a low of 3494 for the early trial at location 2 to a high of 88,245 for the late trial at location 6. Bartlett's test of homogeneity of variance for these data produced a chi-square value of 30.283. The χ^2 at $\alpha = 0.05$ for 15 df is 24.906. Since $24.906 < 30.283$, the indication is that the variances are not homogeneous.

Consideration of Table 6.27 shows that the three smallest variances occurred at locations 1 and 2. These locations were on the principal research station of ICARDA near Aleppo. They were managed by the research staff and were more closely controlled than were the other trials. The other trials were conducted on grower's fields throughout the province. They were managed by the local grower under the supervision of the research scientist. Under these circumstances it seem reasonable to omit locations 1 and 2

Table 6.26 **Mean Grain Yield (kg/ha) of Late-Planted Lentils at Eight Locations in Aleppo Province, Syria, 1985–1986**

	Location			
Treatment *	1	2	3	4
(1)	944.5	1061.5	576.0	1108.5
S	1125.5	1237.0	831.0	1136.5
F	1079.0	1132.5	631.0	1404.5
SF	1236.0	1178.0	1020.0	1471.0
W	959.5	988.0	714.5	1200.5
SW	1076.0	1124.0	950.0	1263.5
FW	898.5	1177.0	838.0	1164.5
SFW	1241.0	1154.5	1186.0	1601.0

	Location			
Treatment	5	6	7	8
(1)	526.0	1773.5	572.0	533.0
S	613.5	1540.5	600.0	915.0
F	820.5	1631.0	471.5	653.5
SF	814.5	1383.5	624.0	910.0
W	1208.0	1572.5	687.5	776.5
SW	1002.0	1258.5	611.0	933.5
FW	1167.5	1700.0	566.5	783.0
SFW	1170.5	1766.0	864.5	971.0

*(1), check; S, 1 kg/ha carbofuran; F, 50 kg/ha P_2O_5; SF, carb. + P_2O_5; W, herbicide; SW, carb. + herb.; FW, P_2O_5 + Herb.; SFW, carb. + P_2O_5 + herb.

Table 6.27 **Analysis of Variance of Grain Yields from Early (E)- and Late (L)-Planted Lentils Grown at Eight Locations in Aleppo Province, Syria, 1985–1986**

		Mean Square at Location			
		1		2	
Source	df	E	L	E	L
Block	1	105	576	4970	79383
Treatment	7	23113*	33533*	8779	11840
Error	7	5260	6673	3494	25276

		3		4	
Source	df	E	L	E	L
Block	1	2525	37733	62750	2704
Treatment	7	102750	83811	221985**	63842
Error	7	34406	56555	30131	33315

		5		6	
Source	df	E	L	E	L
Block	1	100172	338433	176400	581788
Treatment	7	214325**	138202	32464	66458
Error	7	27704	45743	20190	88245

*Significant at 5%; **significant at 1%.

(*continued*)

Table 6.27 **Continued**

		7		8	
Source	df	E	L	E	L
Block	1	3220	47524	155433	35627
Treatment	7	39984	26252	44523	37571
Error	7	21276	17473	32759	17003

and to consider a combined analysis of the data from the remaining six locations. Bartlett's test of homogeneity of variance on the data from these locations (locations 1 and 2 omitted) produced a chi-square value of 9.586. The χ^2 for $\alpha = 0.05$ with 11 df is $19.675 > 9.586$. The conclusion is that the variances from these trials are not heterogeneous. We proceed with the combined analysis using these data.

6.6.2 Mean Square Expectations in the Combined Analysis of Variance

Before proceeding with a preliminary analysis we must consider a problem that we did not have to address in previous analyses. This has to do with the proper way to form the F ratio for testing treatment effects. In combined analyses involving several locations or several seasons, we tested the location × treatment or the season × treatment interaction against the pooled error. If the interaction was significant, we used it to test treatments; otherwise, we used the pooled error. To understand the reasoning behind this choice, we should examine the expected values of the mean squares in the combined analysis.

In general, to form an F ratio to test the hypothesis that a particular component of variation is zero, we place the mean square whose expectation contains the component under test in the numerator of the ratio. The denominator of F is the mean square whose expectation contains all of the components of the mean square in the numerator except the component under test.

To illustrate, suppose that we have an experiment with t treatments replicated r times at each of l locations. We can describe an observation from this type of data with the model.

$$\bar{y}_{ij} = \mu + \pi_i + \tau_j + (\pi\tau)_{ij} + \bar{\epsilon}_{ij}$$

where

$\bar{y}_{ij}$ = observed mean of the j-th treatment at the i-th location
μ = grand mean
π_i = effect of the i-th location
τ_j = effect of the j-th treatment
$(\pi\tau)_{ij}$ = interaction of the j-th treatment with the i-th location
$\bar{\epsilon}_{ij}$ = average of the errors on the r plots that receive the j-th treatment at the i-th location

It is reasonable to assume that the treatments are fixed while the other components are random. Hence we assume that

$$\sum_j \tau_j = 0$$
$$\pi_i \sim N(0,\sigma_l^2)$$
$$(\pi\tau)_{ij} \sim N(0,\sigma_{lt}^2)$$
$$\bar{\epsilon}_{ij} \sim N(0,\sigma_e^2)$$

where $\sim N(0,\sigma^2)$ means "is the normally distributed with mean zero and variance σ^2." Under the preceding conditions, the expected values of the mean squares in the analysis of variance are as shown in Table 6.28.

The ratio F = MSLT/MSE estimates the ratio $(\sigma_e^2 + r\sigma_{lt}^2)/\sigma_e^2$. This ratio follows the F distribution only if $\sigma_{lt}^2 = 0$. Hence F = MSLT/MSE is a test statistic for the significance of the location × treatment interaction.

Table 6.28 **Expected Values of the Mean Squares, E(MS), in the Combined Analysis of Variance of Data from the Same Experiment Conducted at Several Locations**

Source	df	MS	E(MS)
Treatment	(t-1)	MST	$\sigma_e^2 + r\sigma_{lt}^2 + rl\sum_j \tau_j^2/(t-1)$
Loc. by Treat	(l-1)(t-1)	MSLT	$\sigma_e^2 + r\sigma_{lt}^2$
Pooled error	l(r-1)(t-1)	MSE	σ_e^2

Now consider the ratio F = MST/MSLT. This estimates the ratio

$$\frac{\sigma_e^2 + r\sigma_{lt}^2 + rl\ \Sigma_j\tau_j^2/(t-1)}{\sigma_e^2 + r\sigma_{lt}^2}$$

which follows the F distribution only if rl $\Sigma_j\ \tau_j^2/(t-1)$ is zero. This condition is met only if H_o: $\tau_1 = \tau_2 = \cdots = \tau_t = 0$ is true. Hence this is the proper ratio for testing the significance of the differences among treatment means.

On the other hand, suppose that an F test leads us to the conclusion that H_0: $\sigma_{lt}^2 = 0$ is true. In this case, both MSE and MSLT estimate σ_e^2, and MST estimates σ_e^2 + rl $\Sigma_j\ \tau_j^2/(t-1)$. Under this condition either MSE or MSLT could serve as the denominator of an F test for treatments. Since MSE usually carries more degrees of freedom than does MSLT. MSE is usually a better estimate of σ_e^2 than is MSLT. For this reason, if we conclude that the location × treatment interaction is not significant, we use F = MST/MSE to test the significance of differences among treatment means.

Suppose that we now consider the combined analysis of data from an experiment repeated at several locations over several seasons. Assume that there are l locations and s seasons. Assume, further, that a new site is chosen at each location each season and that a new randomization is used in each repetition. A treatment mean, in this case, is described by the model

$$\bar{y}_{ijk} = \mu + \pi_i + \tau_j + (\pi\tau)_{ij} + \gamma_k + (\pi\gamma)_{ik} + (\tau\gamma)_{jk} + (\pi\tau\gamma)_{ijk} + \epsilon_{ijk}$$

in which

$\bar{y}_{ijk}$ = observed mean of the j-th treatment at the i-th location during the k-th season
γ_k = effect of the k-th season
$(\pi\gamma)_{ik}$ = location × season interaction
$(\tau\gamma)_{jk}$ = treatment × season interaction
$(\pi\tau\gamma)_{ijk}$ = location × treatment × season interaction

and where μ, π_i, τ_j, $(\pi\tau)_{ij}$, and $\bar{\epsilon}$ are defined as for the multilocation model discussed previously. If all the effects are random except the treatments, we have the following additional assumptions about the terms in the model:

$$\gamma_k \sim N(0,\sigma_y^2)$$
$$(\pi\gamma)_{ik} \sim N(0,\sigma_{ly}^2)$$
$$(\tau\gamma)_{jk} \sim N(0,\sigma_{ty}^2)$$
$$(\pi\tau\gamma)_{ijk} \sim N(0,\sigma_{lty}^2)$$

Table 6.29 **Expected Mean Squares, E(MS), in the Combined Analysis of Variance of Data from an Experiment Conducted at Several Locations Over Several Seasons**

Source	df	MS	E(MS)
Treatment	t-1	MST	$\sigma_e^2 + r\sigma_{lty}^2 + rs\sigma_{lt}^2 + rl\sigma_{ty}^2 + \frac{rl\sum_j \tau_j^2}{t-1}$
Season by Treatment	(s-1)(t-1)	MSST	$\sigma_e^2 + r\sigma_{lty}^2 + rl\sigma_{ty}^2$
Loc.by Treatment	(l-1)(t-1)	MSLT	$\sigma_e^2 + r\sigma_{lty}^2 + rs\sigma_{lt}^2$
Sea by Loc by Treat	(s-1)(l-1)(t-1)	MSSLT	$\sigma_e^2 + r\sigma_{lty}^2$
Pooled Error	sl(r-1)(t-1)	MSE	σ_e^2

Under these assumptions the expected values of the mean squares, E(MS), in the combined analysis of variance are as shown in Table 6.29. On considering the E(MS) in Table 6.29, we see that F = MSSLT/MSE is the test statistic for testing the significance of the season × location × treatment interaction. Similarly, MSSLT is the proper denominator in the F ratio for testing both the season × treatment and the location × treatment interactions.

A problem arises, however, in finding an F ratio for testing treatments. None of the other mean squares have expectations that contain all of the components of the expectation of MST except the component involving $\sum_j \tau_j^2$. Thus, in general, there is no ratio of mean squares to use as an F statistic for testing treatments. If either the season × treatment or the location × treatment interactions is negligible (not significant when tested against MSSLT), the other interaction mean square may be used as the denominator of the F for testing treatments.

When both of the two-factor interactions are significant, we must construct a test ratio by combining mean squares in both the numerator and denominator of the ratio. On considering Table 6.29 we see that E[MSSLT + MST] contains $\sum_j \tau_j^2$, while [MSST + MSLT] contains all of the components of the preceding combination of mean squares except $\sum_j \tau_j^2$. This suggests that we could form a test statistic for H_o: $\tau_1 = \tau_2 = \cdots = \tau_t = 0$ using the ratio

$$\frac{\text{MSSLT} + \text{MST}}{\text{MSST} + \text{MSLT}}$$

This ratio does not follow the F distribution when H_o: $\tau_j = 0$ for all j is true. However, it does follow an approximation proposed by Satterthwaite (1946). With this ratio we can use the F tables with n_1' df for the numerator and n_2' df for the denominator. We compute n_1' as

$$n_1' = \frac{[MST + MSSLT]^2}{[MST^2/(t-1)] + [MSSLT^2/(s-1)(l-1)(t-1)]}$$

and n'_2 as

$$n_2' = \frac{[MSST + MSLT]^2}{[MSST^2/(s-1)(t-1)] + [MSLT^2/(l-1)(t-1)]}$$

6.6.3 Preliminary Combined Analysis

A preliminary combined analysis of variance of lentil grain yields from an experiment conducted for two seasons at six off-station locations is given in Table 6.30. Significance testing in this table was a sequential process. The three-factor location × season × treatment interaction was tested against the pooled error and found not to be significant. Hence the pooled error was also used to test the two-factor interactions. With this test the location × treatment interaction was found to be nonsignificant, while the season × treatment interaction was found to be significant at the 5% level. For this reason, the season × treatment interaction mean square was used to test the significance of the treatment effects. Under this test the treatment

Table 6.30 **Preliminary Combined Analysis of Variance of Grain Yields of Lentils Planted in Two Seasons, Early and Late, at Six Locations in Aleppo Province, Syria, 1985–1986**

Source	df	MS	F
Treatment	7	484,521	4.97*
Location by Treatment	35	68,373	1.93NS
Season by Treatment	7	97,420	2.75*
Loc by Seas by Treat	35	29,678	.84NS
Pooled Error	84	35,400	

*Significant at 5%; NS, not significant.

effects were found to be significant at the 5% level. Since there is evidence that the treatment effects are not the same from one season to another, we should examine the data in more detail.

6.6.4 Detailed Combined Analysis of Variance

A detailed analysis of variance of the treatment effects and the location × treatment, season × treatment, and location × season × treatment interactions is presented in Table 6.31. The decomposition of the variance sources in Table 6.31 is based on the factorial nature of the treatments. The treatment source is partitioned into the standard factorial main effects and interactions, each carrying 1 df. The season and location interactions with treatment are partitioned into components corresponding to the effects in the treatment partition.

Significance testing was begun by testing the second-order (season × location × treatment) interaction components against the pooled error. None of the second-order components were significant. Hence the pooled error was also used to test the first-order interaction components.

Both the season × sitona and the location × sitona components were significant at the 1% level. The effect of sitona weevil control varied from location to location and from season to season. This is probably the result of differential sitona weevil infestation in the study site. None of the other first-order interaction components were significant.

As a consequence of the preceding tests the pooled error was used to test all of the treatment components except for the main effect of sitona weevil control. Under these tests the main effect of fertilizer (F) and the main effect of herbicide (W) were found to be significant at the 1% level. None of the interactions among the treatment factors were judged to be significant.

Since both L by Si and Se by Si are significant, we test the main effect of sitona control using a ratio of combinations of mean squares. This ratio is computed as

$$\frac{\text{MS(Si)} + \text{MS(SeLSi)}}{\text{MS(SeSi)} + \text{MS(LSi)}} = \frac{1{,}510{,}350 + 71{,}251}{340{,}118 + 191{,}288}$$

$$= \frac{1{,}581{,}601}{531{,}406} = 2.98$$

Degrees of freedom for determining the significance of this ratio using the tables of the F distribution are obtained using Satterthwaite's approximation.

Table 6.31 **Detailed Combined Analysis of Variance of Grain Yields of Lentils Planted in Two Seasons at Six Locations in Aleppo Province, Syria, 1985–1986**

Source	df	MS	F
Treatment	7		
Sitona(Si)	1	1,510,350	2.98 a
Fertilizer(F)	1	366,888	10.36**
Si F	1	10,165	.29
Weed Control(W)	1	1,468,200	41.48**
Si W	1	23,122	.65
F W	1	18,116	.51
Si F W	1	16,188	.46
Season by Treatment	7		
Se Si	1	340,118	9.61**
Se F	1	20,152	.57
Se Si F	1	103,556	2.92
Se W	1	66,120	1.87
Se Si W	1	26,767	.76
Se F W	1	64,131	1.81
Se Si F W	1	61,097	1.78
Location by Treatment	35		
L Si	5	191,288	5.40**
L F	5	47,497	1.34
L Si F	5	17,024	.48
L W	5	15,542	.44
L Si W	5	15,554	.44
L F W	5	42,016	1.19
L Si F W	5	98,061	2.77

Table 6.31 **Continued**

Source	df		MS	F
Seas by Loc by Treat	35			
Se L Si		5	71,251	2.01
Se L F		5	26,916	.76
Se L Si F		5	51,485	1.45
Se L W		5	24,105	.68
Se L Si W		5	28,672	.81
Se L F W		5	40,918	1.16
Se L Si F W		5	10,735	.30
Pooled Error	84		35,400	

**Significant at 1%; a, Satterthwaite's approximation.

We have numerator df:

$$n_1' = \frac{[MS(Si) + MS(SeLSi)]^2}{\{[MS(Si)^2/df(Si)] + [MS(SeLSi)^2/df(SeLSi)]\}}$$

$$= \frac{(1{,}510{,}350 + 71{,}251)^2}{[(1{,}510{,}350)^2/1] + [(71{,}251)^2/5]}$$

$$= 1.10 \cong 1$$

$$n_2' \cong \frac{[MS(SeSi) + MS(LSi)]^2}{\{[MS(SeSi)^2/df(SeSi)] + [MS(LSi)^2/df(LSi)]\}}$$

$$= \frac{(340{,}118 + 191{,}288)^2}{[(340{,}118)^2/1] + [(191{,}288)^2/5]}$$

$$= 2.30 \cong 2$$

Since $F_{0.05(1.2)} = 18.51 < 2.98$, we conclude that the effect of sitona weevil control is not significant.

Report of Statistical Analysis

An experiment was designed to measure the effect of sitona weevil control, fertilizer, and herbicide on the grain yield of lentils was conducted at six locations on two planting dates in Aleppo Province, Syria, during 1985–1986. The results are summarized in Tables 6.32 to 6.35.

Table 6.32 **Yield of Lentil Grain (kg/ha) With and Without 50 kg/ha Added P_2O_5[a]**

Added P_2O_5	None	50 kg/ha	Standard Error
Mean Yield	884.10	971.53	13.58

[a]Data from two planting dates at six locations.

Consideration of Table 6.32 indicates that the application of 50 kg/ha of phosphate increased the yield of lentil grain by an average of more than 85 kg/ha for the 1985–1986 growing season in Aleppo Province. Similarly, during the same period the use of herbicide to control weeds resulted in an average yield increase throughout the province of nearly 175 kg/ha (see Table 6.33).

Response to the use of carbofuran to control sitona weevils varied from location to location and from early to late planting. This was probably the result of irregular weevil infestation throughout the province. In Table 6.34 it is seen that, on average, weevil control was relatively more effective for early-planted lentils than for the late planting. Similarly, consideration of Table 6.35 shows that the response to sitona control varied from an increased grain yield of about 360 kg/ha at location 4 to a decrease in yield of about 50 kg/ha at location 6.

In summary, lentil grain yield, which averaged about 710 kg/ha for both planting times at all locations under no improved practices, was increased by the addition of phosphate fertilizer and by the use of herbicide to control weeds. On the other hand, response to weevil control was spotty, with no

Table 6.33 **Yield of Lentil Grain (kg/ha) With and Without Herbicide Spray[a]**

Weed control	None	Spray	Standard Error
Mean Yield	841.01	1014.63	13.58

[a]Data from two planting dates at six locations.

Table 6.34 **Yield of Lentil Grain (kg/ha) With and Without Sitona Weevil Control on Two Planting Dates**[a]

Sitona Control	Planting Date	
	Early	Late
None	715.75	962.50
Carbofuran	977.31	1055.71

[a]Standard error = 19.20. Data from six locations.

Table 6.35 **Yield of Lentil Grain (kg/ha) With and Without Sitona Weevil Control at Six Locations in Aleppo Province, Syria**[a]

Location	Weevil Control	
	None	Carbofuran
3	649.25	976.88
4	969.94	1327.94
5	672.12	800.31
6	1281.63	1231.19
7	639.12	724.06
8	822.69	1038.69

[a]Standard error = 33.26. Data from two planting dates.

consistent effect from season to season or from location to location within the province.

6.7. COMBINED ANALYSIS OF ON-STATION YIELD TRIALS

Replicated yield trials form an integral part of most plant breeding programs. They provide a way to assess the yield potential of new selections by comparing them to one or more checks and by comparing them with other new selections. The typical procedure is to conduct a number of trials on a given speciès in a given year on an experiment station. In the usual practice the trials are identical in a number of ways. All of the trials are run in proximity to one another. The same experimental design, usually a randomized block design, is used for all trials. The same checks are used in each trial. A different set of new selections is used in each trial, but the number of new selections is the same for all trials.

It is customary to analyze the results of each trial separately. This permits comparison of the new selections with the checks and with other new selections in the same trial. It does not, however, permit comparison of new selections in one trial with those in another trial. Further, the experimental error common to a group of similar trials is not as well estimated from the individual trials as it would be from a combined analysis of data from all the trials.

The combined analysis presented here includes the data from all the yield trials conducted at the same location during a given year. There are several purposes of a combined analysis:

1. To provide a single estimate of the experimental error that is common to all the trials.
2. To adjust the means of the new selections for yield differences from one trial to another.
3. To provide a way of comparing the adjusted selection means with check means and with adjusted means of other selections in both the same and different trials.

6.7.1 Analysis

The analytical procedure for the combination of data from several on-station yield trials is an extension of the methods developed by Federer and his co-workers (Federer, 1956, 1961; Federer and Ragavarao, 1975) in their work on augmented designs.

Suppose that we have a number of similar yield trials, each conducted at the same location during the same growing season. Suppose, further, that all of the trials use a randomized block design, that there are two or more checks common to all of the trials, and that the same number of new selections, a different set in each trial, is included in each trial. In particular, suppose that

p = number of trials
r = number of blocks (replications) per trial
c = number of check varieties
v = number of new selections per trial

Table 6.36 **Analysis of Variance of Check Entries in a Combined Analysis of On-Station Yield Trials**

ANOVA

Source	df	SS	MS
Total	prc−1	SSTot	
Blocks	pr−1	SSR	
Checks	c−1	SSC	MSC
Error	(pr−1)(c−1)	SSE	MSE_c

The checks constitute, in effect, one experiment conducted as a randomized block design with c treatments and pr blocks. Let x_{ijk} represent the yield of the i-th check in the k-th block of the j-th trial. The analysis of variance of the check entries takes the form shown in Table 6.36. The error mean square, MSE_c, from this analysis is an unbiased estimate of σ^2, the experimental error variance for the yield trials.

The new selections constitute entries in p trials, each conducted as a randomized block design with v treatments and r blocks. Let y_{ijk} represent the yield of the i-th new selection in the k-th block of the j-th trial. The analysis of variance of new selections, ignoring checks, in any of the trials has the format given in Table 6.37.

Table 6.37 **Analysis of Variance of New Selections in an On-Station Yield Trial**

ANOVA

Source	df	SS	MS
Total	rv−1	SSTot	
Blocks	r−1	SSR	
Selections	v−1	SSV	
Error	(r−1)(v−1)	SSE	MSE_j

With p yield trials there will be p repetitions of the ANOVA of Table 6.37. The error mean square, MSE_j, from each of these analyses is also an unbiased estimate of σ^2, the experimental error variance.

6.7.2 Pooled Variance

Although an estimate of variance is available from each trial, a better estimate may be obtained by combining (pooling) the information from all the trials. The pooled estimate, E_p, of σ^2 may be computed as the weighted average of the error mean squares from the individual analyses. We have

$$E_p = \frac{SSE_c + \Sigma_j SSE_j}{(pr - 1)(c - 1) + p(r - 1)(v - 1)}$$

with $[(pr - 1)(c - 1) + p(r - 1)(v - 1)]$ degrees of freedom. This pooled error may be used for significance tests and for interval estimates of means and differences.

6.7.3 Adjusted Selection Means

One of the goals of the combined analysis is to be able to compare the mean of a selection in one trial with the mean of a selection in another trial. To do this the means must be adjusted for differences in the general yield level from one trial to another. An adjustment factor, a_j, for the j-th trial can be computed from the yields of the checks.

The mean of the checks in the j-th trial, $\bar{x}_j$, is computed as

$$\bar{x}_j = \frac{\Sigma_i \Sigma_k x_{ijk}}{rc}$$

The overall mean of checks in all of the trials, $\bar{\bar{x}}$, is found to be

$$\bar{\bar{x}} = \frac{\Sigma_i \Sigma_j \Sigma_k x_{ijk}}{prc}$$

The adjustment factor, a_j, for the j-th trial is then computed as

$$a_j = \bar{x}_j - \bar{\bar{x}}$$

The mean, $\bar{y}_{ij}$, of the i-th selection in the j-th trial is

$$\bar{y}_{ij} = \frac{\Sigma_k y_{ijk}}{r}$$

To adjust a selection mean in the j-th trial for the differential yield level of the trial the adjustment factor is subtracted, giving an adjusted mean, $\hat{\bar{y}}_{ij}$. We have

$$\hat{\bar{y}}_{ij} = \bar{y}_{ij} - a_j$$

6.7.4 Means and Standard Errors

The mean of a check variety, $\bar{x}_i$, is obtained as

$$\bar{x}_i = \frac{\sum_j \sum_k x_{ijk}}{pr}$$

The standard error, $s_{\bar{x}}$, of this mean is

$$s_{\bar{x}} = \sqrt{\frac{E_p}{pr}}$$

The adjusted mean, $\hat{y}_{ij}$, of the i-th selection in the j-th trial is

$$\hat{y}_{ij} = \bar{y}_{ij} - a_j$$

and its standard error, $s_{\hat{y}}$, is

$$s_{\hat{y}} = \sqrt{\frac{E_p(c + 1)}{rc}}$$

There are a number of comparisons that may be of interest in the combined analysis of several yield trials. These comparisons and their standard errors are as follows:

1. Difference between two check means:

   $$\bar{d}_1 = \bar{x}_i - \bar{x}_{i'}$$

   The standard error, $s_{\bar{d}1}$, is

   $$s_{\bar{d}1} = \sqrt{\frac{2E_p}{pr}}$$

2. Difference between adjusted means of two selections in the same trial:

   $$\bar{d}_2 = \hat{y}_{ij} - \hat{y}_{i'j} = \bar{y}_{ij} - \bar{y}_{i'j}$$

   with a standard error, $s_{\bar{d}2}$, of

   $$s_{\bar{d}2} = \sqrt{\frac{2E_p}{r}}$$

3. Difference between adjusted means of two selections in different trials:

   $$\bar{d}_3 = \hat{y}_{ij} - \hat{y}_{i'j'}$$

The standard error, $s_{\bar{d}3}$, is

$$s_{\bar{d}3} = \sqrt{\frac{2E_p(c + 1)}{rc}}$$

4. Difference between a check mean and the adjusted mean of a new selection:

   $\bar{d}_4 = \hat{\bar{y}}_{ij} - \bar{x}_i$

   with a standard error, $s_{\bar{d}4}$, of

$$s_{\bar{d}4} = \sqrt{\frac{E_p(pc + p + c)}{prc}}$$

These standard errors may be used for constructing interval estimates or in testing the significance of the individual differences. The most common comparison sought in a breeding program is the comparison of new selection means with the mean of the checks to determine which, if any, of the new selections outyield the checks. For this purpose we compute the LSI (least significant increase) over the checks. For the combined yield trials the LSI is computed as

$$\text{LSI} = t_\alpha s_{\bar{d}4}$$

where t_α is the one-sided α-level t at $[(pr - 1)(c - 1) + p(r - 1)(v - 1)]$ degrees of freedom, and $s_{\bar{d}4}$ is the standard error of the difference between an adjusted selection mean and a check mean.

6.7.5 Numerical Example

In the 1983–1984 growing season a number of yield trials on spring barley were conducted at the principal research station of the Oregon Agricultural Experiment Station near Corvallis, Oregon. Each trial consisted of 24 entries replicated three times in a randomized block design. To illustrate the computations involved in a combined analysis of data from trials such as this, a subset of data, shown in Table 6.38, was extracted from three of these trials.

The three trials had two checks in common: 'Benton' and 'Steptoe'. The remaining 10 entries in each trial were new selections, and these differed from trial to trial. For this set of data $c = 2$, $p = 3$, $r = 2$, and $v = 10$. To begin the analysis we compute the check means, the adjustment factors, and the analysis of variance of the checks. The means and adjustment factors are given in Table 6.39, and the analysis of variance is presented in Table 6.40.

The adjusted mean yields of the new selections in each trial are now computed. These and the unadjusted means are shown in Table 6.41. The

Table 6.38 **Yields (kg/ha) of Spring Barley from Three Yield Trials Grown by the Oregon Agricultural Experiment Station**

	Trial					
	1		2		3	
Entry	I	II	I	II	I	II
Benton	2550	4241	3259	3375	3366	3283
1	2433	3833	3075	3625	3200	3591
2	2150	4250	3291	3741	3141	3233
3	2950	3225	2941	2750	2975	2291
4	4125	3925	4283	3891	3708	2708
5	3441	3916	2958	3233	3950	3383
6	3066	3883	2875	2791	2458	2666
7	3433	3725	2708	2641	2575	2350
8	3233	3966	2933	2941	3583	2900
9	3008	2600	3425	4041	3766	3800
10	2766	2708	3633	3091	2783	2600
Steptoe	3500	3375	3633	2866	3058	2875

analyses of variance of the new selections are now computed. These are presented in Table 6.42. With the error sums of squares from the analyses of variance of the checks and of the new selections, the pooled error mean square is now computed as

$$E_p = \frac{SSE_c + \Sigma_j\, SSE_j}{(pr - 1)(c - 1) + p(r - 1)(v - 1)}$$

$$= \frac{1{,}107{,}364.5 + 2{,}613{,}486.2 + 708{,}393.1 + 930{,}422.1}{5 + (3)(9)}$$

$$= \frac{5{,}359{,}647.9}{32} = 167{,}489.0$$

The pooled error mean square, E_p, may be used to compute the standard

Table 6.39 **Mean Yields (kg/ha) of Checks and Adjustment Factors for Three Yield Trials on Spring Barley**[a]

Variety	Trial 1	Trial 2	Trial 3	Mean
Benton	3395.5	3317.0	3324.5	3345.67
Steptoe	3437.5	3249.5	2966.5	3217.83
Adjustment, a_j	134.75	1.50	-136.25	

[a]The adjustment factors sum to zero.

Table 6.40 **Analysis of Variance of Barley Check Yields**

ANOVA

Source	df	SS	MS
Total	11	2,039,994.2	
Blocks	5	883,623.7	176,724.7
Checks	1	49,024.0	49,024.0
Error	5	1,107,346.5	221,469.3

errors of means and the various differences. The standard errors of the means are:

1. Check means, $s_{\bar{x}}$:

$$s_{\bar{x}} = \sqrt{\frac{E_p}{pr}} = \sqrt{\frac{167,489.0}{6}} = 167.08$$

2. Adjusted selection means, $s_{\hat{y}}$:

$$s_{\hat{y}} = E_p\sqrt{\frac{c+1}{rc}} = \sqrt{\frac{(167,489.0)(3)}{4}} = 354.42$$

The standard errors of differences are:

Table 6.41 **Unadjusted and Adjusted Mean Yields (kg/ha) of New Barley Selections in Three Yield Trials**

Trial	1		2		3	
Selection	$\bar{y}$ *	$\hat{y}$	$\bar{y}$	$\hat{y}$	$\bar{y}$	$\hat{y}$
1	3133.0	2998.2	3350.0	3348.5	3395.5	3531.8
2	3200.0	3065.2	3516.0	3514.5	3178.0	3323.2
3	3087.5	2952.8	2845.5	2844.0	2633.0	2769.2
4	4025.0	3890.2	4087.0	4085.5	3208.0	3344.2
5	3678.5	3543.8	3095.5	3094.0	3666.5	3802.8
6	3474.5	3339.8	2833.0	2831.5	2562.0	2698.2
7	3579.0	3444.2	2574.5	2673.0	2642.5	2598.8
8	3599.5	3464.8	2937.0	2935.5	3241.5	3377.8
9	2804.0	2669.2	2733.0	3731.5	3783.0	3919.2
10	2737.0	2602.2	3362.0	3360.5	2691.5	2827.8
a_j	134.75		1.50		-136.25	

*$\bar{y}$, unadjusted mean; $\hat{y}$, adjusted mean; a_j, adjustment factor.

1. Difference between two check means, $s_{\bar{d}1}$:

$$s_{\bar{d}1} = \sqrt{\frac{2E_p}{pr}} = \sqrt{\frac{(2)(167{,}489.0)}{6}} = 236.28$$

2. Difference between adjusted means of two selections in the same trial, $s_{\bar{d}2}$:

$$s_{\bar{d}2} = \sqrt{\frac{2E_p}{r}} = \sqrt{\frac{(2)(167{,}489.0)}{2}} = 409.25$$

3. Difference between adjusted means of two selections in different trials, $s_{\bar{d}3}$:

Table 6.42 **Analysis of Variance of Yields (kg/ha) of New Barley Selections in Three Yield Trials**

Source	df	Sums of squares for trial 1	2	3
Total	19	7,091,153.2	4,326,438.5	5,256,706.9
Block	1	1,472,073.8	19,406.4	342,434.4
Selection	9	3,005,593.2	3,643,529.0	3,983,850.4
Error	9	2,613,486.2	708,393.1	930,422.1

$$s_{\bar{d}3} = \sqrt{\frac{2E_p(c + 1)}{rc}} = \sqrt{\frac{(2)(167{,}489.0)(3)}{4}} = 501.23$$

4. Difference between an adjusted selection mean and a check mean, $s_{\bar{d}4}$:

$$s_{\bar{d}4} = \sqrt{\frac{E_p\,(pc + p + c)}{prc}} = \sqrt{\frac{(167{,}489.0)(6 + 3 + 2)}{12}}$$

$$= 391.83$$

For this example the 5% LSI is

$$\text{LSI} = t_{0.05}s_{\bar{d}4} = (1.695)\,(391.83) = 664.15$$

Report of Statistical Analysis

A group of yield trials on new spring barley selections was conducted on the principal research farm of the Oregon Agricultural Experiment Station near Corvallis. A combined analysis was conducted on a subset of data from three of these trials. There were two checks in common in each trial. These were ‘Steptoe’, a long-term baseline variety, and ‘Benton’, a recent release. Each trial contained 10 new selections, which differed from trial to trial. The entries in each trial were replicated twice in a randomized block design.

The combined analysis produced adjustment factors that could be used to adjust the mean yields of the new selections for yield differences from one trial to another. The adjusted means of the new selections in the three trials are shown in Table 6.43.

The mean yield of the baseline, long-time check, 'Steptoe', was 3217.8 kg/ha, while that of the recent release, 'Benton', was 3345.67 kg/ha. For these trials the 5% LSI (least significant increase) for comparing an adjusted selection mean with a check mean was 664.2. Hence any adjusted selection mean greater than 3217.8 + 664.2 = 3882.0 significantly outyields 'Steptoe', while any adjusted selection mean greater than 3345.7 + 664.2 = 4009.9 significantly outyields 'Benton'. On considering Table 6.43 it is seen that only selection 4 in trial 2 outyields 'Benton', while selection 4 in trial 1, selection 4 in trial 2, and selection 9 in trial 3 outyield 'Steptoe'.

Table 6.43 **Adjusted Mean Yields (kg/ha) of New Barley Selections from Three Yield Trials Grown Near Corvallis, Oregon**[a]

	Trial		
Selection	1	2	3
1	2998.2	3348.5	3531.8
2	3065.2	3514.5	3323.2
3	2952.8	2844.0	2769.2
4	3890.2	4085.5	3344.2
5	3543.8	3094.0	3802.8
6	3339.8	2831.5	2698.2
7	3444.2	2673.0	2598.8
8	3464.8	2935.5	3377.8
9	2669.2	3731.5	3919.2
10	2602.2	3360.5	2827.8
Adjustment	134.75	1.50	-136.25

[a]Standard error = 354.42.

6.8 COMBINED ANALYSIS FOR OFF-STATION TRIALS

Much of the basic research in a program to determine the way that various factors of production operate to affect yield and quality should be done on an experiment station. Here it is possible to control tightly treatment application and the operations required to produce the crop. On-station it is also convenient to take observations on a number of variables, such as time of emergence, stand density, tillering, and lodging. There are, however, two stages in a research program that are more meaningfully conducted off-station than on-station. These are the initial exploratory stages and the final field verification stages.

No matter how much care has gone into the selection of its location, an experiment station can represent, at best, only a very limited sample of soil and climatic conditions. Often, because conditions on the station have been altered by many seasons of previous experimentation, the station really represents only the limited area it occupies. Although this usually does not affect the results of experiments designed for basic research, it cannot be expected that experiments conducted on a station can produce results on which broad recommendations can be made. This type of information can only be obtained by off-station experimentation to sample a wide variety of growing conditions.

6.8.1 Exploratory Research

In the exploratory stages of a research program we want to examine a number of factors over a wide variety of conditions in order to separate the factors that are important from those that need not be considered further. This points us to the need to conduct two-level factorial experiments with a large number of factors. Because we would like to sample a variety of conditions, it would seem reasonable to use location as a blocking factor and conduct one replication of the experiment at each of a number of locations. With this approach we could examine a number of factors over a range of conditions using experiments of manageable size at each location. For example, we could study five factors each at two levels, usually absence and presence, in blocks of 32 plots. That is, with the full factorial set of treatments we would have $2^5 = 32$ treatment combinations. If we conducted this 32-plot experiment at each of r locations our analysis of variance would take the form shown in Table 6.44.

We assume that there is no real location $\times$ treatment interaction and take this mean square to be estimating only experimental error. This is a reasonable assumption if the locations are not too divergent. If there is a location $\times$ treatment interaction, the F tests will be conservative because

Table 6.44 **Analysis of Variance for a 2^5 Factorial Trial with Single Replications Conducted at Each of r Locations**

ANOVA

Source	df
Total	32r - 1
Location	r - 1
Main effects	5
Two-factor interactions	10
Three-factor interactions	10
Four factor interactions	5
Five factor interaction	1
Error (location by treatment)	31(r-1)

the denominator of the F will be too large relative to the true error. However, any effects that test significant under this test will certainly be significant under a proper test.

6.8.2 Field Verification Trials

Field verification trials are also, of necessity, conducted off-station. The aim of these trials is to verify, or confirm, levels of input factors that are believed to be optimum in some sense. Further, this verification should be done throughout the area in which recommendations are to be made. A second and sometimes more important aim is to demonstrate new practices or recommended levels to the growers, who must ultimately decide whether or not to adopt the practices.

In general, the number of treatments in these trials should be kept small. At most three factors should be included in a two-level factorial experiment. Thus there would be a maximum of 8 plots per replication. The low level of a factor would usually be absence of the factor, and the high level would be presence at the assumed optimum rate. Because of the demonstration aspects of these trials, the plots should be larger (0.1 to 1.0 ha) than those used in on-station research.

Again, the aim is to sample as wide a set of conditions as possible within the area for which recommendations are to be made. For this reason the best strategy is to select a large number of locations and conduct one replication of the trial at each location. Statistical analysis of the resulting data would involve consideration of locations as blocks and the use of the location × treatment interaction as error.

6.8.3 Numerical Example

A number of field verification trials were conducted throughout southeastern Iowa several years ago. The purpose of these trials was to evaluate the effect of phosphorus and potassium fertilizers and to determine the need for the minor element boron on alfalfa. A second aim was to demonstrate the use of these fertilizers to farmers throughout the area. Although trials on the experiment station had demonstrated that alfalfa might benefit from additional phosphorus and potassium, the effect of additional boron had not been studied. However, a survey of the area revealed visual symptoms of what appeared to be boron deficiency in some of the alfalfa in the area.

It was decided to use a two-level factorial combination of the three factors with the following treatments:

(1) = no added fertilizer
P = 20 kg/ha P_2O_5
K = 20 kg/ha K_2O
PK = 20 kg/ha P_2O_5 + 20 kg/ha K_2O
B = 10 kg/ha borax ($Na_2B_4O_7$)
PB = 20 kg/ha P_2O_5 + 10 kg/ha borax
KB = 20 kg/ha K_2O_0 + 10 kg/ha borax
PKB = 20 kg/ha P_2O_5 + 20 kg/ha K_2O_0 + 10 kg/ha borax

Treatments were applied in late March and early April.

A single replication of the eight treatment combinations was applied to established stands of alfalfa in grower's fields at each of five locations. Plots 0.1 ha in size were used. Although the experiments were conducted off-station, they were managed by the research scientists and their technicians.

Yields (kg/ha) of air-dried alfalfa from the first cutting after fertilizer application are summarized in Table 6.45. The first cutting was made in late May and early June. The preliminary analysis of variance of the yields in Table 6.45 is given in Table 6.46. Note that when tested against the treatment × location mean square, the F ratio for treatments is 9.02. Since $9.02 > 3.36$, the differences among the treatment means are significant at the 1% level.

Table 6.45 **Yield (kg/ha) of First-Cutting Alfalfa from Five Demonstration Trials in Iowa**

Treat.	Trial 1	2	3	4	5	Sum	Mean
(1)	1770	2170	4030	2375	2240	13125	2625
P	2350	2780	4055	2735	3070	14990	2998
K	1905	2710	4210	2420	2780	14025	2805
PK	2420	2890	4435	2710	3360	15815	3163
B	1680	2400	3965	2175	2260	12480	2496
PB	2195	2735	4080	2575	3115	14700	2940
KB	1635	2400	3965	2440	2510	12950	2590
PKB	2375	2780	4300	2710	3765	15930	3186
Sum	16330	21405	33040	20140	23100	114015	2850

Table 6.46 **Analysis of Variance of First-Cutting Alfalfa Yields (kg/ha) from Iowa Demonstration Trial**

ANOVA

Source	df	SS	MS	F
Total	39	22,989,969		
Location	4	19,478,760	4,869,690	
Treatment	7	2,432,169	347,453	9.02**
Error	28	1,079,040	38,537	

**Significant at the 1% level.

The partition of the treatment sum of squares into its factorial components was done by using contrasts of totals according to the methods presented in Section 4.6. This is shown in Table 6.47. The final detailed analysis of variance is presented in Table 6.48. Note that the effect of added phosphorus is highly significant (1% level) and that for potassium was significant (5% level). The effect of boron was not significant, nor were there any real interactions among the fertilizer elements.

Report of Statistical Analysis

A series of demonstration trials on alfalfa fertilization was conducted in southeastern Iowa. The purpose of the trials was to demonstrate the effect of the application of phosphate and potash fertilizers and to determine whether or not alfalfa in southeastern Iowa would benefit from addition of the minor element boron. Pertinent results of the analysis of first-cutting yields from these trials are summarized in Tables 6.49 and 6.50.

Note that the addition of 20 kg/ha of phosphate fertilizer increased the air-dry yield of first-cutting alfalfa by about 440 kg/ha. Similarly, the addition of 20 kg/ha of potash increased first-cutting alfalfa yield by about 220 kg/ha. The addition of fertilizer boron had no real effect on the yield of first-cutting alfalfa whether or not phosphate or potash fertilizer had

Table 6.47 **Partition of Treatment SS from Alfalfa Trial into Factorial Components**

Treat.	(1)	P	K	PK	B	PB	KB	PKB	
Sum	13125	14990	14025	15815	12480	14700	12950	15930	L
Contrast									
P	−	+	−	+	−	+	−	+	8855
K	−	−	+	+	−	−	+	+	3425
PK	+	−	−	+	+	−	−	+	685
B	−	−	−	−	+	+	+	+	−1895
PB	+	−	+	−	−	+	−	+	1545
KB	+	+	−	−	−	−	+	+	−25
PKB	−	+	+	−	+	−	−	+	835

Table 6.48 **Final Analysis of Variance of Alfalfa Yields from Iowa Demonstration Trial**

ANOVA

Source	df	SS	MS	F
Total	39	22,989,969		
Treat.	7	2,432,169	347,453	9.02**
P	1	1,960,276	1,960,276	50.87**
K	1	293,266	293,266	7.16*
PK	1	11,730	11,730	NS
B	1	89,775	89,775	2.32NS
PB	1	59,676	59,676	1.55NS
KB	1	16	16	NS
PKB	1	17,431	17,431	NS
Error	28	1,079,040	38,537	

**Significant at 1%; *significant at 5%; NS, not significant.

Table 6.49 **Effect of Phosphate Fertilizer on the Dry Matter Yield (kg/ha) of First-Cutting Alfalfa in Southeastern Iowa**

Phosphate	None	20 kg/ha	Standard Error
Mean Yield	2,629	3,072	43.90

Table 6.50 **Effect of Potash Fertilizer on the Dry Matter Yield (kg/ha) of First-Cutting Alfalfa in Southeastern Iowa**

Potash	None	20 kg/ha	Standard Error
Mean Yield	2,765	2,986	43.90

been used. Further, the response to added phosphate took place regardless of whether or not potash was added.

REFERENCES

Bartlett, M. S., 1937. Some examples of statistical methods of research in agriculture and biology. *J. Roy. Stat. Soc. Suppl.* 4:137–183.

Cochran, W. G., 1954. The combination of estimates from different experiments. *Biometrics* 10:101–129.

Cochran, W. G., and G. M. Cox, 1957. *Experimental Designs*, 2nd ed. Wiley, New York, pp. 553–555.

Eberhardt, S. A., and W. A. Russell, 1966. Stability parameters for comparing varieties. *Crop Sci.* 6:36–40.

Federer, W. T., 1956. Augmented (or Hoonuiaku) designs. *Hawaii. Plant. Rec.* 55:191–208.

Federer, W. T., 1961. Augmented designs with one-way elimination of heterogeneity. *Biometrics* 17:447–475.

Federer, W. T., and D. Ragavarao, 1975. On augmented designs. *Biometrics* 31:29–35.

Satterthwaite, F. E., 1946. An approximate distribution of estimates of variance components. *Biom. Bull.* 2:110–114.

Snedecor, G. W., and W. G. Cochran, 1980. *Statistical Methods*, 7th ed. Iowa State University Press, Ames, Iowa, pp. 252–253.

Yates, F., and W. G. Cochran, 1938. The analysis of groups of experiments. *J. Agric. Sci.* 28:556–580.

FURTHER READING

Complete citations for books for further reading are given in the bibliography at the end of the book.

1. Cochran and Cox, Chap. 14.
2. John and Quenouille, Chap. 13.
3. LeClerg, Leonard, and Clark, Chap. 14.
4. Steel and Torrie, Chap. 16.

7

Experiments with Perennial Crops

7.1. INTRODUCTION

The majority of field experiments deal with annual crops. These are crops whose life cycle is completed in a single growing season. Examples are cereals, such as wheat and rice, and food legumes, such as lentils and chick-peas. Experiments with annuals are initiated, conducted, and terminated in a single growing season. If information is wanted for more than one season, a new trial must be conducted each season. Experimental results from several seasons may be combined and analyzed using the methods described in Chapter 6.

On the other hand, a significant amount of research is conducted on perennial crops. These are crops that live through several growing seasons before they die. Examples include such horticultural crops as fruit trees, berries, and ornamental shrubs, and forage crops such as alfalfa, perennial grasses, and forbs. Trials involving perennials involve the same plant grown at the same location for several seasons. Once a trial has begun, data are taken from the same plant each season until the trial is completed. Because of this continuity, the methods of Chapter 6 cannot be used to analyze the combined seasonal data from a perennial trial.

7.1.1 Perennial Crops as Research Subjects

As research plants, perennials differ in some respects from annuals. Because perennials live longer than annuals, they may be exposed to more problems during the course of the trial than is the case with annuals. Perennials are usually larger than annuals, and hence interest is more apt to be directed toward individual plants in perennial trials, whereas with trials on annuals, groups of plants are of interest. This means that in perennial trials, in contrast to annual trials, variation among individual plants contributes more to experimental error than does positional variation in the experimental site. Another difference is that because the same plant remains at the same location from season to season, observations made on a plant in one season are likely to be correlated with those made in another season.

We might consider some of these points in detail. Since perennial trials, once begun, extend over several seasons, they can be affected by the loss of plants before the trial has been completed. This means that care must be taken in the choice of an experimental design so that loss of plants during the trial does not greatly affect analysis of the data from the plants remaining in the trial. It also means that provision should be made to replace plants that are lost early in the trial or to adjust the results for the effect of the missing plants.

In contrast to annuals, the large size of most perennials affects perennial trials in a couple of ways. Construction of the experimental plots, the basic experimental unit, is somewhat more restricted for perennial trials than it is for annuals. Further, while the experimental unit in a trial on annuals is usually a plot containing hundreds of plants, a plot in a perennial trial contains only one or just a few distinct plants. In a plot of annuals the plants are of varying sizes, which tend to compensate for each other in making measurements on the plot. This leaves positional variation as the principal component of experimental error in annual trials. On the other hand, Pearce (1953) has indicated that the experimental error of perennial trials is almost entirely the result of plant-to-plant variation. This is due to:

1. Variation among young plants that have not been subjected very long to the effects of their planting position.
2. Variation due to genetic differences among plants

Although variation among young plants will be reduced, to some extent, as they mature, genetic variation will persist throughout the life of the trial. Thus the use of a complex experimental design to control positional variation in perennial trials is largely unnecessary.

Some perennials behave in some respects like both perennials and annuals; an example is forage crops. They are long-lived and are exposed to the possibility of damage for a longer period than are annuals. However, plants damaged during one growing season often repair themselves in the following season, but the effects of the damage may show up in the trial results. Forage plants are small compared to trees and shrubs, and plots in trials with forages contain many plants that are measured in aggregate during the trial. Hence although care must be taken to allow for longevity, positional variation must also be controlled with this type of crop.

7.2 SELECTION AND CALIBRATION OF PERENNIAL PLANTS

As already pointed out earlier, with perennial crops a large part of the experimental error comes from plant-to-plant variation. This should be controlled as much as possible when selecting plants for a trial. Since some of the variation is genetic, an ideal source of plants, if available, is clonal material. If clonal material is not available, variability can sometimes be reduced by using inbred stock. In any case, the source of each plant should be known, and care should be taken to select all the plants from the same source. When this is not possible, the plant source can serve as a blocking criterion for the trial.

It is usually a good idea to standardize the initial size of plants used in a trial. While some plants tend to end up the same size, others tend to maintain size differences throughout a trial. Where variety is one of the treatments, however, standardization on initial size should be avoided. Some varieties are naturally small, while others are naturally large. Standardization on size would result in some varieties being represented by either abnormally small or abnormally large plants. When a stock of suitable plants has been identified, the plants to be used in a trial should be selected at random from the stock. These should be randomly assigned to field position within the trial.

7.2.1 Calibration

Differences among perennial plants are present at the beginning of a trial and tend to persist throughout the trial. We should try to control this plant-to-plant variability as much as possible. One way to do this is to calibrate the plants to be used in the trial. Calibration involves measuring one or more variates on each plant before the trial begins. Examples of such variates include trunk diameter or prior yield. The purpose of these observations is to provide an objective measure of the plant's inherent ability to produce independent of any effect of treatment. These calibrating measurements can be used to control some of the effect of plant-to-plant differences when the experimental data are analyzed.

One way to make use of the calibrating measurements is as a basis for forming blocks in an experimental design that calls for blocking. Plants with high values of the variable would be put into one block and those with low values of the variable would be put into another. The primary objection to this use is that it prevents the use of blocking to control positional variation, which may also be present and which should be controlled if possible. A second, and preferable use of the calibrating measurement is as a covariable in a covariance analysis of the results of the trial (Pearce, 1953; Rigney et al., 1949). In a covariance analysis the trial results are adjusted for differences in the calibrating variable. The assumption is that this will remove the inherent plant differences and permit the expression of treatment effects clear of plant differences.

The analysis of covariance is a fairly standard statistical technique that is well described in most texts on statistical methods (Snedecor and Cochran, 1980; Steel and Torrie, 1980). We illustrate its application in the numerical example that follows later in this chapter.

The choice of a calibrating variable depends on a number of things. It should be a variable that reflects the ability of the plant to perform. It should also be one that can be expressed as a measurement for the plot, or experimental unit, to be used in the trial. It should be a variable that can be measured before the treatments have been applied, or have had time to have an effect. Pearce (1953) has given a summary of some of the variables that have been used for calibration. These include trunk girth and trunk cross-sectional area for tree crops: average yield in the preceding two or more years, since many tree crops bear on a biennial cycle; and growth increment during a previous season. As calibrating variables, Rigney et al. (1949) used number of shoots per plant for blueberries, and average diameter of eight arms for grapevines. Other possibilities include plant density or ground cover for perennial forages. In any case, the choice should be one that would reflect future performance if the plants were to be left untreated.

7.3 EXPERIMENTAL PLOT

The experimental unit in a trial on perennial crops is usually a plot of ground on which one or more plants are growing. In some trials, however, the unit might be a single branch or a single quadrant on a tree, or a box of fruit in a storage facility. For the most part we will be concerned with units that are field plots in the usual sense. We have a plot of ground on which plants are growing and we evaluate the effect of treatments by measuring yield, or some other similar variable, on a per plot or a per unit area basis.

7.3.1 Plot Size and Shape

For the most part, the basic principles for deciding the size and shape of plots in a perennial trial are those presented in Chapter 2. The general rule is that plots should be small and that for a given size of experimental area, precision is improved more rapidly by increasing the number of replications per treatment than by increasing plot size.

Certain factors set a lower limit on plot size. If the plant under study is a tree, the minimum size must clearly accommodate at least one tree. Other limiting factors include the need to have the plots large enough to accommodate the machinery to be used and to provide sufficient material to produce measurements of acceptable accuracy. Further, plots must be large enough to permit the use of guard material (guard trees or between-plot buffer areas). Pearce (1953) states that as a general rule, small plots provide economy of land use, whereas large plots provide more information per replication, especially when lacking skilled labor to conduct the experiment.

An additional consideration is the possibility of loss of plants during the trial. In very small plots the loss of a plant is equivalent to the loss of the plot, and missing plot procedures must be used in analysis of the data. On the other hand, loss of one or two plants from a plot containing a large number of plants causes little effect if sufficient plants remain to provide a reasonable measure of plot yield. With medium-sized plots, missing plants can cause some difficulties. These can be handled in a number of ways: (1) regard all plots with missing plants as missing plots and proceed accordingly. (2) discard plants randomly until all plots have the same number of plants remaining, and (3) replant all losses at some time in the trial.

No hard and fast rules can be laid down for what constitutes the proper size for a particular crop. Pearce (1953) summarizes the plot size recommendations obtained by a number of workers using uniformity trials. In general, for tree crops, plots of from four to 12 trees are recommended, with eight being the most common. The recommendation for grapes varies between 4 and 16 vines per plot. For other crops the recommendations are:

Raspberries: one row 30 ft long
Strawberries: 25 to 50 plants/plot
Coffee 20 bushes/plot
Tea: $\frac{1}{18}$ acre/plot

For crops such as forage legumes and perennial grasses, the plots resemble those for annual crops rather than those for trees and shrubs. For these crops the principles used for determining plot size for annuals should apply (see Chapter 2).

The preceding recommendations are meant to serve only as a guide. In practice, plot size will be determined by the need to adapt to local conditions. When working with a new crop it is probably better to start with larger plots. Size can then be reduced as experience is gained with the new crop.

As with annuals, the preferential shape for plots in trials with perennial crops is a long, narrow rectangle. The aim should be to make the variability within the plots as small as possible and to make the plots themselves as nearly alike as is possible. These goals are best accomplished with narrow rectangular plots. Plots so shaped also accommodate the machinery used to conduct the experiment better than do square plots.

Orientation of plots should be across the gradient of any expected variation. For example, if disease is present and is expected to progress from north to south, the plots should be oriented with their long axis from east to west.

7.3.2 Guard Plants and Alleys

As is the case with annuals, border effects may be present in the plots of perennials. At the edge of an experimental area, this may be an effect of differential competition for light and nutrients or a differential effect of disease infestation. In the interior of an experiment, roots of plants from one plot may extend into another plot, or there may be carryover of a treatment on one plot into another by machinery drag. In any case, the differential effects at the plot borders should be eliminated from the experimental results. This is commonly done by the use of guard rows or by the use of alleys between the plots.

With trees or shrubs it is common to plant guard rows of similar plants around the exterior of the experiment. The aim is to protect the experiment itself from the effect of disease or insect encroachment into the experiment, and to reduce the effect of the differential edge environment. Usually, a single row is used, but two or three border rows may be required in some circumstances. In the interior of the experiment it is usually sufficient to use a single row of plants between the plots. Treatment effects are evaluated from measurements made on plants in the center of each plot.

For trials involving perennial grasses or legumes, border effects are usually eliminated by harvesting only the center portion of the plot. A common practice is first to remove material from the sides and ends of the plot. Yield is then measured on the remaining plant material.

When the treatments are applied directly to the trees or shrubs themselves, pruning treatments for example, it is common to use guard alleys rather than guard rows between the plots. These are spaces that are constructed by, in effect, omitting rows of trees or shrubs between the plots. With fertilizer trials it may be possible to eliminate guard rows by using guard alleys with a deep furrow in the alley to prevent the passage of roots from one plot to another.

7.3.3 Plot in Relation to Treatment

The size, shape, and composition of the experimental plot in a perennial trial depends both on the plants to be used and the treatment to be applied to them. If the trial involves fertilizer treatments on trees or shrubs, the plot must be constructed to prevent cross feeding from one plot to another. This means that internal guard rows are almost essential for this type of experiment. If the plants are to be fertilized only in the immediate vicinity of the plant, however, it may be necessary to allow only guard alleys between the plots.

For trials involving perennial forage crops the factors that govern the choice of plot size and shape for annuals apply. In these trials the plots contain hundreds of plants, and measurements are made on the aggregate rather than the individual. For these trials, long, narrow plots that can accommodate machinery are preferred. These plots should be oriented, if possible, so that their long axis crosses any gradients that exist.

Trials with cultivation or irrigation treatments also require long, narrow plots. These treatments, of necessity, involve machinery, and the plots must accommodate this machinery. In such a trial, additional information can sometimes be obtained by subdividing the plots onto smaller subplots on which to study the effect of an additional factor, such as fertilizer or variety. Thus cultivation and irrigation trials lend themselves readily to split-plot experimentation.

When an experiment involves spraying or dusting treatments, large plots are essential. It would be difficult to avoid contamination of neighboring plots with unwanted spray if the plots were small. The usual practice with these treatments is to use large plots with single or double guard rows between them. Measurements are then made on plants in the center of the plot. An alternative procedure is to provide guard alleys between the plots and to use screens to prevent the drift of spray or dust.

Trials involving varieties, rootstocks, or pruning treatments can usually be carried out on single-plant plots without guard rows. When the treatments include varieties of different size, however, plots containing varieties of similar size should be grouped together to reduce the differential effect of competition. Similar consideration should be given to grouping plots when pruning treatments are expected to produce plants of divergent size and shape.

7.4 CHOICE OF AN EXPERIMENTAL DESIGN

As we stated earlier, plant-to-plant variation in trials with perennials contributes more to experimental error than does positional variation at the experimental site. One of the goals of complex experimental designs is to control local variation. Hence, with perennials, the use of simple designs should be the rule rather than the exception. Thus the completely randomized design and the randomized block design, with an occasional split-plot experiment, should meet most of the design requirements for experiments on perennial crops.

7.4.1 Completely Randomized Design

The simplest and probably the most underused design for trials with perennials is the completely randomized design. In this design the treatments are applied to plots selected completely at random from among those available. There is no attempt to control positional variation with this design, but positional variation is not of major importance in trials with perennials.

There are a number of advantages for the completely randomized design as far as research with perennial plants is concerned:

1. It provides maximum degrees of freedom when the total number of plots used in the trial is small.
2. Its analysis is not greatly affected by loss of plots during the trial, a frequent occurrence with perennials.
3. It is easy to lay out and conduct because it avoids the necessity of restricted randomization.
4. Its analysis is not materially affected by unequal replication of the treatments.

In summary, the completely randomized design has much to recommend it for use in research on perennial crops. Additional information and details of the analysis of data from this design are provided in Section 3.2.

7.4.2 Randomized Block Design

The most commonly used experimental design in agricultural research is the randomized block design. It is a relatively simple design but it does provide some control of local variation. In the randomized block design the plots are first grouped into blocks of similar plots. The number of blocks is equivalent to the number of replications being used in the trial. The number of plots in each block is equal to the number of treatments in the trial. Treatments are randomly assigned to the plots within the blocks in such a way that each treatment occurs once, and only once, in each block.

There should be a recognizable source of variation upon which to base the construction of the blocks. The effect of blocking is maximized when the variation among blocks is made as large as possible while minimizing the variation among plots within blocks. A number of criteria can be used as a basis for blocking. Probably the most frequently used criterion is proximity; that is, blocks are formed from collections of adjacent plots. Proximity is seldom a good blocking basis variable in perennial trials. Other blocking criteria include position on a slope, exposure to disease or insects, variety, and so on. It is not necessary that all blocks have the same shape, nor is it necessary that all blocks be together in the same location. Additional information may be gained from an experiment by placing blocks at different locations.

Once formed, blocks should be treated as a unit. All operations, such as cultivation, measurement, and harvest, should be completed in one block before moving to another. Again, the aim is to treat the plots in a block as uniformly as possible.

There are a number of reasons for using a randomized block design in a perennial trial:

1. It provides a measure of local control and may reduce experimental error.
2. It is relatively simple to lay out.
3. The analysis of the results is not complicated by the loss of a block or of a treatment during the trial.

In summary, the randomized block is almost always the most complex experimental design that needs to be considered for a trial with perennials. It is fairly simple to lay out and use while allowing control of an unwanted source of variation. This design is described in detail in Section 3.3.

7.4.3 Split-Plot Experiments

In a split-plot experiment the levels of one factor, say cultivation treatments, are assigned to large main plots. These large plots are then partitioned into smaller subplots and the levels of a second factor, variety, for example, are assigned to these. The concept of splitting the plots to which one set of treatments have been assigned to permit the use of a second set of treatments in the same trial may be useful in experiments with long-lived perennials.

There are two primary reasons for using a split-plot experiment:

1. For ease and convenience in applying some types of treatments in the field.
2. To introduce new treatments into an experiment already in progress.

It is the second reason that may be of most value in a trial with perennials.

A trial may be designed to test a set of treatments over a period of years. After the trial has been in progress for some time a question might arise as to the effect of a new set of treatments in combination with the original set. The split-plot concept can be used to introduce the new set by subdividing the original plots and applying the new treatments to the subplots thus formed within the original plots. The split plot and its variants are described in Sections 4.8 and 4.9.

A further use of the split-plot concept is in the analysis of multiyear data from a perennial trial. One way to analyze such data is to add the yearly results and analyze the totals or the means of the aggregate. Another way is to consider years to be a split-plot factor and treat the data as if they were from a split-plot experiment with treatments on the main plots and years on the subplots. There are some possible difficulties with this approach. We address these when we consider the analysis of data from perennial trials.

7.5 ANALYSIS OF DATA FROM PERENNIAL TRIALS

There are two types of analysis that must be considered with data from perennial trials. One is the analysis of data from a single season or of data aggregated over several seasons. The other is the combined analysis of data from several seasons or groups of seasons. We deal with each type of analysis separately.

Before looking at the analyses in detail, however, some preliminary observations are in order. It is well to examine the data carefully before proceeding directly to the analysis of variance. The treatment means should be computed and examined. If the differences among means are very small, they may be of little practical consequence even though they may test statistically significant. In this case it may not be worthwhile to go through the exercise of doing a detailed analysis of variance.

A further type of preliminary examination is to graph the data against time, or, if possible, levels of a quantitative treatment. Such graphs may indicate that differences, if they exist, may occur only at a particular phase of plant growth. Plants may be affected by treatments, for example, only until they become well established, after which time the differences disappear as the plants mature. In this case only the early data need be subject to detailed analysis.

The plan for conducting the analysis of the data should be detailed when the experiment is planned. The analysis should be directed toward answering the questions for which the experiment was planned and for which the treatments were chosen. This means that meaningful sets of contrasts should be constructed at the time the experiment is planned.

7.5.1 Analysis of Single Data Sets

For the most part, the analysis of data from a single season in a perennial trial follows much the same pattern as does the analysis of data from a trial on annuals. The format for the analysis of variance is determined by the experimental design used for the trial. Significance tests, computation and testing of contrasts, standard errors, and so on, are comparable for both types of trial.

Pearce (1953) and Rigney et al. (1949) point out that if the trial involves large plants, covariance analysis is often useful for reducing the effect of plant-to-plant variation. To use covariance, however, the plants should be calibrated before the treatments are applied. The calibrating variable is then used as the covariable in the covariance analysis.

Some additional comments are in order before looking at a numerical example of the analysis of data from a single data set from a perennial trial. When the plot (the experimental unit) contains more than one plant, little is gained by considering the data from the individual plants. Rather, the total, or mean, yield for the plot, on a per unit basis, should be the data used for the analysis. Further, if the data to be analyzed represent the aggregate of several seasons yields, there should be an even number of years in the period over which the data are aggregated. This is because many perennial plants tend to bear on a 2-year cycle, and periods containing an odd number of years may not be comparable.

7.5.2 Numerical Example: Single Data Set

A horticulturist wanted to determine the effect of four different pruning systems on the yield of sweet cherries. He had a newly established orchard in which he could conduct the trial. Because there appeared to be some differences in size, he decided to calibrate the trees before he began. As a calibrating variable he chose trunk cross-sectional area at a height of 1 m above the soil surface. His assumption was that cross-sectional area is a measure of the tree's inherent ability to produce fruit.

The trial was laid out using a randomized block experimental design with a row of four trees in each block. There were six blocks oriented perpendicular to the prevailing wind. Since the pruning treatments were not expected to have a drastic effect on tree growth, no guard rows were used within the trial.

The four pruning treatments were randomly assigned to the individual trees within the blocks. Three years after the pruning treatments had begun, measurement of the yield of cherries from each tree was started. To avoid a possible problem with biennial bearing, the trial results were expressed as the mean yield over 2-year periods. For each tree the cross-sectional area of the trunk and the mean yield of sweet cherries for the first 2-year period, beginning 3 years after treatments were begun, are given in Table 7.1.

In an attempt to reduce the effect of tree-to-tree variation on the yield of cherries, the data were subjected to a covariance analysis. Details of the analysis of covariance are given by Snedecor and Cochran, (1980), Steel and Torrie (1980), and in other standard reference books on statistical methods, so are not presented in detail here. For the present experiment the variable of interest is the average yield over 2 years, y, of sweet cherries, and the covariable is cross-sectional area, x, of the trunk.

Table 7.1 **Cross-Sectional Area, x (cm^2), and Mean 2-Year Yield of Sweet Cherries, y (kg), from a Trial on Pruning Systems**

	Pruning System									
	1		2		3		4		Total	
Block	x	y	x	y	x	y	x	y	x	y
1	140	2.1	139	3.4	203	4.3	110	4.1	592	13.9
2	200	3.5	222	5.9	231	5.0	95	4.1	748	18.5
3	144	1.8	209	3.8	228	5.0	171	4.9	752	15.5
4	127	3.6	179	5.6	114	3.6	199	7.1	619	19.9
5	173	3.8	108	2.7	86	2.3	215	6.5	582	15.3
6	225	5.0	210	5.0	89	1.8	191	6.2	715	18.0
Total	1009	19.8	1067	26.4	951	22.0	981	32.9	4008	101.1
Mean										
$\bar{x}$	168.2		177.8		158.5		163.5		167.0	
$\bar{y}$		3.30		4.40		3.67		5.48		4.21

Table 7.2 **Sums of Squares and Products for Covariance Adjustment of Sweet Cherry Yields (kg) to Trunk Cross-Sectional Area**

		Sum of products		
Source	df	xx	xy	yy
Total	25	55,314	1,139.00	49.5863
Block	5	7,809	70.18	6.5688
Treatment	3	1,219	6.95	16.6846
Error	15	46,286	1,061.87	26.3329
Treat. + error	18	47,505	1,068.82	43.0175

To begin the covariance analysis, the sums of squares and cross products of the variable and covariable are computed. These are shown in Table 7.2. These values enter into most of the subsequent computations in the covariance analysis. The analysis of variance of the cherry yield unadjusted for cross section area is given in Table 7.3. Note that the F test for treatment differences fails to reach significance in this analysis.

To see whether a covariance adjustment might be effective in reducing the variability among tree yields, we can test the significance of the slope, β, of the relationship between cross-sectional area and yield. The entries

Table 7.3 **Analysis of Variance of Sweet Cherry Yield (kg) Unadjusted for Trunk Cross-Sectional Area**[a]

ANOVA

Source	df	MS	F
Block	5	1.3138	
Treatment	3	5.5615	3.168 NS
Error	15	1.7555	

[a]NS, not significant.

needed in this test appear on the error line of Table 7.2. We first obtain an estimate, b, of the slope as

$$\beta \doteq b = \frac{E_{xy}}{E_{xx}} = \frac{1061.87}{46,286} = 0.0230$$

Then to test the significance of the slope, we use as a test statistic

$$F = \frac{bE_{xy}}{(E_{yy} - E_{xy})/[(p-1)(r-1)-1]}$$

$$= \frac{(0.0230)(1061.87)}{[26.3329 - (0.0230)(1061.87)]/[(3)(5)-1]}$$

$$= \frac{24.4230}{0.1364} = 179.03 \qquad \text{with 1,14 df}$$

Since $F_{0.01(1,14)} = 8.86 < 179.03$, the slope of the relationship is significant at the 1% level. Covariance adjustment will be effective in reducing error. To continue the covariance analysis, we compute the entries in the adjusted analysis of variance. This is presented in Table 7.4.

After adjusting for the differences in cross-sectional area, the differences in sweet cherry yields among the four pruning treatments are highly significant. This is the result of eliminating the effect of tree-to-tree differences in trunk area from the experimental error. A measure of the effect of the covariance adjustment on the precision of the trial may be obtained from

Table 7.4 **Analysis of Variance of Sweet Cherry Yields (kg) Adjusted to a Common Trunk Cross-Sectional Area**

ANOVA

Source	df	SS	MS	F
Treat. + Error	17	18.9700 [a]		
Adj. Treat.	3	16.9980 [c]	5.6660	40.22**
Error	14	1.9720 [b]	.1409	

[a]$18.9700 = 43.0175 - [(1068.82^2)/47,505]$.
[b]$1.9720 = 26.3329 - [(1061.87^2)/46,286]$.
[c]$16.9980 = 18.9700 - 1.9720$.
**Significant at the 1% level.

the ratio of the error mean squares before and after the adjustment. We have

$$\text{relative efficiency} = \left[\frac{\text{MSE(unadj.)}}{\text{MSE(adj.)}}\right]100$$

$$= \left(\frac{1.7555}{0.1409}\right)100 = 1246\%$$

In this instance the covariance adjustment resulted in a 1146% increase in efficiency.

The adjusted mean yields, $\hat{y}$, of sweet cherries, adjusted to the grand mean, $\bar{\bar{x}}$, cross-sectional area, are given in Table 7.5.

Report of Statistical Analysis

A study was conducted to determine the effect of four pruning systems on the yield of sweet cherries. The trees to be used in the study were somewhat variable in size, so they were calibrated by measuring the trunk cross-sectional area at a height of 1 m above the ground. Calibration was done before the pruning trial was started.

To compensate for the biennial bearing tendency of fruit trees, yield was expressed as the mean fruit yield over two consecutive seasons. The mean yield of sweet cherries over the third and fourth years after the treatments had been applied was then analyzed using a covariance analysis

Table 7.5 **Mean Cross-Sectional Area, $\bar{x}$, Unadjusted Mean Yield, $\bar{y}$, and Adjusted Mean Yield, $\hat{y}$**

	Means		
Treatment	$\bar{x}$	$\bar{y}$	$\hat{y}$*
1	168.2	3.30	3.27
2	177.8	4.40	4.16
3	158.5	3.67	3.86
4	163.5	5.48	5.56
Standard Error	22.7	.54	.15

*$\hat{y}_j = \bar{y}_j - b(\bar{x}_k - 167.00) = \bar{y}_j - (0.0230)(\bar{x}_j - 167.00)$.

with cross-sectional area as the covariable. The pertinent results of this analysis are presented in Table 7.6.

Covariance adjustment was highly successful for this trial. Precision was increased by over 1100%. At the same time, as shown in Table 7.6, the adjustment for differences in cross section had little effect on the estimate of mean yield of sweet cherries for the various pruning treatments. For this study, treatment 1 produced the smallest 2-year mean yield, about 3.3 kg/tree, and treatment 4 produced the highest 2-year mean yield, about 5.5 kg/tree.

7.5.3 Combined Analysis of Several Data Sets

The basic principles involved in conducting a combined analysis of data from successive seasons, or successive groups of seasons, in a trial with perennials are, in many respects, the same as those used in Chapter 6. There is one difference, however, which leads to questions about the nature of a proper statistical analysis of such data. In the analytical methods discussed in Chapter 6 it was assumed that even though the design and treatments remained unchanged, there was a new trial with a new randomization of treatments to plots for each data set. Hence the data were statistically independent from one data set to another.

In trials involving perennials, however, treatments are assigned to the experimental units (plots, bushes, trees, etc.) at the beginning of the trial

Table 7.6 **Mean Unadjusted and Mean Adjusted (for Trunk Cross Section) Yield (kg) of Sweet Cherries Under Four Pruning Systems**

	Mean yield (kg) of cherries	
Pruning System	Unadjusted	Adjusted
1	3.30	3.27
2	4.40	4.16
3	3.67	3.86
4	5.48	5.56
Std. Error	0.54	0.15

and remain on the same units for the remainder of the trial. The results are obtained by repeatedly measuring the same unit in successive seasons for the duration of the trial without intervening randomization of the treatments to the units. As a consequence, the measurements tend to be positively correlated from one season to the next. Further, this correlation tends to be high for measurements separated by short time intervals and to decrease as the length of the interval increases.

A number of authors (Cochran, 1939; Pearce, 1953) have suggested that the combined seasonal data from perennial trials be analyzed as if they had come from a split-plot experiment with treatments on the whole plots and successive observations as the subplots. The problem with the split-plot approach is the intraplot correlation mentioned previously. The cause of the problem is that it is not possible to randomize time in the trial. As pointed out by Finney (1988), this correlation can lead to different error mean squares for measurements separated by different time intervals. This means that the average error mean square of the usual split-plot analysis is not a proper error.

Some authors have suggested that the sequential data from perennial trials be analyzed using multivariate techniques. Such techniques are beyond the scope of this book and will not be dealt with here. However, both Finney (1988), and Snedecor and Cochran (1980) suggest that a proper analysis of data of this type can be done by fitting a curve describing the relationship between time and the measurement of interest for each treatment. The parameters of these curves are then analyzed using an analysis of variance. With measurements taken at equal intervals, yearly for example, the simplest procedure is to fit curves of increasing order using the method of orthogonal polynomial coefficients (see Section 4.7).

In practice, this type of analysis involves partitioning the time source of variation and the time × treatment interaction into components associated with the parameters of the fitted curve. These partitions are then tested against their interactions with blocks.

7.5.4 Numerical Example: Combined Analysis of Several Data Sets

An agronomist was interested in examining the response of alfalfa to the application of phosphorus and potassium fertilizers. She was also interested in observing the nature of the residual response to fertilizer as a function of time after application.

She established an experimental planting of alfalfa using a randomized block experimental design with four blocks of four plots each. The fertilizer treatments consisted of the following 2 × 2 factorial combinations of phosphorus and potassium:

(1) = no fertilizer
P = 20 kg/ha of phosphorus
K = 20 kg/ha of potassium
PK = 20 kg/ha P + 20 kg/ha K

Table 7.7 **Yearly Yield (kg/plot) of Alfalfa During Four Years of a Phosphate–Potash Fertilizer Trial**[a]

Block	Year	Treatment (1)	P	K	PK	Total
1	1	24.25	43.50	26.00	37.75	124.50
	2	43.50	47.75	43.50	48.25	183.00
	3	38.50	43.50	42.75	46.50	171.25
	4	41.50	42.25	45.50	47.25	176.50
Total		147.75	170.00	157.75	179.75	655.25
2	1	38.75	42.25	43.50	48.50	173.00
	2	47.00	48.50	46.50	50.75	192.75
	3	54.25	56.75	56.50	59.50	227.00
	4	46.00	51.75	45.75	50.25	193.75
Total		186.00	199.25	192.25	209.00	786.50
3	1	22.75	34.00	22.25	37.00	116.00
	2	41.75	48.00	42.00	45.00	176.75
	3	42.25	39.75	42.75	42.25	167.00
	4	40.75	46.25	43.00	49.75	179.75
Total		147.50	168.00	150.00	174.00	639.50

Table 7.7 **Continued**

4	1	33.50	46.50	40.50	49.00	169.50
	2	45.75	48.00	45.00	48.50	187.25
	3	53.25	54.75	53.25	57.75	219.00
	4	42.00	49.25	46.50	52.25	190.00
	Total	174.50	198.50	185.25	207.50	765.75
	Sum	655.75	735.75	685.25	770.25	2847.00

[a](1) check; P, 20 kg/ha phosphorus; K, 20 kg/ha potassium; PK, 20 kg/ha P + 20 kg/ha K.

Fertilizer was applied in the spring of the year following the establishment of the trial. The plots were clipped and weighed three times each year for the duration of the trial. Plot yields were then expressed as total for the three cuttings per year. The trial was continued for 4 years, after which it was abandoned.

The yearly yields (kg/plot) from the trial over the 4-year duration are given in Table 7.7. The analysis of the data was begun by constructing a number of summary tables: Table 7.8, block × treatment summary; Table 7.9, block × year summary; and Table 7.10, treatment × year summary. The entries in these tables are then used to compute the entries in a preliminary analysis of variance table, Table 7.11.

The treatment source of variation and the block × treatment interaction are then partitioned into the factorial main effects and the first-order interaction associated with the phosphorus and potassium treatment factors. The factorial treatment effects are computed using the contrasts shown in Table 7.12. Block × factorial effect sums of squares are computed from the entries in Table 7.13, which, in turn, were computed from the entries in Table 7.8 using the factorial contrast coefficients given in Table 7.12. For example, the sum of squares for the block × P interaction are computed as

$$10.6953 = \frac{44.25^2 + 30.00^2 + 44.50^2 + 46.25^2}{(4)(4)} - \frac{165.00^2}{64}$$

The block × factorial effect mean squares are used as the denominator in the F ratios for testing the corresponding factorial treatment effects.

Table 7.8 **Block × Treatment Summary Table of Total Yields (kg) of Alfalfa**

Block	Treatment (1)	P	K	PK	Total
1	147.75	170.00	157.75	179.75	655.25
2	186.00	199.25	192.25	209.00	786.50
3	147.50	168.00	150.00	174.00	639.50
4	174.50	198.50	185.25	207.50	765.75
Total	655.75	735.75	685.25	770.25	2847.00

Table 7.9 **Block × Year Summary Table of Total Yields (kg) of Alfalfa**

Block	Year 1	2	3	4	Total
1	124.50	183.00	171.25	176.50	655.25
2	173.00	192.75	227.00	193.75	786.50
3	116.00	176.75	167.00	179.75	639.50
4	169.50	187.25	219.00	190.00	765.75
Total	583.00	739.75	784.25	740.00	2847.00

(*Note*: It is usually sufficient to use the block × treatment mean square as the divisor for all the F ratios used to test the components of the treatment source of variation. The partitioned block × treatment mean squares provide a more conservative test. These are presented here to illustrate the procedure.)

The year source of variation is partitioned into components associated with the linear and quadratic relationship between year and yield. This

Table 7.10 **Treatment × Year Summary Table of Total Yields (kg) of Alfalfa**

	Treatment				
Year	(1)	P	K	PK	Total
1	119.25	159.25	132.25	172.25	583.00
2	178.00	192.25	177.00	192.50	739.75
3	188.25	194.75	195.25	206.00	784.25
4	170.25	189.50	180.75	199.50	740.00
Total	655.75	735.75	685.25	770.25	2847.00

Table 7.11 **Preliminary Analysis of Variance of Yearly Yields (kg) of Alfalfa from 4 Years of a Fertilizer Trial**

ANOVA

Source	df	SS	MS
Total	63	3740.7344	
Block	3	1057.2422	352.4141
Treatment	3	489.7813	163.2604
Blk by Treat	9	17.3046	7.7483
Year	3	1463.4297	487.8099
Blk by Yr	9	424.0937	47.1215
Trt by Yr	9	164.9921	18.3325
Blk by Trt by Yr	27	123.8908	4.5885

Table 7.12 **Factorial Treatment Effect Contrasts for Yields from a 4-Year Fertilizer Trial on Alfalfa**[a]

	Treat. Total (T_j)					
	(1)	P	K	PK		
Contrast	655.75	735.75	685.25	770.25	L_i	S_i^2
P	− 1	+ 1	− 1	+ 1	165.00	425.3907
K	− 1	− 1	+ 1	+ 1	64.00	64.0000
PK	+ 1	− 1	− 1	+ 1	5.00	.3906

[a] $L_i = \Sigma_j k_{ij}T_j$ (T_j = total for j-th treatment); $S^2_i = L^2_i/(16)\ \Sigma_j k^2_{ij}$.

Table 7.13 **Block × Factorial Effect Contrasts**

	Contrast		
Block	P	K	PK
1	[a] 44.25	19.75	-.25
2	30.00	16.00	3.50
3	44.50	8.50	3.50
4	46.25	19.75	-1.75
Total	165.00	64.00	5.00

[a] $44.25 = (-1)(147.75) + (1)(170.00) + (-1)(157.75) + (1)(179.75)$

$19.75 = (-1)(147.75) + (-1)(170.00) + (1)(157.75) + (1)(179.75)$

$-0.25 = (1)(147.75) + (-1)(170.00) + (-1)(157.75) + (1)(179.75)$

Table 7.14 **Regression Contrasts Among Years**[a]

Contrast	Year Total (Y_j) 583.00	739.75	784.25	740.00	L_i	S_i^2
Linear	− 3	− 1	+ 1	+ 3	515.00	830.4383
Quadratic	+ 1	− 1	− 1	+ 1	−201.00	631.2656

[a] $L_i = \Sigma_j k_{ij}Y_j$ (Y_j = total for j-th year); $S^2_i = L^2_i/(16)\Sigma_j k^2_{ij}$.

partition is obtained by means of the orthogonal polynomial regression contrasts shown in Table 7.14. (Details of this procedure are presented in Section 4.7.)

We now partition the treatment × year interaction into its components. The first partition produces the treatment × linear, treatment × quadratic, and treatment × deviation components. For these partitions we operate on the entries in Table 7.15, which, in turn, were obtained by using the coeffi-

Table 7.15 **Treatment × Regression on Year Contrasts**

Treatment	Contrast Linear	Quadratic
(1)	163.25 [a]	− 76.75
P	93.25	− 38.25
K	163.75	− 59.25
PK	95.25	− 26.75
Total	515.50	− 201.00

[a] $163.25 = (-3)(119.25) + (-1)(178.00) + (1)(188.25) + (3)(170.25)$

$-76.75 = (1)(119.25) + (-1)(178.00) + (-1)(188.25) + (1)(170.25)$

cients for the regression contrasts shown in Table 7.14 on the entries in Table 7.10. We have

$$\text{treat.} \times \text{lin. SS} = \frac{163.25^2 + 93.25^2 + 163.75^2 + 95.25^2}{(4)(20)} - \frac{515.50^2}{320}$$
$$= 59.9711$$
$$\text{treat.} \times \text{quad. SS} = \frac{(-76.75)^2 + (-38.25)^2 + (-59.25)^2 + (-26.75)^2}{(4)(4)}$$
$$- \frac{(-201.00)^2}{64}$$
$$= 92.4688$$
$$\text{treat.} \times \text{dev. SS} = \text{treat.} \times \text{year SS} - \text{treat.} \times \text{lin. SS} - \text{treat.} \times \text{quad. SS}$$
$$= 164.9921 - 59.9711 - 92.4688$$
$$= 12.5522$$

These sources are further partitioned into the factorial effect × regression effect components. This partition is accomplished by applying the factorial contrast coefficients given in Table 7.12 to the entries in Table 7.15. For example, the P main effect × linear component is obtained as

$$\text{P} \times \text{lin.} = \frac{[(-1)(163.25) + (1)(93.25) + (-1)(163.75) + (1)(95.25)]^2}{(4)(20)(4)}$$
$$= 59.9445$$

Similarly,

$$\text{P} \times \text{quad.} = \frac{[(-1)(76.75) + (1)(93.25) + (-1)(163.75) + (1)(95.25)]^2}{(4)(4)(4)}$$
$$= 78.7656$$

The final stages in the analysis include the computation of the block × treatment × year source of variation and its decomposition into the factorial × regression components. For this purpose we obtain the block × treatment × linear components given in Table 7.16 and the block × treatment × quadratic components shown in Table 7.17. The entries in Table 7.16 are obtained by applying the linear orthogonal polynomial coefficients to the data in Table 7.7. Similarly, Table 7.17 is constructed using the quadratic orthogonal polynomial coefficients on the entries in Table 7.7.

Finally, the factorial contrast coefficients (Table 7.12) are applied to the entries in Tables 7.16 and 7.17 to obtain Tables 7.18 and 7.19, respectively.

Table 7.16 **Block × Treatment × Linear Year Components**

Block	Treatment (1)	P	K	PK	Total
1	46.75 [a]	13.00	57.75	26.75	144.25
2	29.00	36.75	16.75	14.00	96.50
3	54.50	28.50	63.00	35.50	181.50
4	33.00	15.00	26.25	19.00	93.25
Total	163.25	93.25	163.75	95.25	515.50

[a]46.75 = (−3)(24.25) + (−1)(43.50) + (1)(38.50) + (3)(41.50).

Table 7.17 **Block × Treatment × Quadratic Year Components**

Block	Treatment (1)	P	K	PK	Total
1	− 16.25 [a]	− 12.50	− 14.75	− 9.75	− 53.25
2	− 16.50	− 11.25	− 13.75	− 11.50	− 53.00
3	− 20.50	− 7.50	− 19.50	− .50	− 48.00
4	− 23.50	− 7.00	− 11.25	− 5.00	− 46.75

[a]−16.25 = (1)(24.25) + (−1)(43.50) + (−1)(38.50) + (1)(41.50).

The linear components of the three-way interaction are obtained from Table 7.18. For example,

$$\text{block} \times \text{P} \times \text{linear SS} = \frac{(-64.75)^2 + (5.00)^2 + (-53.50)^2 + (-25.25)^2}{(4)(20)}$$

$$- \frac{(-138.50)^2}{320}$$

$$= 36.5227$$

Table 7.18 **Block × Factorial Effects × Linear Components**

Block	Factorial Effect P	K	PK
1	- 64.75 [a]	24.75	2.75
2	5.00	- 35.00	- 10.50
3	- 53.50	15.00	- 1.50
4	- 25.25	- 2.75	10.75
Total	- 138.50	2.50	1.50

[a]−64.75 = (−1)(46.75) + (1)(13.00) + (−1)(57.75) + (1)(26.75).

Table 7.19 **Block × Factorial Effects × Quadratic Components**

Block	Factorial Effect P	K	PK
1	8.75 [a]	4.25	1.25
2	7.50	2.50	- 3.00
3	32.00	8.00	6.00
4	22.75	14.25	- 10.25
Total	71.00	29.00	- 6.00

[a]8.75 = (−1)(−16.25) + (1)(−12.50) + (−1)(−14.75) + (1)(−9.75).

Similarly, the quadratic components of the second-order interaction are computed from Table 7.19. We have, for example,

$$\text{block} \times \text{P} \times \text{quadratic SS} = \frac{(18.75)^2 + (7.50)^2 + (32.00)^2 + (22.75)^2}{(4)(4)} - \frac{(71.00)^2}{64} = 25.8828$$

The results of the preceding computations are combined to produce the final analysis of variance presented in Table 7.20. In this analysis of variance the mean squares for the interaction of the various components with blocks are used as divisors for the F ratios to test the significance of the components themselves. For example, the F ratio for testing the P × linear interaction is

$$F = \frac{\text{MS(P} \times \text{lin.)}}{\text{MS(B} \times \text{P} \times \text{lin.)}} = \frac{59.9445}{12.1742} = 4.92 \qquad \text{with 1,3 df}$$

The tests made in this way are conservative, but they are, in large part, clear of the correlation among observations on the same plot.

Report of Statistical Analysis

A trial was conducted to determine the effect of phosphorus and potassium fertilizers on the yield of alfalfa over the life of the stand. Yield measurements (sum of three cuttings) were made for 4 years following the application of the fertilizers.

The results indicate that both P and K were effective in increasing the yield of alfalfa. This can be seen by considering mean yearly yields, which are presented in Table 7.21. The application of 20 kg/ha of phosphorus resulted in an average increase of about 5 kg/plot, while 20 kg/ha of potassium increased yield by about 2 kg/plot. There was no indication that the fertilizers used in combination produced different effects than when they were used alone.

As seen in Table 7.22, mean annual yield showed an increase through the first 3 years of the trial followed by a decrease in the fourth year. Further analysis indicates that yield as a function of time since fertilizer application may be described by the relationship

$$\hat{y} = 20.73 + 18.92x - 3.14x^2$$

Table 7.20 **Analysis of Variance of Yearly Yield of Alfalfa During 4 Years of a Fertilizer Trial**

Source	df	SS	MS	F
Total	63	3740.7344		
Block (B)	3	1057.2422		
Treatment (Treat)	3	(489.7813)		
P	1	425.3907	425.3907	126.89 **
K	1	64.0000	64.0000	36.41 **
P K	1	.3906	.3906	NS
B by Treat	9	(17.3046)		
B P	3	10.6953	3.5651	
B K	3	5.2734	1.7578	
B P K	3	1.3359	1.4453	
Year	3	(1463.4297)		
Linear (L)	1	830.4383	830.4383	17.62 **
Quadratic (Q)	1	631.2656	631.2656	13.40 **
Deviation (Dev)	1	1.7258	1.7258	NS
Block by Year	9	424.0937	47.1215	
Treat by Year	9	(164.9921)		
Treat by Lin	3	[59.9711]		
P L	1	59.9445	59.9445	4.92 NS
K L	1	.0195	.0195	NS
P K L	1	.0070	.0070	NS
Treat by Quad	3	[92.4688]		
P Q	1	78.7656	78.7656	9.11 *
K Q	1	13.1406	13.1406	7.78 *
P K Q	1	.5626	.5626	NS

Table 7.20 **Continued**

Treat by Dev	3	[12.5522]	4.1841	2.03 NS
B by Treat by Year	27	(123.3809)		
B by Treat by Lin	9	[65.5807]		
B P Lin	3	36.5227	12.1742	
B K Lin	3	26.0477	8.6826	
B P K Lin	3	2.9384	.9795	
Blk by Trt by Quad	9	[39.8627]		
B P Q	3	25.8828	8.6276	
B K Q	3	5.0703	1.6901	
B P K Q	3	8.9141	2.9714	
Blk by Trt by Dev	9	[18.5149]	2.0572	

**Significant at 1%; *significant at 5%; NS, not significant.

Table 7.21 **Mean Yield (kg/plot/year) of Alfalfa Under Different Levels of P and K Fertilizer**

K	P: None	P: 20 kg/ha	Mean	Std. Err.
None	40.98	45.36	43.48	0.25
20 kg/ha	42.83	48.14	45.48	
Mean	41.91	47.06	44.48	
Std. Err.	0.25			

Table 7.22 **Mean Yield (kg/plot/year) of Alfalfa During 4 Years on a Fertilizer Trial**

Year	1	2	3	4	Mean	Std. Err.
Mean yield	36.44	46.23	49.02	46.25	44.48	1.72

where

$\hat{y}$ = expected yearly mean yield

x = time (years) since fertilizer application

There was an indication that the point of maximum yield and the rate of drop-off after reaching maximum differed between phosphorus and potassium. This effect was not very pronounced, however.

REFERENCES

Cochran, W. G., 1939. Long term agricultural experiments. *J. Roy. Stat. Soc. Suppl.* 6:104–148.

Finney, D. J., 1988. Was this in your statistics textbook? III. Design and analysis. *Exp. Agric.* 24:421–432.

Pearce, S. C., 1953. Field experimentation with fruit trees and other perennial crops. *Tech. Commun. 23.* Commonwealth Bureau of Horticulture and Plantation Crops.

Rigney, J. A., E. B. Morrow, and W. L. Lott, 1949. A method for controlling experimental error for perennial horticultural crops. *Am. Soc. Hortic. Sci. Proc.* 54:209–212.

Snedecor, G. W., and W. G. Cochran, 1980. *Statistical Methods*, 7th ed. Iowa State University Press, Ames. Iowa, Chap. 16, pp. 330–332; Chap. 18.

Steel, R. G. D., and J. H. Torrie, 1980. *Principles and Procedures of Statistics*, 2nd ed. McGraw-Hill, New York, Chap. 17.

8

Pasture Trials

8.1 INTRODUCTION

Pasture trials form the basis of a sound research program whose goal is the evaluation of forage crops as a source of feed for grazing animals. In a pasture trial the pastures are subjected to various treatments (types of forage, agronomic treatments, animal management systems, etc.). The grazing animal is then introduced onto the pasture to serve as a forage management tool or, more generally, as a means of measuring the yield and quality of the forage.

Pasture trials involve two biological systems: plant and animal. The result is that the trials are more complex and often less precise than are other types of field experiments. Because animals are involved, pasture trials require large areas and involve such things as access to water, minerals, and fencing or other means of animal control, which are not required in other field experiments. Hence pasture trials are expensive to run and require more than the usual amount of care in defining the objectives of the trial and in the selection of treatments, usually few in number, to meet the objectives.

Because the number of treatments that can be evaluated in a pasture trial is necessarily limited, these trials are usually run near the end of a forage treatment evaluation program. Prior to conducting a full-scale pasture trial the treatments under initial consideration are usually screened using hay or silage trials. In these trials the forage is harvested mechanically and stored either as hay or as silage. Yield is measured in terms of harvested dry matter per unit area. Quality is measured by chemical analysis or, an some instances, by in vitro digestion trials. Since animals are not involved, the experimental units can be small plots, not large pastures, and many more treatments can be evaluated in a hay or silage trial than in a pasture trial. However, although screening can be accomplished with small-plot trials, the final judgment should be based on a pasture trial if the selected treatment is ultimately to be grazed.

There are basically two situations in which the forage is harvested by grazing animals. One is range grazing and the other is improved, humid-region pasture grazing. Range grazing involves the use of animals to harvest native vegetation produced, primarily in a low-rainfall environment. In a range situation some improvements, such as brush control, introduction of improved forage species, and provision of water and minerals, may be used. For the most part, however, the carrying capacity (number of animals that can be supported per unit area) is low, and hence it is difficult to design and conduct properly replicated grazing trials under range conditions. On the other hand, in a high-rainfall, humid-region environment, pasture is a crop just like any other crop. Land is prepared, a forage crop is seeded, and more or less intense management is applied throughout the year. The carrying capacity of improved, humid-region pastures is relatively high. Under these conditions it is feasible to conduct replicated grazing trials. Although some of the methods of pasture evaluation discussed later in this chapter can be used in range research, we confine ourselves to pasture trials conducted on improved, humid-region pastures.

8.2 OBJECTIVES OF GRAZING TRIALS

The primary objective of a grazing trial is to evaluate pasture forage as a source of feed for grazing animals. Although animals are a component of all grazing trials, the objective is to evaluate the pasture forage, not the grazing animal. Hence the pasture, not the grazing animal, is the experimental unit in these studies.

Pasture forage can be evaluated from two points of view:

1. Quality of the forage as an animal feed
2. Total yield of forage or of animal product produced on the pasture

Pasture studies can be conducted to measure either quality alone or yield alone. Frequently, however, grazing trials are conducted so that both quality and yield can be measured in the same trial.

8.3 QUALITY STUDIES

Quality studies are designed to measure the individual animal's reaction to the forage offered. In pure quality studies yield is not a factor, so that maximum utilization of the forage is not required.

8.3.1 Measurement of Quality

Forage quality can be measured in two ways:

1. By chemical analyses or by in vitro studies of forage clipped from areas that have been protected from grazing
2. By such animal measures as gain per head or gain per animal per day (daily gain)

Although measurements made on clipped forage may be quick and inexpensive, they may not be very highly correlated with animal response. For the most part, animal measures of quality, if they are available, are to be preferred.

8.3.2 Experimental Errors for Quality Studies

Pasture experiments are expensive to conduct and it is essential that the resources devoted to the trial be used as efficiently as possible. One way to increase efficiency is to reduce the experimental error associated with the trial. Mott and Lucas (1952) point out that experimental errors in grazing trials arise from two primary sources: (1) variability among pastures treated alike, and (2) variation among animals on the same pasture. The contribution of these two sources of variation to experimental error per pasture depends on what is being measured on the pasture, being different for quality measures than for quantity measures.

Petersen and Lucas (1960) derived an expression for the experimental error per pasture for average daily product per animal (average daily gain). This expression is

$$C_q = \sqrt{C_p^2 + \frac{C_t^2}{t} + \frac{C_a^2}{a} + \frac{C_d^2}{d}}$$

where

C_q = error of a pasture mean for average daily gain
C_p = among-pasture variation in herbage quality
C_t = pasture × time interaction for herbage quality
C_a = among-animal variability for performance
C_d = animal × time interaction for performance

all expressed as coefficients of variation, and

t = length of grazing period in days
a = number of different animals on pasture during the trial
d = number of animal days the pasture is grazed
= (a)(t) if the same animals remain on the pasture throughout the trial

These authors also examined data from nearly 40 grazing trials, primarily with beef cattle, conducted in the southeastern and midwestern United States. They found that for grazing trials lasting for periods of 6 weeks or longer, C_q could be approximately determined by the formula

$$C_q = \sqrt{\frac{24{,}700}{t} + \frac{300}{a} + \frac{50{,}800}{d}}$$

This expression can be used to predict the error in mean daily gain per pasture for a trial involving a given number of animals for a given grazing period.

Alternatively, the expression can be used to estimate the length of period for a given number of animals or the number of animals for a fixed period required to obtain a given error. For example: For a trial running 90 days how many animals would be required to obtain a per pasture error in average daily gain of 20%?

We have t = 90 days. Assume that a is constant for the trial. Then

$$20 = \sqrt{\frac{24{,}700}{90} + \frac{300}{a} + \frac{50{,}800}{90a}}$$

$$400 = 274.44 + \frac{300}{a} + \frac{564.44}{a}$$

$$125.56 = \frac{864.44}{a}$$

$$6.88 = a \cong 7$$

Thus seven animals per pasture would be required to obtain a 20% error in mean daily gain.

8.3.3 Experimental Design

Experimental designs for grazing trials to measure pasture quality should be as simple as possible. When compared to other field experiments, pasture trials usually have relatively few treatments under study. At the same time the size of the experimental unit (the pasture) is relatively large and the total number of pastures that can be placed at a single location is usually fairly small. Under these conditions a randomized block design should provide as much local control as necessary. Although more complex designs might provide a somewhat smaller experimental error, there is a consequent reduction in error degrees of freedom, making the small reduction in experimental error of little consequence.

In some instances it may be possible to conduct the trial at several locations within the region in which the results are to be applied. In this case the proper strategy is to run a single replication of the treatments at each location. The results are then analyzed as if they came from a randomized block design with location as the blocking factor.

A further refinement may be made if it is desired to introduce new treatment factors into an experiment already in progress. In this case it may be possible to superimpose the new factor into the old pasture treatments and, in this way, convert the original design into a split-plot design.

As mentioned previously, the experimental unit in a pasture trial is the pasture. It is of little use to specify pasture size in terms of land area. Rather, it is better to think about pasture size in terms of carrying capacity (number of animals the pasture will support for the duration of the trial) for a given treatment. Optimum carrying capacity varies with the objective of the trial and the measures being used to evaluate the pasture. As a rule, the carrying capacity should be greater for quality studies than for yield studies. Lucas (1952) and Lucas and Mott (1962) have shown that carrying capacity for trials involving quality measures (daily gain, or gain per head) should be 3 to 6 for growing cattle, 4 to 8 for milking dairy cattle, or 4 to 7 for lambs.

Examination of experimental results and consideration of the experimental error relationship presented previously indicate that the precision of quality measures is quite low for short grazing periods. Petersen and Lucas (1960) indicate that for such measures as daily gain the duration of the trial should be at least 6 weeks, preferably longer. Because of the very high errors for short periods, trials that run for less than 4 weeks should be avoided unless they can be highly replicated in time or space.

8.3.4 Animal Management

When studying forage quality it is particularly important that the class and breed of livestock to which the results are to be applied be used in the study. This is because quality studies are designed to evaluate the animal's response to the feed offered. This response will usually differ with the type and condition of the animal used.

In quality studies, particularly, it is useful to test all of the animals under uniform conditions before they are placed on the experimental pastures.

This might be done by dry lot feeding of the same ration to all animals. However, a better procedure is to place all of the animals on the same large pasture for 3 or 4 weeks before the trial begins. During the standardization period, such measurements as condition, rate of gain, weight, and so on, should be made on each animal. These measurements made during the standardization period can be used for two purposes: (1) to serve as a basis for grouping the animals into blocks of uniform animals, or (2) to serve as a covariable for adjusting the final results for differences in this variable by means of a covariance analysis.

If blocking is used, the number of animals per block should be equal to the number of treatments in the trial. One animal from each block should then be assigned, at random, to each treatment. It may also be worthwhile to use blocking to control variation in one variable, such as condition, and use covariance to control variation in another variable, such as pretest daily gain. Such techniques cost little, but particularly for quality measures, may lead to worthwhile reductions in experimental error.

For quality studies it is only necessary to use enough animals per pasture to keep the forage in the desired physiological condition. This may require periodic adjustment of the number of animals on the pasture. Since yield estimation is not the goal of quality studies, precise adjustment of animal numbers to obtain optimum forage consumption is not necessary.

Each animal in the study should be weighed when placed on or taken off of a pasture and periodically, at 28-day intervals, while on pasture. Some people advocate weighing each animal on three successive days and using the mean of these 3-day weights. Studies comparing 3-day with 1-day weights, however, indicate that little is gained by using 3-day weights (Cook, 1962). In practice, the animals should be restricted from water overnight and weighed early in the morning.

8.3.5 Computations and Analysis

A number of records should be kept on each animal in the trial:

1. Animal's identity and description
2. Pasture assignment
3. Initial weight on pasture
4. Periodic weights during the trial
5. Final weight off pasture
6. Days in the interval between weighings
7. Milk and its fat content, or wool produced on pasture, if applicable
8. Starting and ending body condition for growing animals

Using these data, the total number of days on pasture, total gain while on pasture, total amount of milk or wool produced on pasture, average daily gain, and average daily milk or wool production can be computed for each animal.

The most commonly used measure of forage quality is mean daily gain per animal per pasture (ADG). This can be computed by dividing the sum of the total gains for all animals by the total animal days spent on the pasture. If the number of animals has remained constant throughout the trial, ADG is simply the mean of the average daily gains of all animals on the pasture.

Mean daily gain per pasture (ADG) is the observation that measures the the quality of the forage on that pasture. These observations are subjected to an analysis of variance appropriate for the experimental design used to assign the treatments to the pastures. Interpretation of this analysis follows the same pattern as that for any other analysis of variance.

Analogous calculations may be performed if milk or wool production is to be used as an expression of quality. Mean daily product per animal is computed for each pasture. The results are then subjected to an appropriate analysis of variance.

A numerical example of the computations involved in evaluating forage quality will not be presented here. Rather, it will be included later as part of an example to illustrate the computation of all of the measures that might be used for evaluating the results of a pasture trial.

8.4 YIELD STUDIES

Yield studies are designed to measure the output of a pasture for a grazing season or parts of a season. Pasture output may be expressed on a number of ways: (1) total herbage produced by the pasture during the period, (2) total amount of herbage consumed by the grazing animals during the period, (3) number of animal days supported by the pasture, and (4) total amount of animal product (meat, milk, wool) produced on the pasture during the period. Total herbage production during any period is not a useful measure of production because all of the herbage produced is not available for animal consumption, and it is difficult to express this in terms of the grazing animal. For the other forms of yield it is assumed that in a well-conducted yield trial the pastures will be managed in such a way that optimum production will be obtained.

8.4.1 Measurement of Yield

In general, two types of measures are used to evaluate pasture yield:

1. Clipping methods
2. Animal methods

With clipping methods the pasture is grazed, but production or consumption is determined by clipping and weighing the forage. Clipping methods are of two types: (1) difference measures, and (2) direct measures. With difference measures part of the pasture is excluded from grazing by the use of permanent or movable cages. Forage consumption is then measured as the difference between forage present in the cages and forage present in the grazed areas outside the cages. Under the direct methods forage availability is determined from clippings made at the beginning and end of each grazing period.

The usefulness of clipping methods depends on how well their results correlate with animal performance. Linehan (1952) has indicated that this correlation might be quite high for full-season grazing trials. However, the correlation is fairly low for short grazing periods. Further, even for long periods clipping methods tend to overestimate yield as compared to estimates made from grazing animals.

There are, basically, three commonly used measures of yield as determined from the grazing animal: (1) total product per unit of pasture area (meat, milk, wool), (2) animal days of grazing per unit area and, (3) effective feed units (total digestible nutrients, total energy, starch equivalents) estimated to have been required to maintain the animals and to support production on the pasture. Product per unit area is only a partial measure of pasture production because it does take into account the feed required to maintain the animals and support the work of grazing. Animal days per unit area is a more complete measure in that it does reflect the feed required for maintenance and the work of grazing. Like product per unit area, however, animal days per unit area is dependent on the type and body condition of the animals doing the grazing. Effective feed units (EFU) per unit area, on the other hand, appears to be relatively independent of type and condition of the grazing animal. Further, results expressed in terms of feed units are more easily generalized than are other animal measures of yield.

Because the ultimate use of pasture is to produce animal product and to maintain the grazing animal, yield as measured by the animals themselves is to be preferred to yield as measured by clipping methods. Of the animal-based measures, the estimated effective feed units (EFU) would seem to be the most useful. In a later section of this chapter we present a numerical example of the computation of yield using animal measures.

8.4.2 Experimental Errors

As was the case for quality studies, one way to increase the efficiency of pasture trials for evaluating forage yield is through reduction of the experimental errors involved. Although both pasture variability and animal variability contribute to the experimental errors associated with yield studies, the error associated with quantity measures tends to be lower than that for quality studies. Petersen and Lucas, (1960) derived the following expression for the experimental error, C_y, per pasture for estimated effective total digestible nutrients (ETDN) produced by a pasture:

$$C_y = \sqrt{\frac{C_p^2}{\sqrt{s}} + \frac{C_t^2}{t\sqrt{s}} + \frac{C_a^2}{a} + \frac{C_d^2}{d}}$$

in which

C_y = error of a pasture mean for ETDN per acre
C_p = among pasture variability in yield of ETDN
C_t = pasture × time interaction for ETDN
C_a = among-animal variability in predicting ETDN consumption
C_d = within-animal variability in predicting ETDN consumption

all expressed as coefficients of variation, and

s = pasture size in hectares
t = length of trial in days
a = number of different animals appearing on the pasture during the trial
d = number of animal days the pasture is grazed
= (a)(t) if animal number is constant

Using the data described previously in Section 8.3.2, they developed the following equation to approximate the per pasture error in ETDN for periods ranging from 4 weeks to a full season:

$$C = \sqrt{\frac{6500}{t\sqrt{s}} + \frac{14{,}000}{d}}$$

As was the case for the error expression for average daily gain, this expression can be used to predict the error in ETDN for pastures of a given size grazed for a given number of animal days during a period of given length.

It can also be used to estimate pasture size with a given number of animal days or to estimate animal days for a given pasture size to obtain a specified error in ETDN. For example: For a pasture trial running 90 days using four animals per pasture, what must be the pasture size to obtain a per pasture error in ETDN of 10%?

We have t = 90 days. Assume that animal number is constant throughout the trial. Then d = at = (4)(90) = 360.

$$10 = \sqrt{\frac{6500}{(90)\sqrt{s}} + \frac{14{,}000}{360}}$$

$$100 = \frac{72.22}{\sqrt{s}} + 38.89$$

$$61.11 = \frac{72.22}{\sqrt{s}}$$

$$\sqrt{s} = \frac{72.22}{61.11} = 1.18$$

$$s = 1.39 \cong 1.4$$

Under the given conditions a pasture of about 1.4 ha would be required to obtain an error of 10% in ETDN.

Formulas for estimating errors associated with gain per animal and animal days per acre were not derived by Petersen and Lucas. They indicate, however, that errors for gain per acre, an incomplete measure of production, may be approximated with the expression for estimating errors in average daily gain (8.3.2). Similarly, errors associated with animal days per acre, a more complete measure, may be estimated using the equation for errors in ETDN.

8.4.3 Experimental Design

As is the case with trials designed to study quality, experimental designs should be kept as simple as possible for yield trials. In most cases a randomized block design provides as much local control as is worthwhile. More complex designs should be avoided unless it is clear that they will produce sufficient error reduction to compensate for the loss in error degrees of freedom.

It has been suggested that in view of the long-time nature of many grazing trials, the actual experimental trial be preceded by a uniformity trial. The intent is to determine the nature of pasture variation and, possibly, provide data on the untreated pastures that can be used to adjust for pasture variation by means of a covariance analysis. Gross et al. (1965) reported the results of such a uniformity trial conducted in western North Carolina. They found that precision was increased by covariance adjustment only for one or two seasons following the uniformity trial. The magnitude of the increase in precision decreased sharply with increasing seasons following the uniformity trial. Their conclusion was that because of the expense and time involved, a uniformity trial prior to a grazing study is of questionable value.

For yield trials pastures need not be as large as those required for quality studies. Lucas (1952) and Lucas and Mott (1962) indicate that the optimum pasture size for yield trials is one that has a carrying capacity of 1 to 3 growing cattle or sheep and 2 to 4 milking dairy cattle.

In general, pastures should be as small as possible to permit the use of a reasonable number of treatments and a reasonable amount of replication within a trial. In this regard it should be pointed out that all pastures in a given trial need not be of the same size. Some treatments may be expected to produce more forage than others. In this case, pasture size should be adjusted so that all pastures are grazed with about the same pressure. Final results are then expressed on a per unit area basis and the data are analyzed in this form.

For quantitative measures such as ETDN, the grazing period need not be as long as that required for quality measures. Consideration of the composition of the errors involved, however, leads to the conclusion that for yield studies the grazing period should be at least 4 weeks long. For shorter periods the errors are too high to make such trials worthwhile.

8.4.4 Animal Management

For yield studies it is necessary that the grazing pressure (number of animals per unit of forage) be as close to the pasture's carrying capacity (stocking rate at optimum grazing pressure) as possible. Mott (1960) and Petersen et al. (1965) have shown that stocking rates (number of animals per unit area) either below or in excess of carrying capacity will result in less than maximum animal product per unit area. If grazing pressure is less than carrying capacity, the forage will not be fully utilized. On the other hand, with stocking rates in excess of carrying capacity, product per animal is reduced, and this, in turn, leads to reduced product per unit area. Further, both overstocking and understocking can result in unwanted changes in the botanical composition of the forage.

Pasture forages are not produced at a uniform rate throughout the growing season. A typical forage crop will have a period of rapid growth during the spring. It will then go through a period of slow growth during the summer. This will be followed by another period of increased growth during early autumn. Finally, the forage will go through a period of no growth during the winter. This implies that carrying capacity is not a fixed constant but changes throughout the season in response to changes in the supply of forage. To obtain optimum utilization of a pasture, it is necessary to keep the feed supply in equilibrium with the requirements of the animals. This means that adjustments must be made from time to time during the grazing season.

There are a number of management systems for adjusting grazing pressure during the season:

1. Keep animal numbers constant and adjust the size of the pasture periodically to provide a constant feed supply per animal.
2. Keep animal numbers constant, harvest excess forage during periods of rapid growth, and provide supplementary feed during periods of low growth.
3. Keep the animal numbers constant at less than full capacity and estimate the degree of forage utilization.
4. Adjust the stocking rate periodically to compensate for changes in the forage supply.

The extent to which the pastures are used at their optimum will depend on the skill of the research manager to use one or more of these methods of adjustment properly.

Of these management systems the most widely used and, in many respects, most satisfactory is to adjust the stocking rate periodically. This procedure has been called the "put-and-take" method by Lucas and Mott (1962).

Unlike quality studies, pretesting of animals and blocking or covariance with respect to animal measures will usually not result in any reduction in experimental error for yield studies. However, it is probably worthwhile to place all animals on a common pasture for a week or two prior to placing them on the experimental pastures (Lucas, 1952).

8.4.5 Put-and-Take Grazing

The put-and-take grazing system is one in which animals are added to or removed from the pasture in response to changes in the growth of forage. The aim is to keep the grazing pressure as close to the carrying capacity as possible throughout the trial. In its most useful form the put-and-take system involves two classes of animals, called by Petersen and Lucas (1968) "testers" and "regulators."

Testers are animals that are used to evaluate the quality of the pasture. They are the types of animals about which inferences from the trial results are to be made. More than one type of tester may be used in a trial, which makes it possible to use a single trial to provide inferences about more than one type of animal. One or more of each type of tester is assigned, at random, to each pasture. The same number of each type of tester must be used on every pasture in the trial. Once assigned to a pasture, a tester remains on that pasture throughout the course of the trial. Thus pastures must be large enough to support testers during periods of reduced forage production.

Regulators are animals whose purpose is to maintain the grazing pressure at the desired level. Together with testers, they provide data from which the pasture yield is computed. Regulators are added to the pasture during periods of rapid forage growth and removed from the pasture during slack periods. The decision to add or to remove regulators is made independently for each pasture. In this sense, regulators are a management tool whose purpose is to maintain a constant grazing pressure. Any kind of animal may be used as a regulator as long as its grazing habits are similar to those of testers. Further, regulators need not be assigned to a specific pasture but may be moved from one pasture to another as the need arises. During periods when not on test pastures, the regulators should be kept on a common pasture, if possible, whose forage is similar to that of the test pastures.

The testers are weighed at the beginning and end of the trial and at regular intervals, usually every 28 days, throughout the trial. Each regulator is weighed at the same time as the testers and, in addition, each time it is added to or removed from a test pasture. Weighing should be done early in the morning following overnight restriction from water.

8.4.6 Yield Computations

The animal records that are described for quality studies should be kept for yield studies as well. These include:

1. Animal's identity and description
2. Pasture assignment
3. Date of addition to or removal from pasture
4. Initial weight on pasture
5. Periodic weights while on pasture
6. Final weight off pasture
7. Length of interval between weighings
8. Milk, and its fat content, or wool produced while on pasture
9. Starting and ending body condition

With these records a number of measures of pasture yield may be computed. Mott and Lucas (1952) described three methods of yield computation in common use among pasture investigators:

I. Observed yield
II. Yield based on tester performance
III. Yield based on effective feed units (EFU) consumed

Under method I the following measures are computed:

1. Total production by all animals, reduced to a per unit area basis. Depending on the class of animal used, the product might be meat, milk, or wool.
2. Total animal days of grazing supported by the pasture, reduced to a per unit area basis.
3. Mean daily performance computed from all animals.

In method II yield measures are computed from the performance of the testers. They are computed in the following way:

1. Animal days per unit area computed from all animals, as under method I.
2. Average daily performance computed for each class of tester.
3. Product per unit area, computed as product per unit area = (animal days per unit area) × (average daily performance of testers)

If more than one class of tester is used, method II provides a way of estimating the yield of product for each type of tester.

8.4.7 Effective Feed Unit Approach

Yield measures computed under either method I or method II are deficient in at least two respects. They are highly dependent on the kinds of animals, both testers and regulators, used in the trial. Hence it is difficult to use the results to generalize for different classes of livestock. Further, they do not adequately take into consideration the feed required to maintain the animal and to support the work of grazing while the animal is on pasture.

One method of overcoming these difficulties is to, in effect, apply feeding standards in reverse to estimate the energy (total digestible nutrients, total energy, starch equivalents, etc.) that must have been consumed by each animal to support gain, production, work of grazing, and body maintenance while on pasture. The sum of these estimates for all animals appearing on the pasture is a measure of the total energy yield of the pasture. Mott and Lucas (1952) have outlined a method of yield estimation using this approach. This method has been called (Lucas, 1962) the effective feed unit (EFU) approach. Computation of yield measures using the EFU approach has been described in detail by Petersen and Lucas (1968).

To use the EFU approach the energy estimated to have been consumed by each animal while on pasture is computed as

$$EFU_{ij} = mM_{ij} + gG_{ij} + pP_{ij}$$

where

EFU_{ij} = total effective feed units consumed by the i-th animal on the j-th pasture

m = EFU required per unit of maintenance mass (including that required for the work of grazing)

M_{ij} = maintenance mass, expressed as the product of days on pasture, d_{ij}, and a convenient measure of weight, W_{ij}, of the i-th animal on the j-th pasture

g = EFU required per unit of gain

G_{ij} = gain made by the i-th animal on the j-th pasture

p = EFU required per unit of production

P_{ij} = production (milk or wool) by the i-th animal on the j-th pasture

Note: The multipliers m, g, and p depend on the form chosen to express EFU. A number of forms have been used by various pasture investigators. These include, among others, starch equivalent, total digestible nutrients (TDN), metabolizable energy, and net energy. The most common form in use in the United States is TDN. Tables containing factors for converting weight, gain, and production into TDN equivalents for maintenance, gain, and production for cattle and sheep are given in an appendix to this chapter (Tables A8.1 and A8.2).

The total EFU yield per unit area, $TEFU_j$, on the j-th pasture is the sum of the estimated EFU per unit area consumed by all animals on the pasture. $TEFU_j$ is a measure of the total energy yield of the j-th pasture during the trial.

Carrying capacity, gain, and production for a given pasture can be evaluated in terms of each of the classes of tester used in the trial. This involves computing the average daily EFU requirement, EFU_k, average daily gain, TDG_k, and average daily production, TDP_k, for the class, k, of tester under consideration. Then, in terms of the k-th type of tester:

1. Carrying capacity in animal days per unit area, TAD_k, is

$$TAD_k = \frac{TEFU}{EFU_k}$$

2. Gain per unit area, TGA_k, is computed as

$$TGA_k = (TAD_k)(TDG_k)$$

3. Product per unit area, TPA_k, is

$$TPA_k = (TAD_k)(TDP_k)$$

By computing EFU_k, TDG_k, and TDP_k separately for each type, pasture yield can be estimated in terms of each type of tester appearing on the pasture.

8.5 DUAL-PURPOSE TRIALS

Because of the expense and the amount of land involved, grazing trials are frequently conducted with the dual objectives of evaluating both quality and yield in the same trial. This would seem to be a reasonable thing to do since few adjustments need be made to turn a single-objective trial into a dual-purpose trial.

As is the case in either quality or in yield trials the experimental design for dual-purpose trials should be as simple as possible. A randomized block design should be adequate to handle most situations.

Again, as with single-purpose trials, the pasture is the experimental unit. Recall that the optimum size of pasture for quality studies is larger than that required for yield studies. Lucas (1952) and Lucas and Mott (1962) proposed a compromise in pasture size for dual-purpose trials. They suggest a size with a carrying capacity of 2 to 5 animals for growing cattle or sheep, and a carrying capacity of 3 to 6 milking dairy cattle. Lucas (1952) points out that efficiency changes very slowly in the region of the optimum size, so the compromise seems reasonable. Lucas also suggests that with put-and-take grazing the number of testers per pasture should be less than the optimum carrying capacity. Since quality is to be evaluated, the trial should be run for longer than would be the case if only yield were to be measured. As a working rule, 6 weeks should be the minimum length of trial with the dual-purpose objective.

Since yield is to be measured the preferred procedure for managing the animals is to use the put-and-take system. One or more classes of testers should be defined, keeping in mind the quality aspects of the evaluation since testers are the animals on which the evaluation of quality is based. Each class of tester defines a block as far as the animals are concerned. The same number of animals from each class of tester should be assigned, at random, to each pasture. Regulators, whose grazing habits are nearly the same as those of the testers, should be used to adjust the stocking rate as required. All animals should be weighed when they are put on or taken off the pasture. Periodic weighings of all animals while on pasture should be done.

Computation of measures of product per animal and of yield per pasture should be carried out just as they are for single-objective trials. Each of the appropriate quality and yield measures is computed for each pasture and reduced, where appropriate, to a per unit area basis. These pasture values then serve as entries in analyses of variance appropriate for the experimental design used to assign treatments to the pastures.

8.6 NUMERICAL EXAMPLE

To illustrate the computations involved in evaluating grazing trials we use the results from one pasture of an experiment designed to test the effect of different fertilizer treatments on the output of tall fescue pastures. The pasture was 1.5 ha in size. The trial began in early April and ran until late July. Growth of forage was rapid near the beginning of the trial but declined substantially near the end. The put-and-take grazing system was used in an attempt to compensate for differential growth of forage during the course of the trial. Steers were used as grazing animals. There were two types of tester: (1) yearlings (12 months of age) and (2) long yearlings (18 months of age). One of each type of tester was used for the entire trial in each pasture. Steers of varying age and condition served as regulators.

8.6.1 Basic Data

Characteristics of the animals used on the example pasture are given in Table 8.1. Animal weights at each weighing date are presented in Table 8.2. Table 8.2 also includes some of the basic summary information for each animal. This is required in computating the quality and yield measures. Computations will be illustrated for the three methods presented in Section 8.4.6.

For the computations used in this illustration, let

W_{Ii} = initial weight (on pasture) of the i-th animal

W_{Fi} = final weight (off pasture) of the i-th animal

d_i = days spent on pasture by the i-th animal

$G_i = W_{Fi} - W_{Ii}$ = gain made by the i-th animal while on pasture

$\bar{W}_i = (W_{Ii} + W_{Fi})/2$ = mean weight of the i-th animal while on pasture

Table 8.1 **Characteristics of Steers Used on the Example Pasture from a Grazing Trial**

Animal Number	Type of Animal	Body Condition Start	Finish
T1 *	Yearling	SF	SF **
T2	Long yearling	SF	T
R1	2 year old	T	SF
R2	Calf	SF	SF
R3	Yearling	SF	SF

*T, tester; R, regulator.
**T, thin; SF, slightly fat.

8.6.2 Method I: Observed from All Animals

Under method I the following are computed:

1. Mean daily gain, DG, from all animals:

$$DG = \frac{\Sigma_i G_i}{\Sigma_i d_i}$$

2. Animal days, AD, per unit area from all animals:

$$AD = \frac{\Sigma_i d_i}{\text{pasture size}}$$

3. Product (gain), GA, per unit area from all animals:

$$GA = \frac{\Sigma_i G_i}{\text{pasture size}}$$

From Table 8.2 we have, for the example data,

1. DG = (222 kg)/(420 days) = 0.53 kg/day
2. AD = (420 days)/(1.5 ha) = 280 days/ha
3. GA = (222 kg)/(1.5 ha) = 148 kg/ha

Table 8.2 **Weights (kg) at Intervals Throughout the Trial of Steers Used on the Example Pasture**

Interval		28	28	28	28 days			
	Date					Total		Mean
Animal	Apr 3	May 1	May 29	Jun 26	Jul 24	Gain	Days	Weight
T1	240	254	270	283	300	60	112	270
T2	305	315	324	336	347	42	112	326
R1	295	313	331	347	365	70	112	330
R2	190	206	222			32	56	206
R3		300	318			18	28	309
					Total	222	420	

8.6.3 Method II: Based on Tester Performance

Under method II animal days are computed from all animals, daily performance is computed separately for each type of tester, and yield is estimated as the product of animal days and tester performance. We have

1. Animal days per unit area, AD, from all animals (as for method I):

$$AD = \frac{\sum_i d_i}{\text{pasture size}}$$

2. Mean daily gain, TDG_k, for testers of type k:

$$TDG_k = \frac{\sum_i G_{ik}}{\sum_i d_{ik}}$$

where G_{ik} = gain made by the i-th tester of type k
d_{ik} = days on pasture by the i-th tester of type k

3. Gain per unit area, TGA_k, for animals of tester type k:

$$TGA_k = (AD)(TDG_k)$$

From the data in Table 8.2 we compute:

1. AD = (420 days)/(1.5 ha) = 280 days/ha.
2. Tester daily gain:
   a. Type 1 (yearling) testers:

$$TDG_1 = \frac{60 \text{ kg}}{112 \text{ days}} = 0.54 \text{ kg/day}$$

   b. Type 2 (long yearling) testers:

$$TDG_2 = \frac{42 \text{ kg}}{112 \text{ days}} = 0.38 \text{ kg/day}$$

3. Gain per hectare:
   a. Type 1 (yearling) testers:

$$TGA_1 = (280 \text{ days/ha})(0.54 \text{ kg/day}) = 151.2 \text{ kg/ha}$$

   b. Type 2 (long yearling) testers:

$$TGA_2 = (280 \text{ days/ha})(0.38 \text{ kg/day}) = 106.4 \text{ kg/ha}$$

8.6.4 Method III: Effective Feed Unit Approach

Under this approach the effective feed units (EFU) required to support each animal while on pasture are computed. These are summed over all animals to obtain an estimate of the total EFU, produced by the pasture during the trial. Performance is based on the daily EFU requirement of the various types of tester, and yield is computed from animal days, as estimated through EFU, and tester performance. The computing procedure used here is a modification of that presented by Petersen and Lucas (1968).

Although other measures of feed units may be used, we will use total digestible nutrients, TDN, in this example. Factors for converting maintenance, m, and gain, g, to their TDN equivalent are given in Tables A8.1 and A8.2 of the appendix to this chapter. These conversion factors were developed by Petersen and Lucas (1968). The formula for estimating the individual animal EFU requirement is given in Section 8.4.7. The computation of EFU for each animal in the example is given as follows:

1. Animal T1: AD = 112; G = 60; $\bar{W}$ = 270

$$M = \frac{(AD)(\bar{W})}{1000} = \frac{(112)(270)}{1000} = 30.24$$

EFU = (9.4)(30.24) + (2.0)(60) = 404.26 kg TDN *
* 9.4 = m from Table A8.1; 2.0 = g from Table A8.2

2. Animal T2: AD = 112; G = 42; $\bar{W}$ = 326

$$M = \frac{(112)(326)}{1000} = 36.51$$

EFU = (9.0)(36.51) + (1.8)(42) = 404.19 kg TDN

3. Animal R1: AD = 112; G = 70; $\bar{W}$ = 330

$$M = \frac{(112)(330)}{1000} = 36.96$$

EFU = (9.0)(36.96) + (2.4)(70) = 500.64 kg TDN

4. Animal R2: AD = 56; G = 32; $\bar{W}$ = 206

$$M = \frac{(56)(206)}{1000} = 11.54$$

EFU = (9.7)(11.54) + (1.6)(32) = 163.14 kg TDN

5. Animal R3: AD = 28; G = 18; $\bar{W}$ = 309

$$M = \frac{(28)(309)}{1000} = 8.65$$

EFU = (9.2)(8.65) + (2.0)(18) = 115.58 kg TDN

Data for the individual animals used on the example pasture may now be summarized as shown in Table 8.3. The entries in this table provide the basic data for computing the measures of yield for the pasture (the experimental unit of a pasture trial).

To summarize the output of the pasture we first compute the total EFU, TEFU, per unit area produced by the pasture:

$$TEFU = \frac{\Sigma\ EFU}{\text{pasture size}}$$

where Σ is over all animals that appeared on the pasture during the trial.

Table 8.3 **Summary of Data from Individual Steers Used on the Example Pasture**

Animal	Total gain (kg)	Time on pasture (days)	Daily gain (kg/day)	Mean weight (kg)	Maintenance mass (kg)(days)/1000	EFU (kg)
T1	60	112	0.54	270	30.24	404.26
T2	42	112	0.38	326	36.51	404.19
R1	70	112	0.63	330	36.96	500.64
R2	32	56	0.57	206	11.54	163.14
R3	18	28	0.64	309	8.65	115.58

From Table 8.3 we have

$$\text{TEFU} = \frac{404.26 + 404.19 + 500.64 + 163.14 + 115.58}{1.5}$$

$$= \frac{1587.81 \text{ kg TDN}}{1.5 \text{ ha}}$$

$$= 1059 \text{ kg TDN/ha}$$

For the pasture as a unit carrying capacity and yield are computed for each type of tester separately. For this purpose we first compute the daily gain, TDG_k, of the average tester of type k, and the daily EFU requirement. $DEFU_k$, of the average k type tester. From this we then compute the carrying capacity and gain, on a per unit area basis, supported by the pasture for each type of tester. These computations follow.

Type 1 (yearling) testers:

1. Tester daily gain, TDG_1:

   $TDG_1 = 0.54$ kg/day (Table 8.3)

2. Daily EFU, $DEFU_1$:

$$DEFU_1 = \frac{EFU_1}{AD_1} = \frac{404.26 \text{ kg TDN}}{112 \text{ days}} \quad \text{(Table 8.3)}$$

$$= 3.61 \text{ kg TDN/day}$$

3. Tester animal days, TAD_1:

$$TAD_1 = \frac{TEFU}{DEFU_1}$$
$$= \frac{1059 \text{ kg TDN/ha}}{3.61 \text{ kg TDN/day}}$$
$$= 293.35 \text{ days/ha}$$

4. Gain per unit area, TG_1, for animals of tester type 1:

$$TG_1 = (TAD_1)(TDG_1)$$
$$= (293.35 \text{ days/ha})(0.54 \text{ kg/day})$$
$$= 158.41 \text{ kg/ha}$$

Type 2 (long yearling) testers:

1. $TDG_2 = 0.38$ kg/day (Table 8.3)

2. $$DEFU_2 = \frac{404.19 \text{ kg TDN}}{112 \text{ days}} \quad \text{(Table 8.3)}$$
$$= 3.61 \text{ kg TDN/day}$$

3. $$TAD_2 = \frac{1059 \text{ kg TDN/ha}}{3.61 \text{ kg TDN/day}}$$
$$= 293.35 \text{ days/ha}$$

4. $$TG_2 = (293.35 \text{ days/ha})(0.38 \text{ kg/day})$$
$$= 111.47 \text{ kg/ha}$$

The quality and yield measures for the example pasture are summarized in Table 8.4. Of the methods of pasture evaluation presented here, probably the most useful is the method (III) based on effective feed units (EFU). It provides a complete measure of pasture production if put-and-take grazing is done properly to optimize grazing pressure. The total EFU value takes into consideration not only the energy required for production but also the energy needed to maintain the animals and to support the work of grazing. Neither of the other methods accounts for maintenance and grazing energy. The EFU method also provides a way of using a single pasture to draw inferences about different classes of livestock by using different types of testers in the same experiment.

Table 8.4 **Summary of Quality and Yield Measures for the Example Pasture**

Estimation Method	Daily gain (kg/day)	Animal days (days/ha)	Total gain (kg/ha)
I. Observed-all animals	0.53	280	148.0
II. From tester performance			
a. Yearlings	0.54	280	151.2
b. Long yearlings	0.38	280	106.4
III. EFU Method			
a. Yearlings	0.54	293.4	158.4
b. Long yearlings	0.38	293.4	115.4

EFU yield = 1059 kg/ha TDN

The pasture data, not the animal data, are the observations used to evaluate the effect of the pasture treatments. When the measures of quality and yield have been computed by any of the three methods, and then reduced to a per unit area basis, they are subjected to an analysis of variance appropriate for the experimental design used to assign the treatments to the pastures.

APPENDIX

The tables that appear in this appendix are taken from Petersen and Lucas (1968).

Table A8.1 **Factors (m)* for Converting Maintenance Requirement to TDN Equivalent**

A. *Cattle*

Weight		
kg	lb	m*
136–181	300–399	10.0
182–226	400–499	9.7
227–272	500–599	9.4
273–317	600–699	9.2
318–362	700–799	9.0
363–408	800–899	8.8
409–453	900–999	8.7
454–499	1000–1099	8.6
500–544	1100–1199	8.5
545–589	1200–1299	8.4
590–635	1300–1399	8.3

B. *Sheep*

Weight		
kg	lbs	m*
18–22	40–49	1.11
23–28	50–59	1.08
29–32	60–69	1.06
33–36	70–79	1.04
37–40	80–89	1.02
41–45	90–99	1.01
46–54	100–119	0.99
55–64	120–139	0.97
65–72	140–159	0.95
73–81	160–179	0.94
82–90	180–199	0.93

*m = kg TDN/1000 kg weight/day, varied as weight to the $\frac{7}{8}$ power with an amount added for activity.

Table A8.2 **Factors (g)* for Converting Gain Requirement to TDN Equivalent**[a]

Cattle	Sheep	Starting Condition	Ending Condition VT	T	SF	F**
Calves	0–6 Mo.	VT	1.2	1.3	1.5	1.6
		T	1.3	1.4	1.6	1.7
		SF	1.4	1.5	1.6	1.7
		F	1.4	1.5	1.7	1.8
Yearlings	6–9 Mo.	VT	1.4	1.6	1.8	2.0
		T	1.5	1.7	1.9	2.1
		SF	1.6	1.8	2.0	2.2
		F	1.7	1.9	2.1	1.4
2-yr-old	9–12 Mo.	VT	1.6	1.8	2.2	2.5
		T	1.8	2.0	2.4	2.7
		SF	2.0	2.2	2.6	2.9
		F	2.2	2.5	2.8	3.1
3-yr-old	>12 Mo.	VT	1.9	2.1	2.6	3.0
		T	2.1	2.4	2.8	3.2
		SF	2.4	2.7	3.1	3.6
		F	2.7	3.0	3.5	3.9

*kg TDN/kg gain or lb TDN/lb gain.

[a]Milk-producing cattle require 0.31 kg of TDN per kg of 4% fat-corrected milk, or 0.31 lb of TDN per lb of 4% fat-corrected milk.

**VT, very thin: T, thin; SF, slightly fat; F, fat.

REFERENCES

Cook, C. W., 1962. Weighing of animals. In *Pasture and Range Research Techniques*. Comstock Publishing Associates, Ithaca, N.Y., pp. 30–31.

Gross, H. D., L. Goode, and R. G. Petersen, 1965. Use of a uniformity trial in grazing experiments. *Agron. J.* 57:272–273.

Linehan, P. A., 1952. Use of cage and mower strip methods for measuring the forage consumed by grazing animals. *Proc. 7th Int. Grass. Congr.*, pp. 1328–1333.

Lucas, H. L., 1952. Design, conduct and interpretation of grazing trials. Paper read at Black Mountain Statistics Conference, Department of Statistics, North Carolina State College. Mimeo.

Lucas, H. L., 1962. Determination of forage yield and quality from animal response. In *Range Research Methods*. Misc. Publ. 940. U.S.D.A., Forest Service, pp. 43–54.

Lucas, H. L., and G. O. Mott, 1962. Size and number of pastures, number of animals and length of grazing period. In *Pasture and Range Research Techniques*. Comstock Publishing Associates, Ithaca, N.Y., pp. 7–10.

Mott, G. O., 1960. Grazing pressure and the measurement of pasture production. *Proc. 8th Int. Grass. Congr.*, pp. 606–611.

Mott, G. O., and H. L. Lucas, 1952. The design, conduct and interpretation of grazing trials on cultivated and improved pasture. *Proc. 8th Int. Grass. Congr.*, pp. 1380–1385.

Petersen, R. G., and H. L. Lucas, 1960. Experimental errors in grazing trials. *Proc. 8th Int. Grass. Congr.*, pp. 747–750.

Petersen, R. G., and H. L. Lucas, 1968. Computing methods for the evaluation of pastures by means of animal response. *Agron. J.* 60:682–687.

Petersen, R. G., H. L. Lucas, and G. O. Mott, 1965. Relationship between rate of stocking and per animal and per acre performance on pasture. *Agron. J.* 57:27–30.

FURTHER READING

1. Brown, D., 1954. Methods of surveying and measuring vegetation. *Bull. 42*. Commonwealth Bureau of Pasture and Field Crops. 223 pp.
2. Institute Staff, 1961. Research techniques in use at the Grassland Research Institute, Hurley. *Bull. 45*. Commonwealth Bureau of Pasture and Field Crops. 167 pp.
3. Joint Committee Report, 1952. Pasture and range research techniques. *Agron. J.* 44:39–50.
4. Joint Committee Report, 1962. *Pasture and Range Research Techniques*. Comstock Publishing Associates, Ithaca, N.Y. 242 pp.
5. Symposium, 1962. Range research methods. *Misc. Publ. 940*. USDA, Forest Service. 172 pp.

9

On-Farm Trials in Farming Systems Research

9.1 INTRODUCTION

Since the early 1970s farming systems research (FSR) has received an increasing amount of attention by agricultural scientists, particularly by those in the international agricultural research centers working in the developing areas. The basic goals of FSR are to determine the problems facing the small farmer and to develop technologies for improving the return to small farmers. Sands (1986) has pointed out that FSR is farmer oriented. It is interdisciplinary and while it is complementary to mainstream agricultural research, it does not replace it. FSR uses a problem-solving approach to examine the systems under which small farmers operate. It depends on feedback both from farmers to identify problems and on-station research to develop possible solutions. The foundation of FSR is on-farm trials involving participation by the farmer in adapting new technology within the farm enterprise.

According to Collison (1987), FSR involves three successive stages: (1) diagnosis, (2) planning, and (3) experimentation and assessment. The diagnosis stage involves the identification of a target area composed of farmers operating in a fairly homogeneous system with similar problems and similar opportunities for adoption of new technologies. The delineation of the target area is based on both formal and informal surveys of farmers.

The aim is to identify farm enterprises that absorb a high proportion of the factors of production: cash, labor, land, and so on. Research opportunities are then identified and ranked to set priorities in order of their potential for contributing to production in the farming system.

In the planning stage the problems identified in the diagnosis stage are examined to identify the problem properly and to identify possible causes and their solution. At this stage previous research may be brought to bear, or new on-station research may be conducted in an attempt to develop possible alternative solutions.

Once a possible solution has been agreed upon by the FSR team, on-farm trials are begun. The purpose of these on-farm trials is to evaluate the proposed new practice under farm conditions. They also have the goal of demonstrating the new practice to farmers in the target area. At the conclusion of each trial the results are assessed by the FSR team in cooperation with the participating farmer. Assessment is not confined to yield differences but includes economic returns and social effects as well. On-farm trials and their assessment are the cornerstone of farming systems research. (For readers who would like more detailed information on FSR, see the series of papers that began in 1986 in the journal *Experimental Agriculture*.)

9.2 ON-FARM TRIALS

The bulk of this chapter is devoted to the design and analysis of on-farm trials in FSR. A number of features distinguish these trials from the usual scientist-managed off-station research trial. On-farm trials in FSR are conducted with the active participation of the farmer on whose property the trial is conducted. Hence variables that affect the outcome of the trial are not as well controlled as they are in trials managed by the research scientist. Further, while the usual off-station research trial involves multiple treatments, often in factorial arrangements, the on-farm FSR trial usually involves only two treatments: the farmer's practice and a single package of improved practices. In the scientist-managed off-station research, the results are usually expressed only as yield differences. In on-farm FSR trials, interest is not only in yield differences but also in such economic and social differences as net revenue or allocation of family labor.

The basic goal of on-farm trials is to compare the results from the selected package of improved practices with the results of the production system used by the farmers in the target area. Thus the on-farm FSR trial essentially involves only two treatments: farmer's practice (F) and improved practice (N).

In operation, a target area is chosen. This is the area that contains farm conditions which the new practices are intended to improve. A number of farms in the target are chosen for the study. On each of the farms selected, a trial is conducted. This trial includes, basically, two treatments, F and N, both of which are used on each of the farms chosen. The F treatment is managed by the farmer in his usual way. The N treatment is usually managed by the FSR scientists with the cooperation of the farmer.

The increase from the improved practice is determined by subtracting the response under the farmer's practice, F, from the response under the improved practice, N. Response is measured not only in terms of yield but also in economic terms. Records are kept on the cost of the resources, such as seed, fertilizer, personnel, and so on, required by both practices. Returns from both practices are also recorded and the net returns under both practices are compared. Procedures for the economic analysis of farming systems research are detailed in Perrin et al. (1976).

A test of the significance of the increases attributable to the improved practice is based on an analysis of variance of the data from the trials. This analysis provides information that can be used to determine the minimum increase (LSI) required to attain significance. Observed increases may then be compared to the LSI to determine which, if any, are significant.

Because on-farm trials are not as closely managed as are on-station trials, they must be more highly replicated than are on-station trials. Replication of on-farm trials is best accomplished by selecting a number of farms within the target area and by using a single replication of each treatment on each farm. The experimental data are then analyzed as if they were from a randomized block design with farms as blocks. Blocking by farms serves a number of purposes. It provides wider coverage of the target area than would be the case with multiple replication on only a few farms. Further, it provides for widespread demonstration of the new technologies to farmers in the target area who are not actively participating in the program. Suggestions as regards the minimum number of farms to be used with each type of design are included with the discussion of the individual designs.

The form of the experimental design to be used on the farms chosen will depend on whether the trial is concerned with crops or animals. Also, with crops the design will depend on whether the improved practice has one or more than one component. Similarly, analysis of the resulting data will depend on the experimental design used in the trial. The design and analysis details are given separately for each design type presented here.

9.3 BASIC TRIAL WITH CROPS

The basic on-farm trial with crops compares only two treatments: farmer's practice (F) and improved practice (N) A number of farms are selected from within the target area. There should be enough farms in the trial so that there is a reasonable chance that differences in response to F and N will be detected. One suggestion is to select enough farms so that the error degrees of freedom in the analysis of variance of the resulting data will be 10 or more. With this criterion the suggested minimum number of farms for the basic on-farm FSR trial would be 11.

9.3.1 Trial Layout

The basic experimental design consists of two large plots on each of farm selected from the target area. These plots should be laid out in a representative field on the farm. The two plots should be the same size and shape, but they need not be adjacent. It is sufficient that they both represent the same conditions of soil type, topography, drainage, and so on. The plots should be large enough to permit normal farm operations. One of the plots is assigned, at random, to the improved practice, N, while the other is assigned to the farmer's practice, F. An example of the field plan for the basic crop trial is presented in Figure 9.1. The trial is conducted and yield and other data are recorded for each plot at each location.

9.3.2 Data Analysis

Suppose that the basic two-treatment trial has been conducted at each of n locations. Let y_{ij} represent the yield (or other measure such as net return) per unit area of the j-th treatment at the i-th location. The results might be tabulated as shown in Table 9.1. The data analysis takes a relatively simple form:

1. Compute MSE $= [1/(n - 1)][\Sigma_i d_i^2 - G^2/n]$

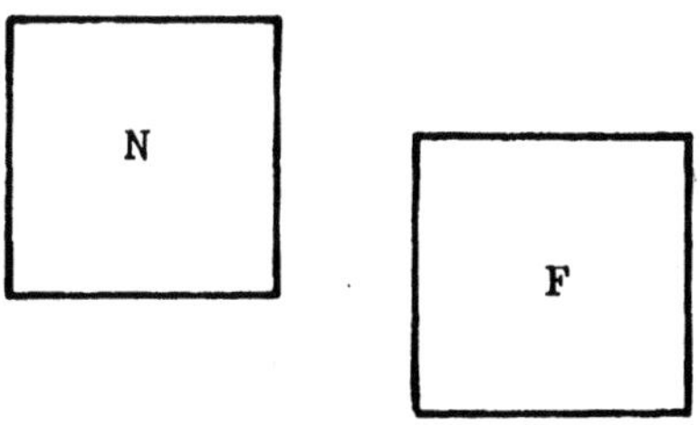

Figure 9.1 Field layout for a basic two-treatment on-farm FSR trial.

Table 9.1 **Data Table for the Basic Two-Treatment On-Farm FSR Trial**[a]

Location	Treatment F	Treatment N	Difference
1	y_{11}	y_{12}	d_1
2	y_{21}	y_{22}	d_2
.	.	.	.
.	.	.	.
.	.	.	.
n	y_{1n}	y_{2n}	d_n
		Sum	G
		Mean	$\bar{d}$

[a]$d_i = y_{i2} - y_{i1}$
= difference between the improved practice and the farmer's practice
= increase from the improved practice
$G = \Sigma_i d_i$ = grand total of differences
$\bar{d} = G/n$ = mean difference

2. The least significant increase (LSI) is computed as

$$LSI_{0.05} = t_{0.05}\sqrt{\frac{MSE}{n}} \quad \text{at the 5\% level}$$

$$LSI_{0.01} = t_{0.01}\sqrt{\frac{MSE}{n}} \quad \text{at the 1\% level}$$

Note: $t_{0.05}$ and $t_{0.01}$ are, respectively, the 5% and 1% one-sided tabular t values with n − 1 df.

If $\bar{d} > LSI_{0.05}$, the increase over the farmer's practice is significant (5%).
If $\bar{d} > LSI_{0.01}$, the increase over the farmer's practice is highly significant (1%).
If $\bar{d} < LSI_{0.05}$, the improved practice is not significantly better than the farmer's practice.

9.3.3 Numerical Example

A trial was conducted in eastern Oregon to compare a new wheat variety, N, with a commonly grown variety, F. The trial was conducted at each of eight locations. Two plots were laid out in a field at each location. The old variety was assigned at random to one of the plots and the new variety was assigned to the other. The resulting yields (kg/plot) are given in Table 9.2.

The computations proceed as follows:

1. $$\begin{aligned} \text{MSE} &= [1/(n-1)][\textstyle\sum_i d_i^2 - G^2/n] \\ &= [1/(8-1)][144.9100 - 31.5^2/8] \\ &= (1/7)\,(144.9100 - 124.0312) \\ &= 20.8788/7 = 2.9827. \end{aligned}$$
2. $\text{LSI}_{0.05} = 1.90\sqrt{2.9827/8} = (1.90)(0.6106) = 1.16.$

   $\text{LSI}_{0.01} = 3.00\ \sqrt{2.9827/8} = (3.00)(0.6106) = 1.83.$

Table 9.2 **Yields (kg/plot) of New (N) and Old (F) Wheat Varieties from an On-Farm FSR Trial**

	Variety		
Location	F	N	d = N - F
1	30.2	32.2	2.0
2	31.5	35.6	4.1
3	37.1	41.9	4.8
4	30.8	35.4	4.6
5	23.2	25.4	2.2
6	24.8	27.1	2.3
7	26.7	33.8	7.1
8	26.7	31.1	4.4
		Sum	31.5
		Mean	3.94

Since 3.94 > 1.83, the increase from the new variety compared to the commonly grown variety is highly significant. The new variety outyields the old by about 4 kg/plot.

9.4 CROP TRIAL WITH SUPPLEMENTARY TREATMENTS

Sometimes it is desired to include other treatments in addition to the farmer's practice and the improved practice in an on-farm FSR trial. For example, it might be of interest to examine some new varieties or some different herbicides that are not included in the improved practice. Further, it might be interesting to see how these new treatments perform under farm conditions.

9.4.1 Trial Layout

The basic design for two treatments, F and N, can be modified to include as many as six supplementary treatments, $S_1, S_2, \ldots, S_6$. As with the basic crop design, a number of farms is selected in the target area. The minimum number of farms to be used depends on the number of supplementary treatments included in the trial. If only one or two extra treatments are used, the minimum number of farms would be seven or eight. If five or six supplementary treatments are included, the minimum number of farms would be four or five.

On each farm the design would consist of two large and several small plots. The farmer's practice, F, would be assigned, at random, to one of the large plots, while the improved practice, N, would be assigned to the other. The supplementary treatments, S_j, would be assigned, at random, to the small plots, one treatment to each plot. An example of the field plan for a trial with four supplementary treatments is shown in Figure 9.2.

The plots should be located in the same field but need not be adjacent to one another. Bear in mind, however, that one of the purposes of an on-farm FSR trial is to demonstrate the improved practice in comparison to

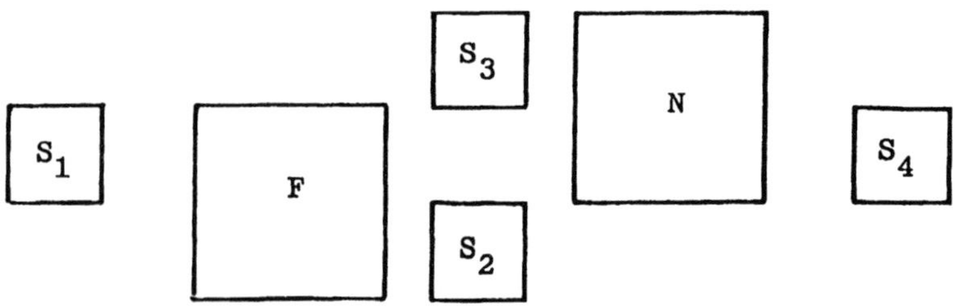

Figure 9.2 Field layout for an on-farm FSR trial with four supplementary treatments.

the farmer's practice. For this reason the large plots should be located relatively close to each other. Keep in mind, also, that all the plots should be selected to be as nearly alike as possible before the treatments are applied.

9.4.2 Data Analysis

Suppose that the trial is conducted at n locations and that the treatments include F, N, and p ($0 < p < 6$) supplementary treatments, S_j. Let y_{ij} represent the yield (or other measure) per unit area from the j-th treatment at the i-th location. (*Note*: Because different sizes of plot are used the yields must be expressed on a per unit area basis to be comparable.) The results of the trial may be tabulated as shown in Table 9.3. The ANOVA format for the analysis of the results is given in Table 9.4.

The entries in Table 9.4 are computed in the following way:

$$\text{SSTot} = \sum_i \sum_j y_{ij}^2 - \frac{G^2}{n(p+2)}$$

$$\text{SSL} = \frac{1}{p+2} \sum_i L_i^2 - \frac{G^2}{n(p+2)}$$

Table 9.3 **Summary Table for the Data from an On-Farm FSR Trial with p Supplementary Treatments**[a]

	Treatment						
Location	F	N	S_1	S_2	...	S_p	Sum
1	y_{11}	y_{12}	y_{13}	y_{14}	...	$y_{1,p+2}$	L_1
2	y_{21}	y_{22}	y_{23}	y_{24}	...	$y_{2,p+2}$	L_2
.	.	.	.	.		.	.
.	.	.	.	.		.	.
.	.	.	.	.		.	.
n	y_{n1}	y_{n2}	y_{n3}	y_{n4}	...	$y_{n,p+2}$	L_n
Sum	T_1	T_2	T_3	T_4	...	T_{p+2}	G
Mean	$\bar{y}_1$	$\bar{y}_2$	$\bar{y}_3$	$\bar{y}_4$	...	$\bar{y}_{p+2}$	

[a]$L_i = \Sigma_j\, y_{ij}$ = total yield at location i
$T_j = \Sigma_i\, y_{ij}$ = total yield of treatment j
$G = \Sigma_i\, L_i = \Sigma_j\, T_j$ = grand total yield
$\bar{y}_j = T_j/n$

Table 9.4 **Format for the Analysis of Variance of Data from an On-Farm FSR Trial with Supplementary Treatments**

ANOVA

Source	df	SS	MS
Total	n(p+2)+1	SSTot	
Location	n−1	SSL	
Treatment	p+1	SST	
Error	(n−1)(p+1)	SSE	MSE

$$SST = \frac{1}{n}\sum_j T_j^2 - \frac{G^2}{n(p+2)}$$

$$SSE = SSTot - SSL - SST$$

$$MSE = \frac{SSE}{(n-1)(p+1)}$$

The standard error, $s_{\bar{y}}$ of a treatment mean is

$$s_{\bar{y}} = \sqrt{\frac{MSE}{n}}$$

and the standard error, $s_{\bar{d}}$, of the difference between two treatment means is computed as

$$s_{\bar{d}} = \sqrt{\frac{2MSE}{n}}$$

Mean increases, $\bar{d}_j$, of the other treatments compared to the farmer's practice are computed by subtracting the mean, $\bar{y}_1$, of the farmer's practice from the means, $\bar{y}_j$, of the other treatments. We have

$$\bar{d}_j = \bar{y}_j - \bar{y}_1$$

The least significant increase (LSI) is found to be

$$LSI_{0.05} = t_{0.05}\sqrt{\frac{2MSE}{n}} \quad \text{at the 5\% level}$$

$$LSI_{0.01} = t_{0.01}\sqrt{\frac{2MSE}{n}} \quad \text{at the 1\% level}$$

where $t_{0.05}$ and $t_{0.01}$, are, respectively, the 0.05 and 0.01 one-tailed tabular t values with (n − 1)(p + 1) degrees of freedom.

If $\bar{d}_j > LSI_{0.05}$, the increase is significant (5%).
If $\bar{d}_j > LSI_{0.01}$, the increase is highly significant (1%).
If $\bar{d}_j < LSI_{0.05}$, the increase is not significant.

9.4.3 Numerical Example

An on-farm FSR trial was conducted to compare a package of improved practices, N, with the usual grower's practice, F, in the rice-growing region of central California. At the same time, four new rice selections, S_1, S_2, S_3, and S_4, were included in the trial to see how they would perform under farm conditions. The trial was conducted at five locations.

Rice yields are tabulated in Table 9.5, and the analysis of variance of these yields is presented in Table 9.6. The standard error, $s_{\bar{y}}$, of a treatment mean is

$$s_{\bar{y}} = \sqrt{\frac{MSE}{n}} = \sqrt{\frac{0.0722}{5}} = 0.1201$$

Table 9.5 **Rice Yields (T/ha) from an On-Farm Trial with Four Supplementary Treatments**

	Treatment						
Location	F	N	S_1	S_2	S_3	S_4	Sum
1	3.21	4.45	4.50	4.46	3.68	4.20	24.50
2	3.31	4.89	5.10	4.79	4.21	4.31	26.61
3	3.86	5.68	6.10	5.67	4.72	4.89	30.92
4	3.74	5.47	6.40	5.21	4.80	4.77	30.39
5	3.99	6.01	7.18	5.79	5.28	5.39	33.64
Sum	18.11	26.50	29.28	25.92	22.69	23.56	146.06
Mean	3.62	5.30	5.86	5.18	4.54	4.71	

Table 9.6 **Analysis of Variance of Rice Yields (T/ha) from an On-Farm FSR Trial with Four Supplementary Treatments**

ANOVA

Source	df	SS	MS
Total	29	24.9997	
Location	4	8.8142	
Treatment	5	14.7419	
Error	20	1.4436	.0722

and the standard error, $s_{\bar{d}}$, of a difference between two treatment means is

$$s_{\bar{d}} = \sqrt{\frac{2MSE}{n}} = \sqrt{\frac{(2)(0.0722)}{5}} = 0.1699$$

The increases of the other treatments over the farmer's practice are given in Table 9.6.

For this set of rice yield data the least significant increases (LSI) are

$$LSI_{0.05} = 1.72\sqrt{\frac{(2)(0.0722)}{5}} = 0.29 \qquad (5\%)$$

$$LSI_{0.01} = 2.53\sqrt{\frac{(2)(0.0722)}{5}} = 0.43 \qquad (1\%)$$

Compared to the mean yield under the grower's practice, all of the increases are greater than 0.43. Hence the improved package of practices and the new selections produced highly significant increases over the grower's practice. Consideration of the means in Table 9.5 and the increases in Table 9.7 leads to the conclusion that selection S_1 shows promise for future study.

9.5 CROP TRIAL WITH TWO-COMPONENT IMPROVED PRACTICE

Often, the package of improved practices consists of two types of input. For example, the inputs might consist of a new crop variety and the use of an herbicide to control weeds. In this case it might be interesting to determine whether any improvement from the new practice is the result of

Table 9.7 **Mean Yield Increase (T/ha) of Other Treatments Compared to the Grower's Practice**[a]

Treatment	N	S_1	S_2	S_3	S_4
Increase	1.68	2.24	1.56	0.92	1.09

[a]Standard error = 0.17.

one or the other of the components or whether both components are acting together. Some information on this question can be obtained by a modification of the basic design. To do this we add two treatments to the basic design and construct, in effect, a 2 × 2 factorial set of treatments. We then look at the main effects and the interaction of the components when we analyze the data.

9.5.1 Trial Layout

Suppose that the improved practice consists of two components: A and B. The A component, for example, might be a new variety, and the B component might be an herbicide. The trial would be conducted using four treatments:

F = farmer's practice
N = improved practice = new variety + herbicide
A = component A alone = new variety without herbicide
B = component B alone = herbicide + farmer's variety

A number of farms are selected within the target area. With this design the minimum number of locations could be reduced to five. At each location the experimental design would consist of two large plots on which treatments F and N would be applied, and two small plots on which treatments A and B would be used. A field plan for the two-component design is illustrated in Figure 9.3.

All of the plots should be located in the same field but they need not be adjacent to each other. The goal in locating the plots is to select plots that are as nearly alike as possible before the treatments are applied.

9.5.2 Data Analysis

Suppose that the two-component trial is conducted at *n* locations. Let y_{ij} represent the yield (or other measure) per unit area from the j-th treatment at the i-th location.

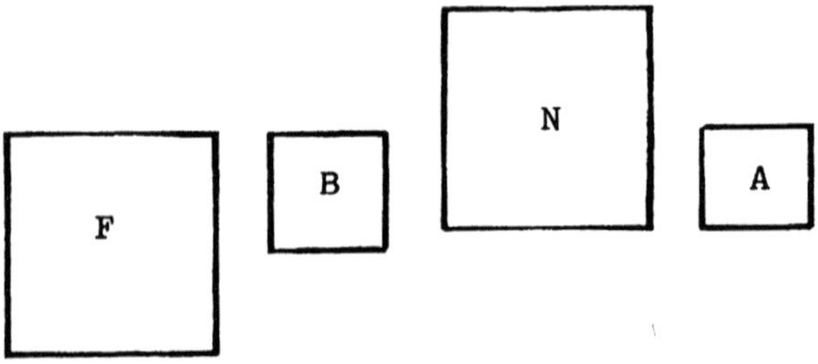

Figure 9.3 Field layout for an on-farm FSR trial with a two-component improved practice.

(*Note*: Because there are different plot sizes in the trial the yields must be expressed on a per unit area basis to be comparable.)

The data may be tabulated as shown in Table 9.8. The data in Table 9.8 are subjected to an analysis of variance that takes the format given in Table 9.9. The entries in this table are computed as follows:

$$\text{SSTot} = \sum_i \sum_j y_{ij}^2 - \frac{G^2}{4n}$$

$$\text{SSL} = \frac{1}{4} \sum_i L_i^2 - \frac{G^2}{4n}$$

$$\text{SST} = \frac{1}{n} \sum_j T_j^2 - \frac{G^2}{4n}$$

$$\text{SSE} = \text{SSTot} - \text{SSL} - \text{SST}$$

$$\text{MSE} = \frac{\text{SSE}}{3(n-1)}$$

(*Note*: The principal reason for completing the ANOVA is to obtain MSE, the experimental error mean square. For this analysis none of the other mean squares need to be computed.)

The standard error, $s_{\bar{y}}$, of a treatment mean is computed as

$$s_{\bar{y}} = \sqrt{\frac{\text{MSE}}{n}}$$

Similarly, the standard error, $s_{\bar{d}}$, of the difference between two treatment means is

$$s_{\bar{d}} = \sqrt{\frac{2\text{MSE}}{n}}$$

The mean increases, $\bar{d}_j$, of the other treatments compared to the farmer's

Table 9.8 **Data Table for an On-Farm FSR Trial with a Two-Component Improved Practice**[a]

Location	Treatment F	N	A	B	Sum
1	y_{11}	y_{12}	y_{13}	y_{14}	L_1
2	y_{21}	y_{22}	y_{23}	y_{24}	L_2
.	.	.	.	.	.
.	.	.	.	.	.
.	.	.	.	.	.
n	y_{n1}	y_{n2}	y_{n3}	y_{n4}	L_n
Sum	T_1	T_2	T_3	T_4	G
Mean	$\bar{y}_1$	$\bar{y}_2$	$\bar{y}_3$	$\bar{y}_4$	

[a] $L_i = \Sigma_j y_{ij}$ = total yield at location i

$T_j = \Sigma_i y_{ij}$ = total yield from the j-th treatment

$G = \Sigma_j L_i = \Sigma_j T_j$ = grand total yield

$\bar{y}_j = T_j/n$ = mean yield of the j-th treatment

Table 9.9 **Format for the Analysis of Variance of Data from an On-Farm FSR Trial with a Two-Component Improved Practice**

ANOVA

Source	df	SS	MS
Total	4n−1	SSTot	
Location	n−1	SSL	
Treatment	3	SST	
Error	3(n−1)	SSE	MSE

practice are computed by subtracting the mean for the farmer's practice, $\bar{y}_1$, from the means of the other treatments, $\bar{y}_j$:

$$\bar{d}_j = \bar{y}_j - \bar{y}_1$$

The smallest increase that will be significant, LSI, is computed as

$$LSI_{0.05} = t_{0.05}\sqrt{\frac{2MSE}{n}} \quad \text{at the 5\% level}$$

$$LSI_{0.01} = t_{0.01}\sqrt{\frac{2MSE}{n}} \quad \text{at the 1\% level}$$

where $t_{0.05}$, $t_{0.01}$ are, respectively, the 0.05, 0.01 tabled, one-tailed t values at 3(n − 1) df.

If $\bar{d}_j > LSI_{0.05}$, the increase is significant (5%).
If $\bar{d}_j > LSI_{0.01}$, the increase is highly significant (1%).
If $\bar{d}_j < LSI_{0.05}$, the increase is not significant.

9.5.3 Components of the Improved Practice

The two components of the improved practice may act independently or they may interact with each other to produce an effect. To determine which is the case, we can partition the treatment source of variation into three parts: the main effect of each component, and their interaction effect. For this purpose we use the concept of contrasts that we introduced in Chapter 4. Recall that the four treatments constitute a 2 × 2 factorial set, and use the coefficients, k_{ij}, in Table 9.10 to accomplish the partition.

Using these coefficients the computations are

$$E_A = -T_1 + T_2 + T_3 - T_4 \qquad S_A^2 = \frac{E_A^2}{4n}$$

$$E_B = -T_1 + T_2 - T_3 + T_4 \qquad S_B^2 = \frac{E_B^2}{4n}$$

$$E_{AB} = +T_1 + T_2 - T_3 - T_4 \qquad S_{AB}^2 = \frac{E_{AB}^2}{4n}$$

We can now test the significance of the main effects and interactions. For this purpose we use an F test.

$F_A = S_A^2/MSE$, with 1, 3(n − 1) df, tests the significance of component A acting alone.
$F_B = S_B^2/MSE$, with 1, 3(n − 1) df, tests the significance of component B acting alone.

Table 9.10 **Contrasts Defining the Partition of Treatment Sums of Squares into Main Effects and Interactions for an On-Farm FSR Trial with a Two-Component Improved Practice**[a]

Effect \ Total	Treatment F	Treatment N	Treatment A	Treatment B		
	T_1	T_2	T_3	T_4	E_i	S_i^2
A	−1	+1	+1	−1	E_A	S_A^2
B	−1	+1	−1	+1	E_B	S_B^2
AB	+1	+1	−1	−1	E_{AB}	S_{AB}^2

[a]T_j = total yield for treatment j
E_i = i-th factorial effect = $\Sigma_j k_{ij}T_j$
S_i^2 = sum of squares (1-df) associated with effect i
= $E_i^2/4n$

$F_{AB} = S_{AB}^2/MSE$, with 1, 3(n − 1) df, tests the significance of the interaction of components A and B.

If F_{AB} is significant, the two components are not acting independently. In this case the increase from the improved practice depends on both components being in the package.

If F_{AB} is not significant, we consider F_A and F_B. If either of these is significant, we can look at the increase from the significant component and we need not be concerned about the presence or absence of the other component.

9.5.4 Numerical Example

An on-farm FSR trial was conducted in the Aleppo Province of Syria to evaluate a two-component improved package in a wheat-growing target area. The two components consisted of (1) a new, improved wheat variety, and (2) herbicide weed control. The trial was conducted on six farms chosen at random in the target area. At each farm there were two large plots, assigned to treatments F and N, and two small plots, assigned to treatments A and B. The treatments were as follows:

F = farmer's variety, no herbicide
N = new variety + herbicide
A = new variety, no herbicide
B = farmer's variety + herbicide

The resulting wheat yields are presented in Table 9.11. The analysis of variance of the yield data in Table 9.11 is given in Table 9.12. The increase from the new practice and its components compared to the farmer's practice is shown in Table 9.13.

For this trial the least significant increase (LSI) is

$$LSI_{0.05} = t_{0.05}\sqrt{\frac{2MSE}{6}} = 1.75\sqrt{\frac{(2)(130{,}524)}{6}} = 365$$

$$LSI_{0.01} = t_{0.01}\sqrt{\frac{2MSE}{6}} = 2.60\sqrt{\frac{(2)(130{,}524)}{6}} = 542$$

Since 758 > 542, the increase from the improved practice is highly

Table 9.11 **Wheat Yields (kg/ha) from an On-Farm FSR Trial Involving a Two-Component Improved Practice**

	Treatment				
Location	F	N	A	B	Sum
1	4100	4030	4365	4030	16525
2	3760	4165	4030	4030	15985
3	3895	3895	4100	3965	15855
4	3030	4570	4100	3360	15060
5	3695	4760	4030	3630	16115
6	2890	4500	3360	3025	14755
Sum	21370	25920	23985	22040	93315
Mean	3562	4320	3998	3673	

Table 9.12 **Analysis of Variance of Wheat Yield (kg/ha) Data from an On-Farm FSR Trial with a Two-Component Improved Practice**

ANOVA

Source	df	SS	MS
Total	23	5,300,691	
Location	5	1,235,697	
Treatment	3	2,107,136	
Error	15	1,957,858	130,524

Table 9.13 **Increase (kg/ha) from the Improved Practice and Its Components Compared to the Farmer's Practice**

Treatment	N	A	B
Increase	758	436	111

significant, and because 542 > 436 > 365, the increase from the new variety alone is significant. Further, since 111 < 365, the increase from herbicide is not significant.

Main effects and interactions of the components of the improved practice are examined using the method of contrasts shown in Table 9.14. As an example of the computations in Table 9.14 we have

$$E_A = (-21{,}370) + 25{,}920 + 23{,}985 + (-22{,}040) = 6495$$

$$S_A^2 = \frac{6{,}495^2}{(6)(4)} = \frac{42{,}185{,}025}{24} = 1{,}757{,}709$$

(*Note*: 1,757,709 + 282,751 + 66,676 = 2,107,136 = SST.) The significance of these effects is tested using an F test.

$$F_A = \frac{1{,}757{,}709}{130{,}524} = 13.47$$

$$F_B = \frac{282{,}751}{130{,}524} = 2.17$$

Table 9.14 **Contrasts Defining Main Effects and the Interaction of New Variety and Herbicide in a Two-Component On-Farm FSR Trial**

Total	Treatment F	N	A	B		
Effect	21370	25920	23985	22040	E	s^2
A	−1	+1	+1	−1	6,495	1,757,709
B	−1	+1	−1	+1	2,605	282,751
AB	+1	+1	−1	−1	1,265	66,676

$$F_{AB} = \frac{66{,}676}{130{,}524} = 0.51$$

$$F_{0.01(1,15)} = 8.68 \qquad F_{0.05(1,15)} = 4.54$$

Since $0.51 < 4.54$, the interaction between variety and herbicide is not significant, and because $2.17 < 4.54$, the main effect of herbicide is not significant. However, since $13.47 > 8.68$, the main effect of variety is highly significant.

The results of the two-component on-farm FSR trial are summarized in Table 9.15. In the table it is seen that on the average, the improved variety outyielded the farmer's variety by about 540 kg/ha. On the other hand, herbicide had no real effect on yield, nor did the increase from the new variety depend on whether or not herbicide was used.

Table 9.15 **Mean Yield (kg/ha) of Wheat Under Farmer's Variety and Improved Variety of Wheat**[a]

Variety	Farmer's	Improved	Std. error
Mean yield	4,159	3,618	104.3

[a]Standard error = MSE/2n = 130,524/12 = 104.3.

9.6 CROP TRIAL WITH THREE-COMPONENT IMPROVED PRACTICE

Occasionally, the package of improved practices will contain three components. For example, the inputs might include (1) an improved variety, (2) an herbicide, and (3) an application of fertilizer. Again, we might be interested in the response to the components individually as well as the interactions among them. As with the two-component design we can obtain information on the three components by using the principle of factorial treatment combinations to add treatments to the basic design.

9.6.1 Trial Layout

Suppose that the improved practice includes three components: A (variety), B (herbicide), and C (fertilizer). The trial would be conducted using eight treatments:

F = farmer's practice
N = improved package of practices
= new variety + herbicide + fertilizer
A = component A alone
= new variety without herbicide or fertilizer
B = component B alone
= herbicide + farmer's variety without fertilizer
AB = both components A and B
= new variety + herbicide without fertilizer
C = component C alone
= fertilizer + farmer's variety without herbicide
AC = both components A and C
= new variety + fertilizer without herbicide
BC = both components B and C
= farmer's variety + herbicide + fertilizer

Several farms are selected in the target area. With this design the minimum number of locations would be four, although more locations should be used, if possible, to provide expanded coverage of the target area. On each farm the design would consist of two large plots, one for treatment F and the other for treatment N, and six small plots, one each for treatments A, B, AB, C, AC, and BC. An example of the field plan for an on-farm FSR trial with a three-component improved practice is shown in Figure 9.4.

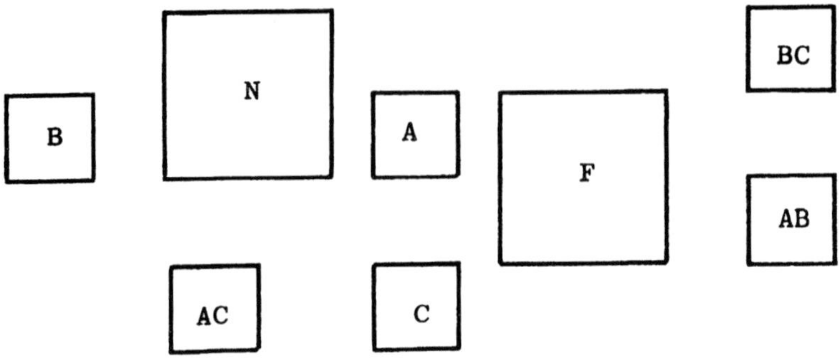

Figure 9.4 Field layout for an on-farm FSR trial with a three-component improved practice.

All of the plots should be located in the same field but need not be placed adjacent to each other. As with any of the on-farm trials, the plots should be as nearly alike as possible before the treatments are applied.

9.6.2 Data Analysis

Suppose that the three-component trial is conducted at n locations. Let y_{ij} represent the yield (or other measure) per unit area from the j-th treatment at the i-th location. (*Note*: Because different plot sizes are used, yields must be expressed on a comparable per unit area basis.) At the conclusion of the trial the results may be summarized as shown in Table 9.16. The analysis of variance of the data from the three-component system has the format given in Table 9.17. The computation of the entries in the analysis of variance, Table 9.17, are as follows:

$$\text{SSTot} = \sum_i \sum_j y_{ij}^2 - \frac{G^2}{8n}$$

$$\text{SSL} = \frac{1}{8} \sum_i L_i^2 - \frac{G^2}{8n}$$

$$\text{SST} = \frac{1}{n} \sum_j T_j^2 - \frac{G^2}{8n}$$

$$\text{SSE} = \text{SSTot} - \text{SSL} - \text{SST}$$

$$\text{MSE} = \frac{\text{SSE}}{7(n-1)}$$

The standard error, $s_{\bar{y}}$, of a treatment mean is obtained as

$$s_{\bar{y}} = \sqrt{\frac{\text{MSE}}{n}}$$

Table 9.16 **Summary Table for an On-Farm FSR Trial with a Three-Component Improved Practice**[a]

Location	Treatment								Sum
	F	N	A	B	AB	C	AC	BC	
1	y_{11}	y_{12}	y_{13}	y_{14}	y_{15}	y_{16}	y_{17}	y_{18}	L_1
2	y_{21}	y_{22}	y_{23}	y_{24}	y_{25}	y_{26}	y_{27}	y_{28}	L_2
.	.	.	.	.	.	.	.	.	.
.	.	.	.	.	.	.	.	.	.
.	.	.	.	.	.	.	.	.	.
n	y_{n1}	y_{n2}	y_{n3}	y_{n4}	y_{n5}	y_{n6}	y_{n7}	y_{n8}	L_n
Sum	T_1	T_2	T_3	T_4	T_5	T_6	T_7	T_8	G
Mean	$\bar{y}_1$	$\bar{y}_2$	$\bar{y}_3$	$\bar{y}_4$	$\bar{y}_5$	$\bar{y}_6$	$\bar{y}_7$	$\bar{y}_8$	

[a]$L_i = \Sigma_j y_{ij}$ = total yield at location i
$T_j = \Sigma_i y_{ij}$ = total yield of treatment j
$G = \Sigma_i L_i = \Sigma_j T_j$ = grand total yield
$\bar{y}_j = T_j/n$

Table 9.17 **ANOVA Format for Data from a Three-Component Improved Practice On-Farm Trial**

ANOVA

Source	df	SS	MS
Total	8n−1	SSTot	
Location	n−1	SSL	
Treatment	7	SST	
Error	7(n−1)	SSE	MSE

and the standard error, $s_{\bar{d}}$, of the difference between two means is

$$s_{\bar{d}} = \sqrt{\frac{2MSE}{n}}$$

The increase, $\bar{d}_j$, of the other treatments over the farmer's practice is obtained by subtracting the mean, $\bar{y}_1$, of the farmer's practice from the mean, $\bar{y}_j$, of the other treatment:

$$\bar{d}_j = \bar{y}_j - \bar{y}_1$$

The least significant increase (LSI) is computed as

$$LSI_{0.05} = t_{0.05}\sqrt{\frac{2MSE}{n}} \quad \text{at the 5\% level}$$

$$LSI_{0.01} = t_{0.01}\sqrt{\frac{2MSE}{n}} \quad \text{at the 1\% level}$$

where $t_{0.05}$ and $t_{0.01}$, are, respectively, the 0.05 and 0.01 one-tailed tabular t values at $7(n - 1)$ degrees of freedom.

If $\bar{d}_j > LSI_{0.05}$, the increase is significant (5%).
If $\bar{d}_j > LSI_{0.01}$, the increase is highly significant (1%).
If $\bar{d}_j < LSI_{0.05}$, the increase is not significant.

9.6.3 Components of the Improved Practice

To determine whether the three components interact or are acting independently, we partition the treatment source of variation into the factorial effects. For this purpose we use a table of contrasts that is an extension of the set of contrasts used in the two-component analysis. The coefficients, k_{ij}, for these contrasts and their associated sums of squares are presented in Table 9.18. The computation of the entries in the last two columns of this table are

$$E_A = -T_1 + T_2 + T_3 - T_4 + T_5 - T_6 + T_7 - T_8 \qquad S_A^2 = \frac{E_A^2}{8n}$$

$$E_B = -T_1 + T_2 - T_3 + T_4 + T_5 - T_6 - T_7 + T_8 \qquad S_B^2 = \frac{E_B^2}{8n}$$

$$\vdots \qquad\qquad \vdots$$

$$E_{ABC} = -T_1 + T_2 + T_3 + T_4 - T_5 + T_6 - T_7 - T_8 \qquad S_{ABC}^2 = \frac{E_{ABC}^2}{8n}$$

Table 9.18 **Contrasts Defining the Partition of Treatment Sum of Squares into Main Effects and Interactions in a Trial with a Three-Component Improved Practice**[a]

Effect \ Total	F	N	A	B	AB	C	AC	BC		
	T_1	T_2	T_3	T_4	T_5	T_6	T_7	T_8	E_i	S_i^2
A	−1	+1	+1	−1	+1	−1	+1	−1	E_A	S_A^2
B	−1	+1	−1	+1	+1	−1	−1	+1	E_B	S_B^2
AB	+1	+1	−1	−1	+1	+1	−1	−1	E_{AB}	S_{AB}^2
C	−1	+1	−1	−1	−1	+1	+1	+1	E_C	S_C^2
AC	+1	+1	−1	+1	−1	−1	+1	−1	E_{AC}	S_{AC}^2
BC	+1	+1	+1	−1	−1	−1	−1	+1	E_{BC}	S_{BC}^2
ABC	−1	+1	+1	+1	−1	+1	−1	−1	E_{ABC}	S_{ABC}^2

Treatment columns: F, N, A, B, AB, C, AC, BC.

[a]T_j = total yield for treatment j
E_i = i-th factorial effect = $\Sigma_j k_{ij} T_j$
S_i^2 = sum of squares (1 df) associated with the i-th factorial effect
= $E_i^2/8n$

We can now test the significance of the main effects and interactions. For this we use an F test of the form

$$F_i = \frac{S_i^2}{MSE} \qquad \text{with } 1,7(n-1) \text{ df}$$

We begin by testing the three-component, ABC, interaction using

$$F_{ABC} = \frac{S_{ABC}^2}{MSE}$$

If this test is significant, the components are not acting independently and we interpret our results on the basis of the individual treatments considered separately.

If F_{ABC} is not significant, we can test the two-component interactions:

$$F_{AB} = \frac{S^2_{AB}}{MSE}$$

$$F_{AC} = \frac{S^2_{AC}}{MSE}$$

$$F_{BC} = \frac{S^2_{BC}}{MSE}$$

If any of these tests is significant, F_{AB} for example, the response to changes in one component, A, depends on the presence or absence of B. In this case the interpretation is based on the AB combination means.

If any of the two-component interactions test nonsignificant, we test the main effects of the two components themselves. For example, if F_{AB} is not significant, we test

$$F_{A} = \frac{S^2_{A}}{MSE}$$

$$F_{B} = \frac{S^2_{B}}{MSE}$$

If any of these tests is significant, that component has a real effect independent of the other components. We can interpret the effect of that component without having to consider whether or not the other components are present.

9.6.4 Numerical Example

An on-farm FSR trial was conducted in Baluchistan to compare yield of lentils under an improved system with the yield under the normal farmer's practices. The improved practice included three components: A = addition of potassium fertilizer (K), B = addition of phosphorus fertilizer (P), and C = reduced seed rate (R).

Because of budgetary restrictions the trial was conducted at only three locations. At each location there were two large plots, one of which was assigned to the new practice and the other of which was assigned to the farmer's practice. In addition, six small plots were included in the field with the large plots. On these small plots the following treatments were used:

A = K without P at the farmer's seed rate

B = P without K at the farmer's seed rate

AB = K + P at the farmer's seed rate

C = reduced seed rate without K or P

AC = reduced seed rate + K without P

BC = reduced seed rate + P without K

Table 9.19 **Yield (T/ha) of Lentils from an On-Farm FSR Trial with a Three-Component Improved Practice**

Location	Treatment								Sum
	F	N	A	B	AB	C	AC	BC	
1	0.6	2.8	1.0	1.8	1.5	1.3	1.7	2.2	12.9
2	1.3	3.2	1.7	1.6	2.0	1.3	1.8	2.0	14.9
3	1.1	3.1	1.8	2.3	2.2	2.2	1.9	2.7	17.3
Sum	3.0	9.1	4.5	5.7	5.7	4.8	5.4	6.9	45.1
Mean	1.00	3.03	1.50	1.90	1.90	1.60	1.80	2.30	

(*Note*: F = farmer's seed rate without K or P, and N = K + P at the reduced seed rate.)

Lentil yields are shown in Table 9.19, and the analysis of variance of the yields is given in Table 9.20. The mean increases, $\bar{d}_j$, from the improved package and its components as compared to the farmer's practice are shown in Table 9.21.

Table 9.20 **Analysis of Variance of Lentil Yields (T/ha) from a Trial with a Three-Component Improved Practice**

ANOVA

Source	df	SS	MS
Total	23	9.5996	
Location	2	1.2133	
Treatment	7	7.5329	
Error	14	.8534	.0610

Table 9.21 **Mean Increase in Yield (T/ha) of Lentils from a Package of Improved Practices and Its Components Compared to the Farmer's Practice**[a]

Treatment	N	A	B	AB	C	AC	BC
Mean increase	2.03	.50	.90	.90	.60	.80	1.30

[a]Standard error = 0.1426.

The least significant increases (LSI) for this trial are:

$$LSI_{0.05} = t_{0.05}\sqrt{\frac{2MSE}{n}} = 1.76\sqrt{\frac{(2)(0.0610)}{3}} = 0.35$$

$$LSI_{0.01} = t_{0.01}\sqrt{\frac{2MSE}{n}} = 2.62\sqrt{\frac{(2)(0.0610)}{3}} = 0.53$$

Increases from all of the treatments except A are greater than 0.53. Hence increases from all treatments except A are highly significant (1%). Since $0.35 < 0.50 < 0.53$, the increase from treatment A is significant at the 5% level.

Main effects and interactions among the components of the improved practice are examined using the table of contrasts and their associated sums of squares as presented in Table 9.22.

On comparing the F value in Table 9.22 with the critical F values, we see that the three-component interaction is significant at the 5% level. None of the two-component interactions is significant, but all of the main effects are highly significant.

Although there is some indication that the three components are not acting independently, it is clear that the individual components of the improved practice are effective. Further, this effect is present regardless of the presence or absence of the other components. The results of the trial may be summarized by the means shown in Table 9.23. From this table it is seen that the addition of potassium (K) fertilizer resulted in an average increase of 0.36 T/ha, the addition of phosphorus (P) produced, on the average, an increase of 0.80 T/ha, and the average increase from the improved seed rate was 0.60 T/ha.

9.7 ANIMAL TRIAL WITH TWO FEEDING PERIODS

The choice of an experimental design for an on-farm trial involving animals depends on the treatments to be applied to the animals. When the treatment

Table 9.22 **Main Effect and Interaction Contrasts for Data from an On-Farm FSR Trial with a Three-Component Improved Practice**[a]

Effect \ Total	F	N	A	B	AB	C	AC	BC	E_i	s_i^2	F_i
	3.0	9.1	4.5	5.7	5.7	4.8	5.4	6.9			
A	−	+	+	−	+	−	+	−	4.3	.7704	12.63
B	−	+	−	+	+	−	−	+	9.7	3.9204	64.27
AB	+	+	−	−	+	+	−	−	.1	.0004	.01
C	−	+	−	−	−	+	+	+	7.3	2.2204	36.40
AC	+	+	−	+	−	−	+	−	1.3	.0704	1.15
BC	+	+	+	−	−	−	−	+	1.9	.1504	2.47
ABC	−	+	+	+	−	+	−	−	3.1	.4004	6.56

[a]$F_{0.05(1,14)} = 4.60$; $F_{0.01(1,14)} = 8.86$.

Table 9.23 **Mean Yields (T/ha) of Lentils With and Without K, P, and Improved Seed Rate in an On-Farm FSR Trial with a Three-Component Improved Practice**[a]

Component	Practice: Farmer	Practice: Improved
K	1.70	2.06
P	1.48	2.28
Seed rate	1.58	2.18

[a]Standard error $= \sqrt{MSE/4n} = \sqrt{0.0610/12} = 0.0713$.

involves a single application of an improved practice, such as a vaccination or a drench, the experimental design is the same as the basic, two-treatment design used for crops. A number of farms on which livestock constitute an enterprise are selected at random from the target area. In this case the suggested minimum number of farms is 10. The animals on each farm are divided at random into two groups. One of the groups, F, is handled by the farmer in his usual way. The other group, N, receives the improved practice. The response to treatment is then expressed on a per animal per group basis on each farm. These average responses, one for group F and one for group N, are analyzed in the same way as the basic yield data from a crops trial. This analysis is described in Section 9.3.2.

When the practice involves treating the animals in a particular way over a period of time, such as the introduction of a new feed, a different type of experimental design is to be preferred to the basic design. This is a particular example of a class of designs called *changeover designs*, designs that are balanced to remove the effect of changes in the animals' performance over time. These designs have been described by Patterson and Lucas (1962) and Petersen (1985), among others.

9.7.1 Description

The experimental design for on-farm FSR trials involving the treatment of animals over time is called a crossover, or simple reversal, design. This is a design that involves two treatments: farmer's practice (F) and improved practice (N). Each trial lasts for two treatment periods. At each farm the animals to be used are divided at random into two groups. In one group the animals receive the improved practice during the first period and the farmer's practice during the second period. In the second group the animals receive the treatments in the sequence farmer's practice followed by the improved practice. The basic pattern is

	Group	
Period	I	II
1	N	F
2	F	N

The observation periods should be of equal length and for most measures should be a minimum of 4 weeks' duration. The same number of animals should be used in each group, although the data analysis is still valid if the groups are of unequal size.

Records are kept for each animal for each period. At the conclusion of the trial the mean performance per animal is determined for each group during each period on every farm. These means are the data to be used in the analysis.

9.7.2 Data Analysis

Suppose that the trial is conducted on n farms. Let y_{ijk} represent the mean performance per animal for the i-th group during j-th period on the k-th farm. The data might be tabulated as shown in Table 9.24. The analysis of variance of data from a crossover design has only two lines. The format

Table 9.24 **Performance per Animal for a Two-Period On-Farm FSR Trial Using Animals**[a]

Group I		Location				
Period	Treat.	1	2	...	n	Sum
1	N	y_{111}	y_{112}	...	y_{11n}	P_{11}
2	F	y_{121}	y_{122}	...	y_{12n}	P_{12}
Difference		D_{11}	D_{12}	...	D_{1n}	G_1

Group II		Location				
Period	Treat.	1	2	...	n	
1	F	y_{211}	y_{212}	...	y_{21n}	P_{21}
2	N	y_{221}	y_{222}	...	y_{22n}	P_{22}
Difference		D_{21}	D_{22}	...	D_{2n}	G_2

[a]D_{ik} = difference between first and second periods for the i-th group on the k-th farm

$= y_{i1k} - y_{i2k}$

P_{ij} = total for the i-th group during the j-th period

$= \Sigma_k\, y_{ijk}$

Gi = total difference between first and second periods for the i-th group

$= P_{i1} - P_{i2} = \Sigma_k\, D_{ik}$

for this analysis is given in Table 9.25. The entries in this table are computed in the following way:

$$SST = \frac{(G_1 - G_2)^2}{4n}$$

$$SSE = \frac{1}{2}\sum_i \sum_k D_{ik}^2 - \frac{1}{2n}(G_1^2 + G_2^2)$$

$$MSE = \frac{SSE}{2(n-1)}$$

The standard error, $s_{\bar{y}}$, of a treatment mean is

$$s_{\bar{y}} = \sqrt{\frac{MSE}{2n}}$$

The standard error, $s_{\bar{d}}$, of the difference between the two treatment means is computed as

$$s_{\bar{d}} = \sqrt{\frac{MSE}{n}}$$

Treatment means are computed in the following way:

$$\bar{y}_N = \bar{\bar{y}} + d = \text{mean of the improved practice}$$
$$\bar{y}_F = \bar{\bar{y}} - d = \text{mean of the farmer's practice}$$

where

$$\bar{\bar{y}} = \frac{1}{4}\sum_i \sum_j P_{ij}$$

$$d = \frac{G_1 - G_2}{4n}$$

Table 9.25 **Analysis of Variance of Data from an On-Farm FSR Crossover Trial**

ANOVA

Source	df	SS	MS
Treatment	1	SST	
Error	2(n−1)	SSE	MSE

The mean increase, I_N, from the improved practice compared to the farmer's practice is

$$I_N = \bar{y}_N - \bar{y}_F = 2d$$

The minimum difference required for significance, LSI, is found to be

$$LSI_{0.05} = t_{0.05} \sqrt{\frac{MSE}{n}}$$

$$LSI_{0.01} = t_{0.01} \sqrt{\frac{MSE}{n}}$$

in which $t_{0.05}$ and $t_{0.01}$ are, respectively, the 0.05 and 0.01 one-tailed tabular t values at $2(n - 1)$ degrees of freedom.

If $I_N > LSI_{0.05}$, the increase is significant (5%).
If $I_N > LSI_{0.01}$, the increase is highly significant (1%).
If $I_N < LSI_{0.05}$, the increase is not significant.

9.7.3 Numerical Example

A feeding trial was conducted to measure the gain produced by feeding a protein supplement along with the farmer's usual feed. The two treatments in the trial were

F = farmer's ration
N = farmer's ration + protein supplement

Flocks of sheep at four farms were divided into two equal groups on each farm. One group was fed the farmer's ration, F, while the other group received the farmer's ration plus the protein supplement, N. The trial lasted for two periods of 30 days each. A crossover experimental design was used. On each farm the average gain per animal was measured at the end of each period for both groups.

The mean gains (kg) per animal per period are presented in Table 9.26. The analysis of variance of these mean gains is given in Table 9.27. The computations for the entries in this table are

$$SST = \frac{(G_1 - G_2)^2}{4n} = \frac{[92 - (-69)]^2}{(4)(4)} = 1620.06$$

$$SSE = \frac{1}{2} \sum_i \sum_k D_{ik}^2 - \frac{1}{2n} (G_1^2 + G_2^2)$$

$$= \frac{1}{2} [28^2 + 17^2 + \cdots + (-21)^2] - \frac{1}{(2)(4)} [92^2 + (-69)^2]$$

$$= 1742.50 - 1653.12 = 89.38$$

$$\text{MSE} = \frac{\text{SSE}}{2(n-1)} = \frac{89.38}{(2)(3)} = 14.90$$

The standard error, $s_{\bar{y}}$, of a treatment mean for this trial is

$$s_{\bar{y}} = \sqrt{\frac{\text{MSE}}{2n}} = \sqrt{\frac{14.90}{(2)(4)}} = 1.36$$

while the standard error of the difference, $s_{\bar{d}}$, between treatment means is

$$s_{\bar{d}} = \sqrt{\frac{\text{MSE}}{n}} = \sqrt{\frac{14.90}{4}} = 1.93$$

To compute the treatment means we first find

$$\bar{\bar{y}} = \frac{1}{4n} \sum_i \sum_j P_{ij} = \left[\frac{1}{(4)(4)}\right](354 + \cdots + 381) = 81.81$$

and

$$d = \frac{G_1 - G_2}{4n} = \frac{92 - (-69)}{(4)(4)} = 10.06$$

Then the mean gain, $\bar{y}_N$, for the farmer's ration plus protein supplement is

$$\bar{y}_N = \bar{y} + d = 81.81 + 10.06 = 91.87$$

while the mean gain, $\bar{y}_F$, for the farmer's ration without the supplement is

$$\bar{y}_F = \bar{\bar{y}} - d = 81.81 - 10.06 = 71.75$$

The mean increase, I_N, from the addition of protein to the ration is

$$I_N = \bar{y}_N - \bar{y}_F = 2d = (2)(10.06) = 20.12$$

The least significant increases (LSI) are

$$\text{LSI}_{0.05} = t_{0.05}\sqrt{\frac{\text{MSE}}{n}} = 1.94\sqrt{\frac{14.90}{4}} = 3.74 \quad (5\%)$$

$$\text{LSI}_{0.01} = t_{0.01}\sqrt{\frac{\text{MSE}}{n}} = 3.14\sqrt{\frac{14.90}{4}} = 6.06 \quad (1\%)$$

Since 20.12 > 6.06, the increase from added protein is highly significant. The addition of a protein supplement to the farmer's ration increased sheep gain by about 20 kg per animal over a 30-day feeding period.

Table 9.26 **Mean Gain (kg) per Animal Fed Rations With and Without Protein Supplement in a Crossover FSR Trial**[a]

Group I		Farm				
Period	Treat.	1	5	3	4	Sum
1	N	98	89	85	82	354
2	F	70	72	61	59	262
Difference		28	17	24	23	92

Group II		Farm				
Period	Treat.	1	2	3	4	Sum
1	F	72	83	90	67	312
2	N	96	96	101	88	381
Difference		−24	−13	−11	−21	−69

[a]For example,

$D_{11} = y_{111} - y_{121} = 98 - 70 = 28$

$P_{11} = \Sigma_k y_{11k} = 98 + 89 + 85 + 82 = 354$

$G_1 = P_{11} - P_{12} = 354 - 262 = 92$

Table 9.27 **Analysis of Variance of Daily Gains by Sheep in a 30-Day Period Crossover FSR Trial**

ANOVA

Source	df	SS	MS
Treatment	1	1620.06	
Error	6	89.38	14.90

REFERENCES

Collinson, M. P., 1987. Farming systems research: procedures for technology development. *Exp. Agric.* 23: 365–386.

Patterson, H. D., and H. L. Lucas, 1962. Change-over trials. *N.C. Agric. Exp. Stn. Tech. Bull. 147.*

Perrin, R. K., D. L. Winkleman, E. R. Moscardi, and J. R. Anderson, 1976. From agronomic data to farmer recommendations. *Inf. Train. Bull. 21.* CIMMYT, El Bataan, Mexico.

Petersen, R. G., 1985. *Design and Analysis of Experiments.* Marcel Dekker, New York, Chap. 12.

Sands, D. M., 1986. Farming systems research: clarification of terms and concepts. *Exp. Agric.* 22:87–104.

10

Intercropping Research

10.1 INTRODUCTION

Intercropping is a farming system that involves growing two or more crops together on the same area of land. The crops need not be planted at exactly the same time and their harvest times might be quite different, but they occupy the same land area together for a significant part of the growing season (Willey, 1979a). The crops are usually harvested separately and the yields of the crops are kept separate.

Intercropping is practiced primarily by subsistence farmers in the tropics and subtropics. It is used in Africa, India, and other parts of Asia and in South and Central America. Mead and Riley (1981) in a review of intercropping research point out that 98% of the cowpeas in Africa, 90% of the beans in Columbia, 73% of the beans in Guatemala, 80% of the beans in Brazil, and 60% of the maize in Latin America are produced in an intercropping enterprise.

The usual intercrop is a cereal–legume mixture. Maize, pearl millet, and sorghum are widely used as the cereal component of the intercrop, while cowpea, chickpea, and pigeon pea constitute the legume component. Other crops, including groundnut, coconut, castor, wheat, alfalfa, potatoes, radishes, cotton, sugarcane, cassava, oats, and clover, have been intercropped as well.

Although intercropping has been practiced for many years, there has been very little research on this farming system until relatively recently. This is probably because of the complexity of handling more than one crop simultaneously, the difficulty of introducing machinery into the system, and the difficulty of assessing the results of the research. With the advent of the Consultative Group on International Agricultural Research (CGIAR) and the development of the system of International Agricultural Research Centers, research on intercropping has expanded. This is particularly true of the International Crops Research Institute for the Semi-Arid Tropics (ICRISAT), based in Hyderabad, India.

10.1.1 Terminology

A number of terms, "intercropping," "mixed cropping," and "relay cropping," have been used to describe a farming system in which two or more crops are grown together in the same place for some part of their life cycle (Willey, 1979a). *Relay cropping* is often used to describe a situation in which one crop closely follows another, with possibly some brief overlap, on the same piece of land. This is not the system described in this chapter, so we will not consider it further. *Mixed cropping* usually refers to a system in which the crops are intermingled, and *intercropping* refers to a system in which the crops are grown in separate rows or other regular spacing arrangements. The material in this chapter is restricted to intercropping as defined here, and we will not consider mixed cropping.

We use the term *component crop* to refer to the individual crops making up the intercrop mixture. The term *sole crop* refers to a component crop grown alone. When sole crops are used in this chapter it is assumed that they are grown at their optimum spacing and population density. The yield of a component crop in an intercrop mixture is called the *component crop yield.* The sum of the component crop yields is often called the *intercrop yield,* and it is expressed over the total intercrop area. The yield of a component crop when grown as a sole crop is called the *sole crop yield.* Although intercropping systems with more than two components have been grown, the more common system is one in which only two crops are included. In this chapter it is assumed, for simplicity and clarity of presentation, that only two-component crops are included in the intercropping system.

10.1.2 Reasons for Intercropping

A number of reasons have been advanced for the use of intercropping in place of sole cropping. The primary reason is that in many instances, intercropping provides greater total yield from a given land area than does a proportional allocation of sole crops to that area. This is particularly true in the subsistence agriculture practiced in the area where intercropping is prevalent. It has also been suggested that intercropping provides better control of weeds, diseases, and pests than does sole cropping. Another reason is the perceived stability of intercropping as compared to sole cropping over a number of seasons. Finally, intercropping might enable the farmer to optimize the input of such things as labor and fertilizer.

Of the advantages attributed to intercropping, perhaps the most important is total yield advantage. In a review of intercropping research, Willey (1979a) indicated that the primary cause of this advantage is the increased use of growth resources by intercrops as compared to sole crops. The resource that seems to be most important in this regard is light. Differences in the canopies of the component crops appear to provide more efficient light use by intercropping than by sole cropping. Other resources that may be more efficiently utilized by intercropping include belowground resources, such as nutrients and water, and nitrogen in combinations that include both a legume and a nonlegume.

Intercropping is said to provide a measure of yield stability that is not present in sole cropping. If one crop fails, or grows poorly, the other component crops can compensate for the failure. Little research is available on the stabilizing effect of intercropping. In addition to a way to measure stability, such research requires experiments at multiple locations over several seasons.

10.1.3 Practical Objectives

Willey (1985) points out that the farmer considering intercropping might have one of a number of practical objectives behind his choice to use an intercropping enterprise:

1. He might wish to maximize total yield. In this case any proportional yield of the components of the intercrop is acceptable as long as the total yield of the intercrop is greater than the highest sole crop yield.
2. He might wish to obtain a given ratio of yields of the component crops. In this case he expects the intercrop to outyield the combination of sole crops that provides the required ratio. He is also prepared to allot part of his land to a sole crop, if necessary, to obtain the required ratio.
3. He might wish to obtain some fixed minimum yield of one or both component crops. In this he might require some of his land to be intercropped while the remainder is sole cropped.
4. As an extension of the previous objective, he might require a full yield of a main crop and some additional yield of another crop. In this

case the component main crop must yield as much in intercropping as it does in sole cropping.

Evaluation of the outcome of an intercropping experiment depends, to some extent, on which of these practical objectives is under consideration.

10.2 ASSESSING THE ADVANTAGES OF INTERCROPPING

Assessing the advantages of intercropping and comparing the yield of intercrops to that of sole crops is not at all straightforward. There is the difficulty of constructing a meaningful total yield from the yields of two different component crops. An alternative to yield for this purpose is to use units, such as monetary units or nutritional values, which may be applied equally to both crops (Willey, 1985). However, this alternative is usually not very satisfactory from the standpoint of the farmer using intercropping. For the farmer, for example, one crop might be a staple food crop while the other is a cash crop, and neither will substitute for the other. Competition among the interplanted crops is another problem. Because of competition the proportion of the total yield from the components in an intercropping system is usually different from that of equivalent sown proportions of sole crops.

10.2.1 Land Equivalent Ratio

A number of measures have been suggested for assessing the output of intercropping. Among these the measure that has received the widest adoption is the land equivalent ratio (LER), proposed by Willey and his associates (Willey and Osiru, 1972; Willey, 1979a; Mead and Willey, 1980). The land equivalent ratio is a standardized index that is defined to be the relative land area required by sole crops to produce the same yield as intercrops.

To clarify the presentation of LER and its variants, as well other measures of intercropping results, it is convenient to define a number of symbols. Suppose that the intercrop consists of two component crops, a and b. Let

Z_a = sown proportion of a

Z_b = sown proportion of b

s_a = yield (per unit area) of a as a sole crop

s_b = yield (per unit area) of b as a sole crop

y_a = yield (per unit area) of a as an intercrop

y_b = yield (per unit area) of b as an intercrop

$$L_a = \frac{y_a}{s_a} = \text{partial LER from a}$$

$$L_b = \frac{y_b}{s_b} = \text{partial LER from b}$$

$$\text{LER} = L_a + L_b = \text{land equivalent ratio}$$

$$P_a = \frac{L_a}{\text{LER}} = \text{proportion of intercrop yield from a}$$

$$P_b = \frac{L_b}{\text{LER}} = 1 - p_a = \text{proportion of intercrop yield from b}$$

Mead and Riley (1981) point out a problem in the use of LER to compare different intercrop treatments. Since LER is defined as the ratio of intercrop to sole crop yields, large values of LER may arise not only from large intercrop yields but also from small sole-crop yields. To address this problem, Mead and Willey (1980) suggest that s_a and s_b be defined as the average or maximum of the sole-crop yields for the intercrop components. Further, Oyejola and Mead (1982) reported that using a single sole-crop yield for each crop as a divisor for computing all LER values in an experiment results in greater precision and a more nearly normal distribution of the LER values than does the use of a separate divisor for each block or treatment. Because of its simplicity and its statistical properties, our preference is to use the sole crop mean for an experiment to compute all LER values for that experiment.

We can illustrate the computations involved using some data cited by Mead and Willey (1980). The data are from an intercropping trial on maize, a, and pigeon pea, b, conducted by ICRISAT in India. These data are given

Table 10.1 **Yields (kg/ha) from an Intercropping Trial on Maize and Pigeon Pea conducted by ICRISAT**

Yields	Maize (a)	Pigeonpea (b)
Sole crop yield, s	3398	1035
Intercrop yield, y	2234	896

in Table 10.1. From these results we can compute the partial LERs as

$$\text{Maize: } L_a = \frac{y_a}{s_a} = \frac{2234}{3398} = 0.66$$

$$\text{Pigeon pea: } L_b = \frac{y_b}{s_b} = \frac{896}{1035} = 0.87$$

Then

$$\text{LER} = L_a + L_b = 0.66 + 0.87 = 1.53$$

The proportions of the total intercrop yield from each of the component crops are

$$P_a = \frac{L_a}{\text{LER}} = \frac{0.66}{1.53} = 0.43$$

$$P_b = \frac{L_b}{\text{LER}} = 1 - P_a = 1 - 0.43 = 0.57$$

10.2.2 Effective Land Equivalent Ratio

When basing recommendations on the LER, it is assumed that the yield proportions from the trial are those which are acceptable to the farmer. Suppose, however, that the farmer requires that a certain proportion of his total yield, different from that implicit in the LER, be obtained from one of the component crops. Suppose, further, that the required proportion is greater than would be obtained if the entire area were to be intercropped. In this case he is obligated to use part of his area for intercropping and part of his area for sole cropping. The question becomes one of determining the proper allocation of sole crop/intercrop to obtain the required proportion.

Mead and Willey (1980) have presented a means of solving this problem. Let a be the crop under consideration and p_a' be the required proportion of a. Assume that $p_a' > p_a$. Let k_I be the proportion of the total area to be in intercrop. Then

$$k_s = 1 - k_I$$

is the proportion of the total area required to be in sole crop.

Under these conditions

$$k_I = \frac{1 - p_a'}{p_a'L_b - (1 - p_a')L_a + (1 - p_a')}$$

where L_a and L_b are the proportions of a and b in the intercrop yield from which the data are derived. Once k_I is known, an effective LER (ELER) can be obtained to express the biological efficiency of the combined sole crop and intercrop. The calculation is as follows:

$$\text{ELER} = \frac{L_b}{(1 - L_a) + (\text{LER} - 1)p_a'}$$

The assumption made in using the ELER approach is that the land allocated to the intercrop will produce at the same level as the sole crop yield used in the calculating L_a. It is also assumed that a is the crop for which the proportion must be increased and hence there must be some sole crop of a.

We can use the ICRISAT data from Table 10.1 to illustrate the computations involved in the ELER. Suppose that the farmer requires that 70% of his total yield be from maize. From Section 10.2.1 we have $p_a = 0.43$, $L_a = 0.66$, $L_b = 0.87$, and LER = 1.53. Under the farmer's requirement 70% of the yield must be from maize. That is, $p_a' = 0.70 > 0.43$ and the proportion of land allocated to intercropping must be reduced to k_I, computed as

$$\begin{aligned} k_I &= \frac{1 - p_a'}{p_a'L_b - (1 - p_a')L_a + (1 - p_a')} \\ &= \frac{1 - 0.70}{(0.70)(0.87) - (1 - 0.70)(0.66) + (1 - 0.70)} \\ &= 0.42 \end{aligned}$$

and

$$k_s = 1 - k_I = 1 - 0.42 = 0.58$$

With this allocation of land the effective land equivalent ratio (ELER) is

$$\begin{aligned} \text{ELER} &= \frac{L_b}{(1 - L_a) + (\text{LER} - 1)p_a'} \\ &= \frac{0.87}{(1 - 0.66) + (1.53 - 1)(0.70)} \\ &= 1.22 \end{aligned}$$

Because of the requirement that 70% of the total yield be from maize the biological efficiency of the cropping system has been reduced from 1.53 to 1.22.

10.2.3 Staple Land Equivalent Ratio

Suppose that a farmer must produce at least a fixed minimum yield of a staple crop but would also like to produce additional yield from a second crop. That is, he might require a minimum percentage of the sole crop yield of a crop, such as a cereal, to provide food for his family while growing another crop as a cash crop.

Suppose that a is the staple crop under consideration and y'_a is the required minimum yield of a. Assume that $y_a < y'_a < s_a$. Then $L'_a = y'_a/s_a$ = required minimum yield of a as a percentage of its sole crop yield. To obtain y'_a the farmer must allocate a portion, k_I, of his land to intercropping and the remainder, k_s, to the sole crop a. Note that

$$k_I + k_s = 1$$

Reddy and Chetty (1984) have shown that

$$k_s s_a + k_I y_a = y'_a$$

and

$$k_I = \frac{s_a - y'_a}{s_a - y_a} = \frac{1 - L'_a}{1 - L_a}$$

where

s_a = sole crop yield of a
y_a = intercrop yield of a
L_a = proportion of intercrop yield from a

Under these conditions Reddy and Chetty (1984) developed an index to measure the biological efficiency of the system. They defined this index to be the staple land equivalent ratio (SLER). They showed that SLER may be computed as

$$\text{SLER} = \text{LER} - \frac{(\text{LER} - 1)(L'_a - L_a)}{1 - L_a}$$

Again, we can use the ICRISAT data in Table 10.1 and the results from Section 10.2.1 to illustrate the SLER computations. Suppose that the farmer must produce at least 75% of the sole-crop maize yield to provide adequate food for his family. Suppose, further, that he would like to produce some pigeon pea as a cash crop. We have a = maize, b = pigeon pea, $s_a = 3398$ kg/ha, $y_a = 2234$ kg/ha, $s_b = 1035$ kg/ha, $y_b = 896$ kg/ha, $L_a = 0.66$, $L_b = 0.87$, and LER = 1.53.

The farmer requires that $L'_a = 0.75$. We have

$$L'_a = 0.75 = \frac{y'_a}{s} = \frac{y'_a}{3398}$$

Hence

$$y'_a = (0.75)(3398) = 2548 \text{ kg/ha}$$

Then

$$k_I = \frac{1 - L'_a}{1 - L_a} = \frac{1 - 0.75}{1 - 0.66} = 0.74$$

Thus 74% of his land should be maize intercropped with pigeon pea while the remaining 26% should be in maize only. Under this cropping system he might expect to obtain a maize yield of about 2548 kg/ha and about (0.74)(896) = 663 kg/ha of pigeon pea.

With this land allocation to sole-crop maize and intercrop of maize/pigeon pea, the staple land equivalent ratio (SLER) would be

$$\text{SLER} = \text{LER} - \frac{(\text{LER} - 1)(L'_a - L_a}{1 - L_a} = 1.53 - \frac{(1.53 - 1)(0.75 - 0.66)}{1 - 0.66} = 1.39$$

10.2.4 Other Measures for Assessing Intercropping

The land equivalent ratio (LER) and related measures have been the most widely accepted measures for assessing the benefits of intercropping. In this book we consider only intercropping systems that involve one intercrop area and one sole-crop area. Riley (1984) has generalized the LER concept to provide a combined LER for systems involving more than one intercrop and more than one sole-crop area. Consideration of this generalization is beyond the scope of this book.

Because intercropping involves growing two crops together on the same land area, the question of competition between the two crops arises. Although LER provides a basis for comparing intercrop yields with sole-crop yields, it does not address the problem of measuring the degree of competition between the components of the intercrop. This problem was examined by Willey and Rao (1980) among others. They propose a measure that expresses the ultimate yields of the components corrected for the proportional areas on which the crops were sown. This measure is called the *competitive ratio* (CR) and is defined as

$$CR = \frac{L_a}{L_b}\frac{Z_a}{Z_b}$$

where Z_a and Z_b are the sown proportions of a and b, and L_a and L_b are partial LERs.

If it can be assumed for the ICRISAT data in Table 10.1 that maize and pigeon-pea were sown in alternate, equally spaced rows, we have

$$\begin{aligned} CR &= \frac{L_a}{L_b}\frac{Z_a}{Z_b} \\ &= \frac{0.66}{0.87}\left(\frac{0.50}{0.50}\right) \\ &= 0.76 \end{aligned}$$

Willey and Rao would interpret this to mean that maize is only about three-fourths as competitive as pigeon pea in this intercropping system.

Another index for assessing intercrops is the system productivity index (SPI), presented by Odo (1991). The SPI is defined as

$$SPI = \frac{S_a}{S_b}y_b + y_a$$

The aim of the SPI is to standardize the yield of the secondary crop, b, in terms of the primary crop, a.

For the ICRISAT data in Table 10.1 the system productivity index is

$$\begin{aligned} SPI &= \left(\frac{3398}{1035}\right)(896) + 2234 \\ &= 5176 \text{ kg/ha} \end{aligned}$$

10.3 TREATMENTS

As with any other type of experiment, the treatments to be used in an intercropping trial are determined largely by the trial's objectives. These may include one, or more, of the following (Mead and Riley, 1981):

1. Assessment of yield advantages of intercropping compared to sole cropping
2. Identification of suitable genotypes for intercropping

3. Consideration of the level of agronomic factors, such as fertilization, to maximize crop yield
4. Determination of optimum spatial arrangements and plant densities
5. Assessment of the yield stability imparted by intercropping

There are, no doubt, other possible research objectives as well.

One problem with treatment selection for intercropping research, as compared to research on sole crops, is that intercrops involve two or more crops grown in association with each other. Further, consideration must be given to the inclusion of sole-crop treatments, not only for comparison with the intercrops but also as a basis for standardizing the intercrop yields.

10.3.1 Inclusion of Sole-Crop Treatments

The number and type of sole-crop treatments to be included in an experiment depends, to some extent, on the objectives of the experiment. Mead and Stern (1980) suggest that if the aim of the trial is to evaluate the benefits of growing intercrops as compared to sole crops, it may be necessary to have nearly as many sole-crop plots as there are intercrop plots. On the other hand, if the aim is to determine the best way to grow intercrops, a sole-crop yield is necessary only for the purpose of standardizing the intercrop yields. In this case it may be sufficient to eliminate sole-crop treatments from the experiment and to grow a few large plots of the constituent intercrops in the vicinity of the experiment.

If the sole-crop yields are to be used only for standardization, the average yield or maximum yield of the sole-crop plots included in the experiment should be used (Mead and Willey, 1980). Further, if the aim of the trial is to study plant spacing or population density, Huxley and Maingu (1978) suggest that the sole crops should be grown at their optimum spacing and density.

For genotype screening trials the number and type of sole-crop plots to be included depend on the phase of the trial. In the early phases there are many possible selections of both components from which to choose. Here it is best to use a single genotype as one of the components and multiple selections of the other. At this stage screening can be accomplished using a single representative genotype of each component as a standardizing sole crop. When the screening program has reached the stage where only four or five selections of each component remain, it is well to include sole plots of each selection in the intercropping trial. Each selection is then standardized against its own sole-crop yield.

10.3.2 Treatment Structures for Agronomy Trials

One characteristic of experiments aimed at determining the best way to grow intercrops is the large number of factors that must be studied. For example, Mead and Stern (1980) point out that a study of the spatial management of two-component intercrops requires the consideration of five factors: population density of each component; spatial arrangement, within and between rows, of each crop; and the intimacy, or relative arrangement, of the two components.

The traditional approach to studying the problems involving several factors is to vary one factor at a time while holding the rest constant. A much more efficient use of experimental resources, however, is to use factorial combinations of treatments. Treatments that meet the factorial criteria are presented in Section 4.3 as well as in standard texts on experimental design (Cochran and Cox, 1957; Cox, 1958; Petersen, 1985). The factorial approach has a number of advantages compared to the one-at-a-time approach. It permits the experimenter to determine the extent to which one factor is affected by changes in another. That is, it provides a measure of the interaction among factors. A second advantage is the increase in precision due to hidden replication when the factors do not interact.

In the early stages of an agronomic study on an intercropping system the number of factors that must be considered may be quite large. In this case the number of treatment combinations required for a full factorial set might be prohibitive, particularly if the treatments are to be replicated. There are a couple of alternatives. The number of factors to be considered in a single experiment might be restricted, and several trials, using different factors in each trial, could be conducted. A second alternative is to use a fraction of the total number of combinations required for the full factorial set in a fractional factorial trial. Details of the construction of fractional factorial trials are beyond the scope of this book. They may be found, however, in such texts on experimental design as Cochran and Cox (1957) and Petersen (1985).

Frequently in agronomic trials the treatments consist of several levels of one or more quantitative variables. When this is the case, it is useful to try to describe the plant response as a function of the level of the quantitative factor. Often, in the absence of information about the true nature of the response function, a polynomial equation is used for this purpose. In most cases the response can be described by, at most, a cubic polynomial. More often, a quadratic polynomial is sufficient.

When considering an experiment that involves a quantitative factor, the question arises about the number of levels of the factor to be included. To fit a response function requires only as many levels as there are parameters in the function to be fitted. For example, to fit a quadratic polynomial requires only three levels. To test for lack of fit requires one additional level. In practice, then, the inclusion of more than five levels of a quantitative factor is seldom justified. More frequently, four levels are sufficient.

When choosing the number of levels to be included it would appear that they should cover as wide a range as possible. Further, the levels should be equally spaced across the range. Such a distribution is usually optimum for most response functions. In addition, equal spacing permits the use of orthogonal polynomial coefficients for fitting and testing the terms of a polynomial function. (For details, see Section 4.7.)

10.4 EXPERIMENTAL DESIGN

There are a number of things that must be considered in choosing an experimental design for an experiment involving intercrop systems. In many areas where intercropping research is being conducted, especially in the tropics, the land has been developed for research work only recently. Hence the uniformity throughout the experimental site which is common in experiment stations in the temperate regions is likely to be absent. It is seldom possible to find blocks of plots that are uniform in size and shape and are sufficiently large to accommodate the large number of treatments usually used in intercrop trials. It is still important, however, to identify groups of homogeneous plots which can serve as blocks in an intercropping trial. Often, these groups of plots will be irregular in shape and might be located some distance from each other. This should not be a factor in forming the blocks for an intercropping trial. Blocks should be formed regardless of size or shape if a suitable blocking criterion can be found. It should also be pointed out that when the treatments are factorial in nature, block size can be reduced by using fractional replications in place of full replications or by confounding high-order interactions with block effects (Cochran and Cox, 1957).

Many intercropping experiments conducted in recent years have used split-plot designs (Mead and Stern, 1980; Mead and Riley, 1981). The reason for this choice is not clear. Perhaps it is because of the relative ease of allocating treatments to the plots in a split-plot design, or because the researcher does not fully understand the implications regarding the complexity of treatment comparisons, or because of the possible increase in precision with which the interactions are estimated in a split-plot design. Any increase in precision in interaction estimation is more than offset by the decrease in precision with which the main effects, particularly for the factors on the whole plots, are estimated. Further, the split plot involves different treatment comparisons and multiple standard errors for these comparisons, an additional complicating factor.

The only legitimate reason for using a split-plot design is that some treatments must be applied to large plots (main plot factor) while other treatments can be applied to smaller plots (subplot factor). If this is not the case, a split-plot design should not be used unless it is determined that reasonable comparisons among the main plot treatments are possible.

10.4.1 Systematic Designs

One of the objectives of intercropping research is to determine how the yields of the components and of the intercrop as a whole are affected by changes in the density or spatial configuration of the component crops. In particular, determination of the optimum inter- and intrarow spacing of the components, as well as their optimum planting configuration, is the goal of some intercropping research. As mentioned previously, spacing experiments with intercrops involve five factors: density and spatial arrangement of the component crops and the arrangement of the components. The use of factorial treatment combinations to separate the main effects and interactions of five factors would lead to so many treatments as to be impractical in the field. Further, if the aim is to determine optimum levels for the factors, each factor must be included at many levels over a wide range.

Some researchers (Huxley and Maingu, 1978; Wahua and Miller, 1978; Willey, 1979b; Mead and Riley, 1981; Mead and Stern, 1980) have suggested that systematic designs be used for spacing research on intercrops in place of the usual random assignment of treatments to plots. In systematic designs the crop density changes by a small but constant amount from one row to another. There are two advantages of the systematic design over random treatment allocation. A systematic design covers a wide range of densities in a relatively small plot. In addition, because each row is flanked by rows of nearly the same density, guard areas around each row can be avoided. For intercropping experiments involving a wide range of spacings combined factorially, in some cases, with other factors, systematic designs provide much greater efficiency of land use than do conventional designs. The primary disadvantage of systematic designs is that because the treatments are not randomized the usual analysis of variance is not strictly valid.

A number of types of systematic design have been proposed. Nelder (1962) proposed a fan-shaped design in which the rows radiate from a central point. Two alternatives of this design have been proposed for use in intercropping trials. One alternative used by Wahua and Miller (1978) has alternating radii as rows for each component. A second alternative, shown in Figure 10.1, has the two-component crops in alternating arcs of the fan. Note that a fan-shaped sole-crop plot is included as a part of this design. In this design the arcs are harvested separately, and the component crops are harvested separately within each arc. Details of the construction of these designs is given by Bleasdale (1967). In practice the designs are laid out with the apex of the fan pointing in opposite directions from one

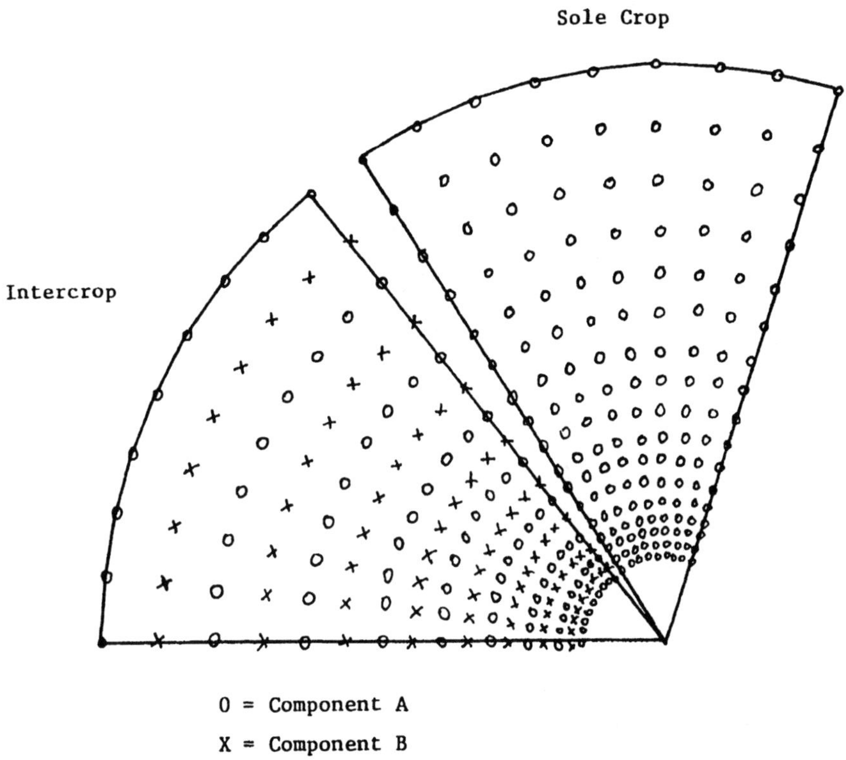

Figure 10.1 Nelder's fan-shaped systematic design.

block to another. Similarly, the sole-crop plot in each couplet is alternated from one component crop to the other.

A second type of systematic design is Bleasdale's row modification of the fan design. This design is shown in Figure 10.2. In the row modification the separation between rows is increased from row to row and the separation between plants within rows is also increased from row to row. Details of the construction of this design are given by Bleasdale (1967). In the field this design should be laid out with the wide part of the plot facing in alternate directions from one plot to another. Systematic sole-crop plots may be included as a part of this design by alternating sole-crop plots with intercrop plots. In this design each row is harvested separately.

A third type of systematic design permits the independent variation in spacing of two crops. This design, an adaptation of Bleasdale's row design, was developed by Mead (1979). A plan for this design is given in Figure

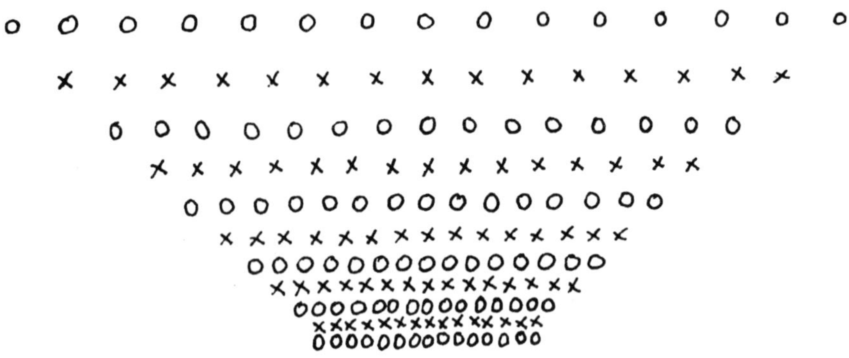

0 = Component A

X = Component B

Figure 10.2 Bleasdale's row modification of the fan-shaped systematic design.

10.3. In practice the orientation of the design should be chosen randomly for each block. At harvest a number of subsamples are taken throughout the plot. Figure 10.3 shows the plot divided into nine subplots, and a sample of several plants of each component would be taken from each subplot.

Because the treatments are not randomly assigned to plots in a systematic design, the analysis of variance of data from these designs is not strictly valid. An appropriate way of examining the data would be first to graph the yields of the component crops and their LER as a function of density or spacing. Then, depending on the outcome of the graphing procedure, an appropriate response function with density or spacing as the independent variable and yield or LER as the dependent variable should be tried. Finally, depending on the experimental design, an analysis of variance of the parameters of the fitted curves could be run.

10.5 DATA ANALYSIS

There is no single, universally accepted procedure for analyzing the data from an intercropping trial. The basic problem has to do with jointly evaluating two or more crops grown together at the same time on the same plot. Pearce and Gilliver (1978, 1979) suggest that since there are multiple crops, a multivariate analysis should be used. Others, however, have attempted to reduce the problem to a univariate one by developing a single

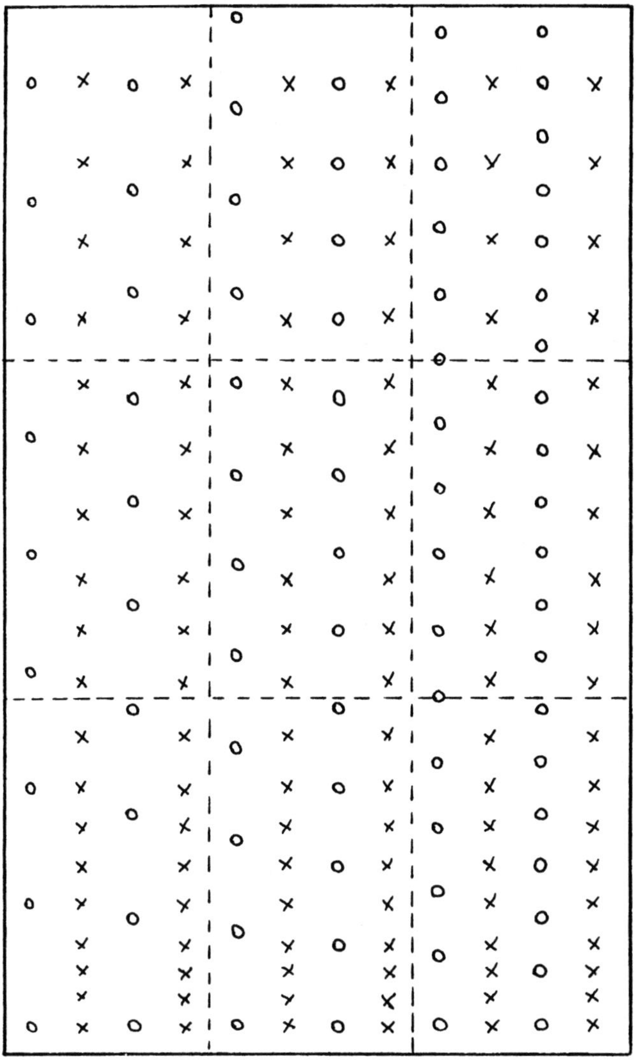

Figure 10.3 Mead's adaptation of Bleasdale's row design; o = component A, x = component B.

index to express the combined effect of all the components in the intercropping system.

Of these indexes, the most widely accepted is the LER discussed previously. Finney (1990) has suggested that other measures might be more meaningful to the farmer seeking recommendations on whether to use intercropping, how much of his land to devote to sole cropping, and how best to grow intercrops. Among alternatives to the LER that have been suggested include value of the produce, cost of the cropping plans, nutritional value of the total crop, and so on. All of these indexes have their advantages and disadvantages. Because of its widespread use and because it seems to be widely accepted as a means of comparing intercropping systems among themselves and comparing intercropping with sole cropping, the LER would appear to be the index of choice.

When beginning the analysis it is best to keep it simple. The first step should be to compute the partial LERs for the components and the LER for the intercrop. It is also useful at this stage to graph the component yields, the total intercrop yield, the partial LERs, and the full LER against spacing, density, amount of fertilizer, or any other quantitative treatment variable.

Mead and Riley (1981) suggest that separate analyses of variance be conducted for the sole-crop yields of each crop, for the sole- and component-crop yields of the intercrop plots, and for the total intercrop yield. Oyejola and Mead (1982) reported that if a single sole-crop yield has been used to compute LER, the LER values approximately conform to the assumptions underlying the analysis of variance. Hence the analysis of variance is appropriate for LERs computed in this way.

Since the treatment levels are not randomly assigned in the systematic designs, it is probably best not to subject the yields from these designs to the analysis of variance. Often, however, the systematic spacing or density levels are applied within the whole plots in a type of split-plot experiment. In this case the data from the whole plots should be analyzed by an analysis of variance appropriate for the design of the whole plots. Further, if response equations are fitted to the systematic data within the whole plots, it would be useful to do analyses of variance on the parameters of these fitted equations.

In place of doing separate analyses of variance on the components of an intercrop, Pearce and Gilliver (1978, 1979) propose a single bivariate analysis of the two yields. They give two different but equivalent methods for dealing with the yields and the correlation between them. In the first approach each variate is adjusted for the other by use of their error variances and covariance. Two new variates, with error variance equal to 1 and covariance equal to zero, are then constructed. These new variates, which are statistically independent, can be plotted on rectangular axes.

The second method uses skew axes based on the correlation between the two variates and their error variance. Transformed yields, adjusted for error variances and covariance, can then be plotted on the skew axes. The transformed variates are also statistically independent.

While the bivariate method provides valid tests of significance, there are a number of difficulties associated with it. The computations involved with this method are more complex than they are with other analytical methods. The analysis is based on the assumption that the correlation between the yield of the two crops is constant and independent of treatment. This assumption is difficult to check. Further, bivariate analysis does not permit a comparison of sole crop and intercrop yields. Finally, bivariate analysis can be used only if the intercrop involves two component crops. If three or more crops comprise the intercrop, a more general, more complicated multivariate analysis procedure must be used. In this case, the graphical presentation provided by the bivariate method is no longer available.

Although the bivariate method provides significance tests, it does not give much assistance in estimation. For the grower who wants to make some decisions regarding intercropping, bivariate analysis does not provide very much help, and the computations involved would not seem to be worth the effort.

10.6 NUMERICAL EXAMPLE

A trial was conducted in Tanzania to determine the effect of different row ratios on the production of intercropped sorghum and cowpea. The treatments consisted of four ratios:

1. One row sorghum to four rows cowpea, 1:4
2. Two rows sorghum to three rows cowpea, 2:3
3. Three rows sorghum to two rows cowpea, 3:2
4. Four rows sorghum to one row cowpea, 4:1

The trial was laid out using a randomized block design with four blocks of four plots each. Soil variation was used as the blocking factor. Pure stands of sorghum and of cowpea were grown as sole crops on separate plots adjacent to, but not part of, the intercropping trial.

Air-dry yields of sorghum and cowpea from the trial are given in Table 10.2. Analyses of variance of these yields are presented in Table 10.3. Note that differences among row ratios for both sorghum and cowpea were highly significant.

The total intercrop yields (sum of the sorghum and cowpea yields) are shown in Table 10.4. The analysis of variance of these yields is given in Table 10.5. Again, the row ratio effect is highly significant.

Table 10.2 **Yield (kg/ha) of Sorghum and Cowpea from a Sorghum/Cowpea Row Ratio Intercrop Trial**[a]

Block	Crop	Rows Sorghum:Rows Cowpea 1:4	2:3	3:2	4:1	Sum
I	Sorghum	956	2144	2635	3255	8990
	Cowpea	1444	1114	818	296	3672
II	Sorghum	1498	2370	3286	3458	10612
	Cowpea	1357	1601	835	400	4193
III	Surghum	1073	2261	2695	3318	9347
	Cowpea	1444	1462	957	400	4263
IV	Sorghum	1622	2466	3252	3508	10828
	Cowpea	1357	1514	940	331	4142
Sum	Sorghum	5149	9221	11868	13539	39777
	Cowpea	5602	5691	3550	1427	16270
Mean	Sorghum	1287	2305	2967	3385	2486
	Cowpea	1400	1423	888	357	1017

[a]Sole crop yield: sorghum = 3585 kg/ha; cowpea = 1740 kg/ha.

The component crop yields and the intercrop yield are graphed in Figure 10.4a. It appears that there is a fairly smooth increase in sorghum yield as the ratio of sorghum rows to cowpea rows is increased. As the sorghum/cowpea ratio increases cowpea yield holds fairly steady until a ratio of 3:2 is reached, at which point cowpea yield falls off rapidly. Total intercrop yield reaches a maximum at a sorghum/cowpea ratio of 3:2.

To further examine the effect of changing row ratio on yield we could try to find an equation to describe the relationship between yield and row ratio. To illustrate the process we will attempt to describe sorghum yield

Table 10.3 **Analyses of Variance of Sorghum and Cowpea Yields (kg/ha) from a Sorghum/Cowpea Row Ratio Intercrop Trial**

		Sorghum		Cowpea	
Source	df	SS	MS	SS	MS
Total	15	10,808,245		3,225,426	
Block	3	623,551	207,850	53,985	17,995
Row ratio	3	10,035,139	3,345,046**	3,057,622	1,019,207**
Error	9	149,555	16,617	113,819	12,647

**Significant at the 1% level.

Table 10.4 **Total Intercrop Yield (kg/ha) from a Sorghum/Cowpea Row Ratio Intercrop Trial**

	Rows Sorghum:Rows Cowpea				
Block	4:1	2:3	3:2	4:1	Sum
I	2400	3258	3454	3551	12662
II	2855	3971	4121	3858	14805
III	2517	3723	3652	3718	13610
IV	2979	3960	4192	3839	14970
Sum	10751	14912	15418	14966	56047
Mean	2688	3728	3854	3741	3503

as a function of the number (out of five) of sorghum rows. (*Note*: The difference between treatment levels is constant, a difference of one row between treatment levels, and treatments are equally replicated, four replications. Since this is the case, we can use the method of orthogonal polynomial coefficients, given in Section 4.7.2, to try to fit a polynomial equation to the data.)

Table 10.5 **Analysis of Variance of Intercrop Yield (kg/ha) from a Sorghum/Cowpea Row Ratio Intercrop Trial**

Source	df	SS	MS
Total	15	4,588,379	
Block	3	882,679	294,726
Row Ratio	3	3,582,768	1,194,256**
Error	9	122,932	13,659

**Significant at the 1% level.

The polynomial contrasts for partitioning the treatment sum of squares are given in Table 10.6. Coefficients for these contrasts were obtained from Table 4.37. Note that

$$9{,}672{,}319 + 360{,}300 + 2520 = 10{,}035{,}139$$

Hence sums of squares for the contrasts add up to the treatment sum of squares.

To determine the best-fitting polynomial equation to describe the data, we can add terms sequentially and test lack of fit after each addition. This procedure is shown in Table 10.7. Note that after the quadratic term has been added lack of fit is no longer significant. Hence the relationship between sorghum yield and the number of sorghum rows may be described by a quadratic polynomial equation.

We can develop this polynomial equation using the method described in Section 4.7.2. Let

y = yield of sorghum
x = number of rows (out of five) of sorghum

Then

$$\bar{y} = 2486$$

$$b_1 = \frac{L_1}{rD_1} = \frac{27{,}817}{80} = 347.71$$

$$b_2 = \frac{L_2}{rD_2} = \frac{-2401}{16} = -150.06$$

$$\bar{x} = \frac{1 + 2 + 3 + 4}{4} = \frac{10}{4} = 2.5$$

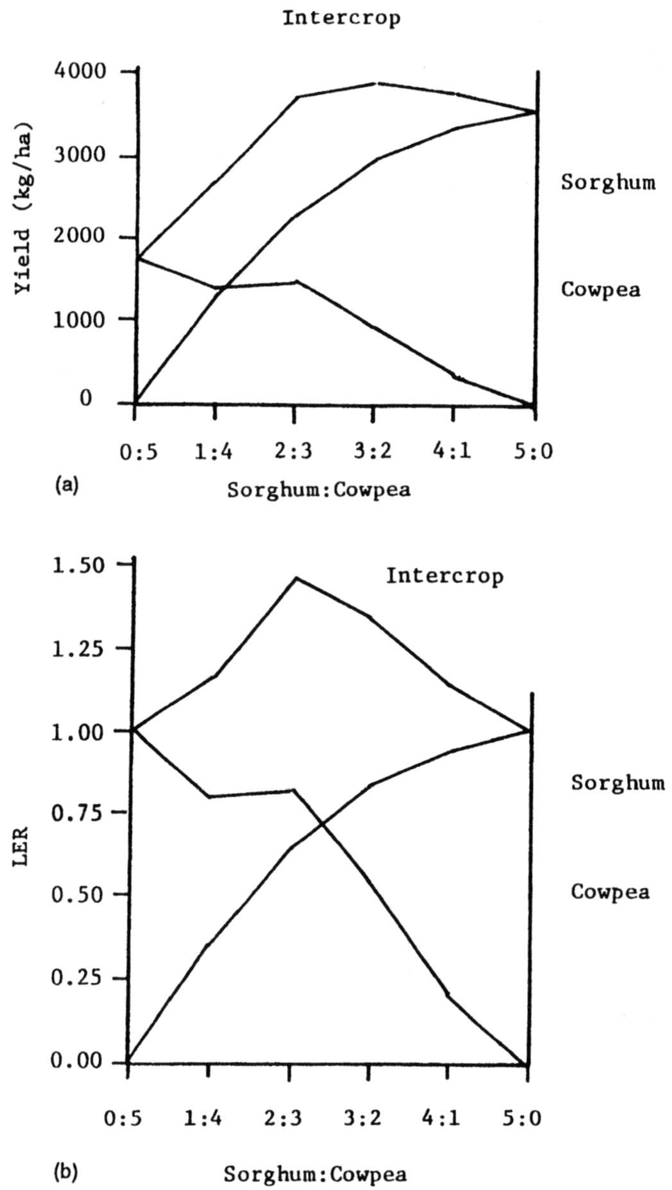

Figure 10.4 (a) Yields of sorghum, cowpea, and total intercrop as a function of sorghum/cowpea row ratio. (b) Land equivalent ratios (LERs) of sorghum, cowpea, and intercrop as a function of sorghum/cowpea row ratio.

Table 10.6 **Orthogonal Partition of Row Ratio Effects for Sorghum Yield (kg/ha) in an Intercrop Trial**[a]

Contrast \ Yield	Sorghum:Cowpea 1:4	2:3	3:2	4:1			
	5149	9221	11868	13539	L_i	rD_i	s_i^2
Linear	−3	−1	+1	+3	27,817	80	9,672,319
Quadratic	+1	−1	−1	+1	−2,401	16	360,300
Cubic	−1	+3	−3	+1	449	80	2,520

[a] $L_i = \Sigma_j k_{ij} T_j$
$D_i = \Sigma_j k^2_{ij}$
$S_i^2 = (L_i)^2/rD_i$

Table 10.7 **Sequential Test of Terms of a Polynomial Equation Describing Relationship Between Number of Sorghum Rows per Five Rows and Sorghum Yield (kg/ha)**

Source	df	MS	F
Row Ratio	15	3,345,046	201.30**
Linear	1	9,672,319	582.07**
Deviation from Linear	2	181,410	10.92**
Quadratic	1	360,300	21.68**
Dev. from Lin. + Quad.	1	2,520	.15NS
Error	9	16,617	

**Significant at the 1% level; NS, not significant.

and from Table 4.37,

$$\lambda_1 = 2$$
$$\lambda_2 = 1$$

From this we have

$$\bar{\bar{y}}_j = \bar{y} + a_1\xi_1 + a_2\xi_2$$
$$= \frac{\bar{y} + \lambda_1 b_1 (x_j - \bar{x})}{\Delta} + \lambda_2 b_2\left[\left(\frac{x_j - \bar{x}}{\Delta}\right)^2 - \frac{p^2 - 1}{12}\right]$$
$$= \frac{2486 + (2)(347.71)(x_j - 2.5)}{1} + (1)(-150.06)$$
$$\left[\left(\frac{x_j - 2.5}{1}\right)^2 - \frac{4^2 - 1}{12}\right]$$
$$= -2.85 + 1445.72x_j - 150.06x^2_j$$

where

$\hat{y}_j$ = expected sorghum yield
x_j = number of sorghum rows (out of five)

We could repeat this process for cowpea yields and for intercrop yields. However, we will not do so.

The sole-crop yield of sorghum from the adjacent plots was 3585 kg/ha, while that of cowpea was 1740 kg/ha. Using these data the partial land equivalent ratios, L_i, of sorghum and cowpea were computed as

$$L_i = \frac{y_i}{s_i}$$

where

y_i = yield of component i in the intercrop
s_i = yield of component i as a sole crop

The partial LERs are presented an Table 10.8. The analyses of variance of these partial LERs are given in Table 10.9. Row ratio effects for both sorghum and cowpea were highly significant.

The land equivalent ratios (LER) were computed as the sum of the partial LERs. These are shown in Table 10.10 and their analysis of variance is given in Table 10.11. Again, the differences in LER among the row ratios were highly significant.

Table 10.8 **Partial Land Equivalent Ratio (L_i) from a Sorghum/Cowpea Row Ratio Intercrop Trial**

		Rows Sorghum:Rows Cowpea				
Block	Crop	1:4	2:3	3:2	4:1	Sum
I	Sorghum	.27	.60	.74	.90	2.51
	Cowpea	.83	.64	.47	.17	2.11
II	Sorghum	.42	.66	.92	.96	2.96
	Cowpea	.78	.92	.48	.23	2.41
III	Sorghum	.30	.63	.75	.92	2.60
	Cowpea	.83	.84	.55	.23	2.45
IV	Sorghum	.45	.68	.91	.98	3.02
	Cowpea	.78	.87	.54	.19	2.38
Sum	Sorghum	1.44	2.57	3.32	3.76	11.09
	Cowpea	3.22	3.27	2.04	.82	9.35
Mean	Sorghum	.36	.64	.83	.94	.69
	Cowpea	.80	.82	.51	.20	.58

The partial LERs and the full LERs are graphed against the sorghum/cowpea row ratio in Figure 10.4b. It will be seen that the highest efficiency of land use in this trial was obtained at a row ratio of 2:3. At this ratio the LER was 1.46.

Mead and Willey (1981) consider the partial LERs, L_i, for the components and the LER for the total intercrop to be standardized yields (standardized against the sole-crop yields of the components). Accordingly, the ratio

$$p_i = \frac{L_i}{LER}$$

is the proportion of the standardized intercrop yield provided by component

Table 10.9 **Analyses of Variance of Partial LERs (L_i) from a Sorghum/Cowpea Row Ratio Intercrop Trial**

		Sorghum		Cowpea	
Source	df	SS	MS	SS	MS
Total	15	.8329		1.0654	
Block	3	.0488	.0163	.0179	.0060
Row Ratio	3	.7729	.2576**	1.0099	.3366**
Error	9	.0112	.0012	.0376	.0042

**Significant at the 1% level.

Table 10.10 **Land Equivalent Ratios (LERs) from a Sorghum/Cowpea Row Ratio Intercrop Trial**

	Rows Sorghum:Rows Cowpea				
Block	1:4	2:3	3:2	4:1	Sum
I	1.10	1.24	1.21	1.07	4.62
II	1.20	1.58	1.40	1.19	5.37
III	1.13	1.47	1.30	1.15	5.05
IV	1.23	1.55	1.45	1.17	5.40
Sum	4.66	5.84	5.36	4.58	20.44
Mean	1.16	1.46	1.34	1.14	1.28

i. Yield proportions, p_i, for this trial are given in Table 10.12. Note that with a row ratio of 2:3 a yield proportion of 0.44 was obtained from sorghum while a yield proportion of 0.56 was obtained from cowpea.

The results of this study are useful if the farmer is only interested in the most efficient land utilization. In this case he would plant his entire area with an intercrop of two rows of sorghum alternating with three rows

Table 10.11 **Analysis of Variance of Land Equivalent Ratios (LERs) from a Sorghum/Cowpea Row Ratio Intercrop Trial**

Source	df	SS	MS
Total	15	.3941	
Block	3	.0989	.0330
Row Ratio	3	.2697	.0899**
Error	9	.0255	.0028

**Significant at the 1% level.

of cowpea. With this intercropping system he would expect to produce about 2300 kg/ha of sorghum and about 1425 kg/ha of cowpea (see Table 10.2). However, he might have different requirements and he might be willing to devote some of his land to sole-cropping as well as to intercropping to satisfy these requirements.

Suppose that the farmer is considering two row ratios, 2:3 or 3:2 rows sorghum to rows cowpea. Suppose, further, that he would like a yield proportion, p'_s, of 0.70 from sorghum. What is his best alternative?

Note from Table 10.12 that with the 2:3 ratio the yield proportion, p_s, from sorghum is 0.44, while with the 3:2 ratio the yield proportion, p_s, from sorghum is 0.62. The required proportion from sorghum, 0.70, cannot be achieved by either row ratio unless some sole-crop sorghum is included

Table 10.12 **Proportional Yield of Sorghum, p_s,* and Cowpea, P_c,* in a Row Ratio Intercrop Trial**

Row Ratio	P_S	P_C
1:4	.31	.69
2:3	.44	.56
3:2	.62	.38
4:1	.82	.18

*$P_s = L_s/LER$; $P_c = L_c/LER$.

in the system. We need to calculate the proportion of total area that must be in sole-crop sorghum and the proportion of area to be in intercrop to meet the farmer's requirements. For this purpose we can use the ELER approach of Mead and Willey discussed in Section 10.2.2.

Let k_I represent the proportion of area to be in intercrop. Then $k_s = 1 - k_I$ is the proportion of area required to be in sole-crop. Further, under the farmer's restriction the required yield proportion, p'_s, of sorghum must be 0.70.

For a row ratio of 2:3 we have

$$L_s = 0.64 = \text{partial LER for sorghum}$$
$$L_c = 0.82 = \text{partial LER for cowpea}$$
$$\text{LER} = 1.46$$

Then

$$k_I = \frac{1 - p'_s}{p'_s L_c - (1 - p'_s)(L_s) + (1 - p'_s)}$$
$$= \frac{1 - 0.7}{(0.7)(0.82) - (1 - 0.7)(0.64) + (1 - 0.7)}$$
$$= \frac{0.3}{0.68} = 0.44$$

With a row ratio of 2:3 the farmer would require 44% of his land to be in intercrop and 56% to be in sole-crop sorghum.

With such an allocation he would have an effective LER (ELER) of

$$\text{ELER} = \frac{L_c}{(1 - L_s) + (\text{LER} - 1)p'_s}$$
$$= \frac{0.82}{(1 - 0.64) + (1.46 - 1)(0.7)}$$
$$= 1.20$$

For a row ratio of 3:2 we have

$$L_s = 0.83$$
$$L_c = 0.51$$
$$\text{LER} = 1.34$$

Then

$$k_I = \frac{1 - 0.7}{(0.7)(0.51) - (1 - 0.7)(0.83) + (1 - 0.7)}$$
$$= \frac{0.3}{0.41} = 0.73$$

With a row ratio of 3:2 the farmer could have 73% of his land in intercrop and 27% in sole-crop sorghum. With this allocation he would have an effective LER of

$$\begin{aligned} \text{ELER} &= \frac{0.51}{(1 - 0.83) + (1.34 - 1)(0.7)} \\ &= 1.24 \end{aligned}$$

With the requirement that the yield proportion from sorghum be 0.70, the farmer would make slightly more efficient use of his land with a row ratio of 3:2 and an ELER of 1.24. This compares with an ELER of 1.20 for a row ratio of 2:3.

Alternatively, suppose that the farmer is considering two row ratios: 2:3 or 3:2 rows sorghum to rows cowpea. Suppose, in addition, that his farming system must produce a cowpea yield at least 90% of what he could produce if the entire area were to be sole-cropped to cowpea. That is, he requires that the cowpea yield be at least

$$(0.9)(1740) = 1566 \text{ kg/ha}$$

With the entire area under intercrop he obtains a cowpea yield of 1423 kg/ha at the 3:2 row ratio and a cowpea yield of 888 kg/ha at the 3:2 row ratio (Table 10.2). To meet his requirements he must grow sole-crop cowpea on a part of his land. To determine his land allocation we can use the approach of Reddy and Chetty described in Section 10.2.3. Let

k_I = proportion of the total area to be intercropped
$k_s = 1 - k_I$ = proportion of area to be sole-cropped to cowpea
$L'_c = 0.90$ = required proportion of the sole-crop yield of cowpea

For a row ratio of 2:3 we have

$$L_c = 0.82$$
$$\text{LER} = 1.46$$

Then

$$\begin{aligned} k_I &= \frac{1 - L'_c}{1 - L_c} \\ &= \frac{1 - 0.90}{1 - 0.82} \\ &= \frac{0.10}{0.18} = 0.55 \end{aligned}$$

He would require 55% of his land to be intercropped and 45% to be sole-cropped with cowpea.

With this allocation his staple LER (SLER) would be

$$\begin{aligned} \text{SLER} &= \text{LER} - \frac{(\text{LER} - 1)(\text{L}_c' - \text{L}_c)}{(1 - \text{L}_c)} \\ &= 1.46 - \frac{(1.46 - 1)(0.90 - 0.82)}{1 - 0.82} \\ &= 1.26 \end{aligned}$$

He would expect to produce

$$\begin{aligned} (0.45)(1740) + (0.55)(1423) &= 1566 \text{ kg of cowpea} \\ (0.45)(2305) &= 1037 \text{ kg of sorghum} \end{aligned}$$

for each hectare of land under this system.

With a row ratio of 3:2 we have

$$\begin{aligned} \text{L}_c &= 0.51 \\ \text{LER} &= 1.34 \end{aligned}$$

Then

$$\begin{aligned} k_I &= \frac{1 - 0.90}{0.90 - 0.51} \\ &= 0.20 \end{aligned}$$

With a 3:2 sorghum/cowpea row ratio he would have to use 26% of his land for intercrop and 74% for cowpea alone.

With this arrangement his staple LER (SLER) would be

$$\begin{aligned} \text{SLER} &= 1.34 - \frac{(1.34 - 1)(0.90 - 0.51)}{1 - 0.51} \\ &= 1.07 \end{aligned}$$

He would expect to produce

$$\begin{aligned} (0.74)(1740) + (0.26)(888) &= 1518 \text{ kg of cowpeas} \\ (0.26)(2976) &= 774 \text{ kg of sorghum} \end{aligned}$$

on each hectare under this system.

If the farmer is required to produce 90% of his sole-crop yield of cowpeas, the intercrop with a row ratio of 2:3 is more efficient. With a 2:3 row ratio the SLER is 1.26, compared to an SLER of 1.07 with the 3:2 row ratio intercrop.

REFERENCES

Bleasdale, J. K. A., 1967. Systematic designs for spacing experiments. *Exp. Agric.* 3:73–85.

Cochran, W. G., and G. M. Cox, 1957. *Experimental Designs*, 2nd ed. Wiley, New York, Chaps. 6, 6a.

Cox, D. R., 1958. *Planning of Experiments*. Wiley, New York, Chaps. 11, 12.

Finney, D. J., 1990. Intercropping experiments, statistical analysis, and agricultural practice. *Expl. Agric.* 26:73–81.

Huxley, P. A., and Z. Maingu, 1978. The use of systematic spacing designs as an aid to the study of intercropping: some general considerations. *Exp. Agric.* 14:49–56.

Mead, R., 1979. Competition experiments. *Biometrics* 35:41–54.

Mead, R., and J. Riley, 1981. A review of statistical ideas relevant to intercropping research (with discussion). *J. Roy. Stat. Soc. A* 144:462–509.

Mead, R., and R. D. Stern, 1980. Designing experiments for intercropping research. *Exp. Agric.* 16:329–342.

Mead, R., and R. W. Willey, 1980. The concept of "land equivalent ratio" and advantages in yield from intercropping. *Exp. Agric.* 16:217–228.

Nelder, J. A., 1962. New kinds of systematic designs for spacing experiments. *Biometrics* 18:283–307.

Oyejola, B. A., and R. Mead, 1982. Statistical assessment of different ways of calculating land equivalent ratios (LER). *Exp. Agric.* 18:125–138.

Pearce, S. C., and B. Gilliver, 1978. The statistical analysis of data from intercropping experiments. *J. Agric. Sci. Camb.* 91:625–632.

Pearce, S. C., and B. Gilliver, 1979. Graphical assessment of intercropping methods. *J. Agr. Sci. Camb.* 93:51–58.

Petersen, R. G., 1985. *Design and Analysis of Experiments*. Marcel Dekker, New York, Chaps. 8, 9.

Reddy, M. N., and C. K. R. Chetty, 1984. Staple land equivalent ratio for assessing yield advantage from intercropping. *Exp. Agric.* 20:171–177.

Riley, J., 1984. A general form of the "land equivalent ratio." *Exp. Agric.* 20:19–29.

Wahua, T. A. T., and D. A. Miller, 1978. Relative yield totals and yield components of intercropped sorghum and soybeans. *Agron. J.* 70:287–291.

Willey, R. W., 1979a. Intercropping: its importance and research needs: Part 1. Competition and yield advantages. *Field Crop Abstr.* 32:1–10.

Willey, R. W., 1979b. Intercropping: its importance and research needs: Part 2. Agronomy and research approaches. *Field Crop Abstr.* 32:73–85.

Willey, R. W., 1985. Evaluation and presentation of intercropping advantages. *Exp. Agric.* 21:119–133.

Willey, R. W., and D. S. O. Osiru, 1972. Studies on mixtures of maize and beans (*Phaseolus vulgaris*) with particular reference to plant populations. *J. Agric. Sci. Camb.* 79:117–125.

Willey, R. W., and M. R. Rao, 1980. A competitive ratio for quantifying competition between intercrops. *Exp. Agric.* 16:117–125.

Appendix

Table A1 **2500 Randomly Assorted Digits**

	00–04	05–09	10–14	15–19	20–24	25–29	30–34	35–39	40–44	45–49
00	70896	44520	64720	49898	48088	76740	47460	83150	78905	59870
01	56809	42909	25853	47624	29486	14196	75841	00494	42390	24847
02	66109	84775	07515	49949	61482	91835	48126	80778	21302	24075
03	18071	36263	14053	52526	44347	04924	68100	57805	19521	15345
04	98732	15120	91754	12657	74675	78500	01247	47919	47635	55514
05	83746	47694	96143	42741	38338	97694	69300	99864	19648	15983
06	27998	42562	63404	10056	81668	48744	08400	83124	19896	18805
07	82685	32323	74625	14510	85927	28017	80588	14756	54937	76369
08	18386	13862	10988	04197	18770	72757	71418	81133	69503	44037
09	21717	13141	68165	58440	19187	08421	23872	03036	22707	34208
10	43674	47103	48614	70823	78252	82406	93424	05236	54588	27757
11	68597	68874	35567	98463	99671	05634	81533	47406	17228	44455
12	91874	70208	06308	40719	02772	69589	79936	07514	44950	35190
13	73854	19470	53014	29375	62256	77488	53949	74388	49607	18916
14	65926	34117	55344	68155	38099	56009	03513	05926	35584	42328
15	93282	62518	17752	53163	63852	44840	02592	88572	03107	90169
16	16215	50809	49326	77232	90155	69955	93892	70445	00906	57002
17	09342	14528	64727	71403	84156	34083	35613	35670	10549	07468
18	38148	79001	03509	79424	39625	73315	18811	86230	99682	82896
19	23689	19997	72382	15247	80205	58090	43804	94548	82693	22799
20	97844	62947	62230	30500	92816	85232	28222	91701	11057	83257
21	07611	71163	82212	20653	21499	51496	40715	78952	33029	64207
22	47744	04603	44522	62783	39347	72310	41460	31052	40814	94297
23	54293	43576	88116	67414	34908	15238	40561	73940	56850	81078
24	67556	93979	73363	00300	11217	74405	18937	79000	68834	48307
25	59391	58030	52098	82718	87024	82848	04190	96754	90464	29065
26	99567	76364	77204	04615	27062	96621	43918	01896	83991	51141
27	10363	97518	51400	25670	98342	61891	27101	37855	06235	33163
28	86859	19558	64432	16706	99612	59798	32803	67708	15297	28612
29	11258	24591	36863	55368	31721	94335	34936	02566	80972	08188
30	31432	96156	89177	75541	81355	24480	77243	76690	42507	84362
31	66890	61505	01240	00660	05873	13568	76082	79172	57913	93448
32	41894	57790	79970	33106	86904	48119	52503	24130	72824	21627
33	11303	87118	81471	52936	08555	28420	49416	44448	04269	27029
34	54374	57325	16947	45356	78371	10563	97191	53798	12693	27928
35	99116	75486	23476	52967	67104	39495	39100	17217	74073	84989
36	15696	10703	65178	90637	63110	17622	53988	71087	84148	11670
37	97720	15369	51269	69620	03388	13699	33423	67453	43269	56720
38	11666	13841	71681	98000	35979	39719	81899	07449	37985	46967
39	71628	73130	78783	73691	41623	09847	61547	18707	85498	69944
40	79788	68243	59732	04257	27084	14743	17520	95401	55811	76099
41	40538	79000	89559	25026	42474	23489	34502	75508	06059	86682
42	64016	73598	18609	73150	62463	33102	45205	87440	96767	67042
43	49767	12691	17903	93871	99721	79109	09425	26904	07419	76013
44	76974	55108	29795	08404	82684	00497	51126	79935	57450	55671
45	57430	82279	10421	00540	43648	75888	66049	21511	47676	33444
46	73528	39559	34434	88596	54086	71693	43132	14414	79949	85193
47	25991	65959	70769	64721	86413	33475	42740	06175	82758	66248
48	78388	16638	09134	59980	63806	48472	39318	35434	24057	74739
49	12477	09965	06657	59439	76330	24596	57994	77515	09577	91871

Table A2 **Student's t Distribution**[a]

d.f.	A .600	.700	.800	.900	.950	.975	.990	.995
	B .400	.300	.200	.100	.050	.025	.010	.005
	C .200	.400	.600	.800	.900	.950	.980	.990
	D .800	.600	.400	.200	.100	.050	.020	.010
1	.325	.727	1.376	3.078	6.314	12.706	31.821	63.667
2	.289	.617	1.061	1.886	2.920	4.303	6.965	9.925
3	.277	.584	.978	1.638	2.353	3.182	4.541	5.841
4	.271	.569	.941	1.533	2.132	2.776	3.747	4.604
5	.267	.559	.920	1.476	2.015	2.571	3.365	4.032
6	.265	.553	.906	1.440	1.943	2.447	3.143	3.707
7	.263	.549	.896	1.415	1.895	2.365	2.998	3.499
8	.262	.546	.889	1.397	1.860	2.306	2.896	3.355
9	.261	.543	.883	1.383	1.833	2.262	2.821	3.250
10	.260	.542	.879	1.372	1.812	2.228	2.764	3.169
11	.260	.540	.876	1.363	1.796	2.201	2.718	3.106
12	.259	.539	.873	1.356	1.782	2.179	2.681	3.055
13	.259	.538	.870	1.350	1.771	2.160	2.650	3.012
14	.258	.537	.868	1.345	1.761	2.145	2.624	2.977
15	.258	.536	.866	1.341	1.753	2.131	2.602	2.947
16	.258	.535	.865	1.337	1.746	2.120	2.583	2.921
17	.257	.534	.863	1.333	1.740	2.110	2.567	2.898
18	.257	.534	.862	1.330	1.734	2.101	2.552	2.878
19	.257	.533	.861	1.328	1.729	2.093	2.539	2.861
20	.257	.533	.860	1.325	1.725	2.086	2.528	2.845
21	.257	.532	.859	1.323	1.721	2.080	2.518	2.831
22	.256	.532	.858	1.321	1.717	2.074	2.508	2.819
23	.256	.532	.858	1.319	1.714	2.069	2.500	2.807
24	.256	.531	.857	1.318	1.711	2.064	2.492	2.797
25	.256	.531	.856	1.316	1.708	2.060	2.485	2.787
26	.256	.531	.856	1.315	1.706	2.056	2.479	2.779
27	.256	.531	.855	1.314	1.703	2.052	2.473	2.771
28	.256	.530	.855	1.313	1.701	2.048	2.467	2.763
29	.256	.530	.854	1.311	1.699	2.045	2.462	2.756
30	.256	.530	.854	1.310	1.697	2.042	2.457	2.750
40	.255	.529	.851	1.303	1.684	2.021	2.423	2.704
50	.255	.528	.849	1.298	1.676	2.009	2.403	2.678
60	.254	.527	.848	1.296	1.671	2.000	2.390	2.660
80	.254	.527	.846	1.292	1.664	1.990	2.374	2.639
100	.254	.526	.845	1.290	1.660	1.984	2.365	2.626
∞	.253	.524	.842	1.282	1.645	1.960	2.326	2.576

Source: Abstracted from Table III of Fisher and Yates, *Statistical Tables for Biological, Agricultural and Medical Research*, published by Longman Group Ltd. London (1974), 6th ed. (previously published by Oliver & Boyd, Ltd., Edinburgh) by permission of the authors and publishers.

[a]Column headings = tail probabilities: A, one-sided confidence interval; B, one-tailed significance test; C, two-sided confidence interval; D, two-tailed significance test.

Table A3 Percentage Points of the F Distribution, 5% Points

Denom. d.f.	Numerator d.f. 1	2	3	4	5	6	7	8	9	10	12	15							
1	161.45	199.50	215.71	224.58	230.16	233.99	236.77	238.88	240.54	241.88	243.91	245.95	248.01	249.05	250.09	251.14	252.20	253.25	254.32
2	18.51	19.00	19.16	19.25	19.30	19.33	19.36	19.37	19.38	19.40	19.41	19.43	19.45	19.45	19.46	19.47	19.48	19.49	19.50
3	10.13	9.55	9.28	9.12	9.01	8.94	8.89	8.84	8.81	8.78	8.74	8.70	8.66	8.64	8.62	8.59	8.57	8.55	8.53
4	7.71	6.94	6.59	6.39	6.26	6.16	6.08	6.04	6.00	5.96	5.91	5.86	5.80	5.77	5.74	5.72	5.69	5.66	5.63
5	6.61	5.79	5.41	5.19	5.05	4.95	4.88	4.81	4.77	4.74	4.68	4.62	4.56	4.53	4.50	4.46	4.43	4.40	4.36
6	5.99	5.14	4.76	4.53	4.39	4.28	4.21	4.14	4.10	4.06	4.00	3.94	3.87	3.84	3.81	3.77	3.74	3.70	3.67
7	5.59	4.74	4.35	4.12	3.97	3.87	3.79	3.72	3.68	3.64	3.57	3.51	3.44	3.41	3.38	3.34	3.30	3.27	3.23
8	5.32	4.46	4.07	3.84	3.69	3.58	3.50	3.44	3.39	3.35	3.28	3.22	3.15	3.12	3.08	3.04	3.00	3.97	2.93
9	5.12	4.26	3.86	3.63	3.48	3.37	3.29	3.23	3.18	3.14	3.07	3.01	2.94	2.90	2.86	2.82	2.79	2.75	2.71
10	4.96	4.10	3.71	3.48	3.32	3.22	3.13	3.07	3.02	2.98	2.91	2.84	2.77	2.74	2.70	2.66	2.62	2.58	2.54
11	4.84	3.98	3.59	3.36	3.20	3.09	3.01	2.95	2.90	2.85	2.79	2.72	2.64	2.61	2.57	2.53	2.49	2.45	2.40
12	4.75	3.88	3.49	3.26	3.10	3.00	2.91	2.85	2.80	2.75	2.69	2.62	2.54	2.50	2.47	2.43	2.38	2.34	2.30
13	4.67	3.80	3.41	3.18	3.02	2.92	2.83	2.76	2.71	2.67	2.60	2.53	2.46	2.42	2.38	2.34	2.30	2.25	2.21
14	4.60	3.74	3.34	3.11	2.96	2.85	2.76	2.69	2.64	2.60	2.53	2.46	2.40	2.35	2.31	2.26	2.22	2.18	2.13
15	4.54	3.68	3.29	3.06	2.90	2.79	2.71	2.64	2.59	2.54	2.48	2.40	2.33	2.29	2.25	2.20	2.16	2.11	2.06
16	4.49	3.63	3.24	3.01	2.85	2.74	2.66	2.59	2.54	2.49	2.42	2.35	2.28	2.24	2.19	2.15	2.10	2.06	2.01
17	4.45	3.59	3.20	2.96	2.81	2.70	2.61	2.55	2.49	2.45	2.38	2.31	2.23	2.19	2.15	2.10	2.06	2.01	1.96
18	4.41	3.55	3.16	2.93	2.77	2.66	2.58	2.51	2.46	2.41	2.34	2.27	2.19	2.15	2.11	2.06	2.02	1.97	1.92
19	4.38	3.52	3.13	2.90	2.74	2.63	2.54	2.47	2.42	2.38	2.31	2.23	2.16	2.11	2.07	2.03	1.98	1.93	1.88
20	4.35	3.49	3.10	2.87	2.71	2.60	2.51	2.45	2.39	2.35	2.28	2.20	2.12	2.08	2.04	1.99	1.95	1.90	1.84
21	4.32	3.47	3.07	2.84	2.68	2.57	2.49	2.42	2.37	2.32	2.25	2.18	2.10	2.05	2.01	1.96	1.92	1.86	1.81
22	4.30	3.44	3.05	2.82	2.66	2.55	2.46	2.40	2.34	2.30	2.22	2.15	2.07	2.03	1.98	1.94	1.89	1.84	1.78
23	4.28	3.42	3.03	2.80	2.64	2.53	2.44	2.37	2.32	2.27	2.20	2.13	2.05	2.00	1.96	1.91	1.86	1.81	1.76
24	4.26	3.40	3.00	2.78	2.62	2.51	2.42	2.36	2.30	2.25	2.18	2.11	2.03	1.98	1.94	1.89	1.84	1.79	1.73
25	4.24	3.38	2.99	2.76	2.60	2.49	2.40	2.34	2.28	2.24	2.16	2.09	2.01	1.96	1.92	1.87	1.82	1.77	1.71
26	4.22	3.37	2.98	2.74	2.59	2.47	2.39	2.32	2.26	2.21	2.15	2.07	1.99	1.95	1.90	1.85	1.80	1.75	1.69
27	4.21	3.35	2.96	2.73	2.57	2.46	2.37	2.30	2.25	2.20	2.13	2.06	1.97	1.93	1.88	1.84	1.78	1.73	1.67
28	4.20	3.34	2.95	2.71	2.56	2.45	2.36	2.29	2.24	2.19	2.12	2.04	1.96	1.91	1.87	1.82	1.77	1.71	1.65
29	4.18	3.33	3.93	2.70	2.54	2.43	2.35	2.28	2.22	2.17	2.10	2.03	1.94	1.90	1.85	1.80	1.75	1.70	1.64
30	4.17	3.32	2.92	2.69	2.53	2.42	2.33	2.27	2.21	2.16	2.09	2.01	1.93	1.89	1.84	1.79	1.74	1.68	1.62
40	4.08	3.23	2.84	2.61	2.45	2.34	2.25	2.18	2.12	2.08	2.00	1.92	1.84	1.79	1.74	1.69	1.64	1.58	1.51
60	4.00	3.15	2.76	2.52	2.37	2.25	2.17	2.10	2.04	1.99	1.92	1.84	1.75	1.70	1.65	1.59	1.53	1.47	1.39
120	3.92	3.07	2.68	2.45	2.29	2.18	2.09	2.02	1.96	1.91	1.83	1.75	1.66	1.61	1.55	1.49	1.43	1.35	1.25
∞	3.84	3.00	2.60	2.37	2.21	2.10	2.01	1.94	1.88	1.83	1.75	1.67	1.57	1.52	1.46	1.39	1.32	1.22	1.00

Table A3 F Distribution (*cont'd*), 1% Points

Denom. d.f.	Numerator d.f. 1	2	3	4	5	6	7	8	9	10	12	15	20	24	30	40	60	120	∞
1	4052.2	4999.5	5403.3	5624.6	5763.7	5859.0	5928.3	5981.6	6022.5	6055.8	6106.3	6157.3	6208.7	6234.6	6260.7	6286.8	6313.0	6339.4	6366.0
2	98.50	99.00	99.17	99.25	99.30	99.33	99.36	99.37	99.39	99.40	99.42	99.43	99.45	99.46	99.47	99.47	99.48	99.49	99.50
3	34.12	30.82	29.46	28.71	28.24	27.91	27.67	27.49	27.34	27.23	27.05	26.87	26.69	26.60	26.50	26.41	26.32	26.22	26.12
4	21.20	18.00	16.60	15.98	15.52	15.21	14.98	14.80	14.66	14.55	14.37	14.20	14.02	13.93	13.84	13.74	13.65	13.56	13.46
5	16.26	13.27	12.06	11.39	10.97	10.67	10.46	10.29	10.16	10.05	9.89	9.72	9.55	9.47	9.38	9.29	9.20	9.11	9.02
6	13.74	10.92	9.78	9.15	8.74	8.47	8.26	8.10	7.98	7.87	7.72	7.56	7.40	7.31	7.23	7.14	7.06	6.97	6.88
7	12.25	9.55	8.45	7.85	7.46	7.19	6.99	6.84	6.72	6.62	6.47	6.31	6.16	6.07	5.99	5.91	5.82	5.74	5.65
8	11.26	8.65	7.59	7.01	6.63	6.37	6.18	6.03	5.91	5.81	5.67	5.52	5.36	5.28	5.20	5.12	5.03	4.95	4.86
9	10.56	8.02	6.99	6.42	6.06	5.80	5.61	5.47	5.35	5.26	5.11	4.96	4.81	4.73	4.67	4.57	4.48	4.40	4.31
10	10.04	7.56	6.55	5.99	5.64	5.38	5.20	5.06	4.94	4.85	4.70	4.56	4.40	4.32	4.25	4.16	4.08	4.00	3.91
11	9.65	7.20	6.22	5.67	5.32	5.07	4.89	4.74	4.63	4.54	4.40	4.25	4.10	4.02	3.94	3.86	3.78	3.69	3.60
12	9.33	6.93	5.95	5.41	5.06	4.82	4.64	4.50	4.39	4.30	4.16	4.01	3.86	3.78	3.70	3.62	3.54	3.45	3.36
13	9.07	6.70	5.74	5.20	4.86	4.62	4.44	4.30	4.19	4.10	3.96	3.82	3.66	3.59	3.51	3.42	3.34	3.25	3.16
14	8.86	6.51	5.56	5.04	4.70	4.46	4.28	4.14	4.03	3.94	3.80	3.66	3.50	3.43	3.35	3.26	3.18	3.09	3.00
15	8.68	6.36	5.42	4.89	4.56	4.32	4.14	4.00	3.89	3.80	3.67	3.52	3.37	3.29	3.21	3.13	3.05	2.96	2.87
16	8.53	6.23	5.29	4.77	4.44	4.20	4.03	3.89	3.78	3.69	3.55	3.41	3.26	3.18	3.10	3.02	2.93	2.84	2.75
17	8.40	6.11	5.18	4.67	4.34	4.10	3.93	3.79	3.68	3.59	3.46	3.31	3.16	3.08	3.00	2.92	2.83	2.74	2.65
18	8.28	6.01	5.09	4.58	4.25	4.01	3.84	3.70	3.60	3.51	3.37	3.23	3.08	3.00	2.92	2.84	2.75	2.66	2.57
19	8.18	5.93	5.01	4.50	4.17	3.94	3.76	3.63	3.52	3.43	3.30	3.15	3.00	2.92	2.84	2.76	2.67	2.58	2.49
20	8.10	5.85	4.94	4.43	4.10	3.87	3.70	3.56	3.46	3.37	3.23	3.09	2.94	2.86	2.79	2.69	2.61	2.52	2.42
21	8.02	5.78	4.87	4.37	4.04	3.81	3.64	3.51	3.40	3.31	3.17	3.03	2.88	2.80	2.72	2.64	2.55	2.46	2.36
22	7.94	5.72	4.82	4.31	3.99	3.76	3.59	3.45	3.34	3.26	3.12	2.98	2.83	2.75	2.67	2.58	2.50	2.40	2.30
23	7.88	5.66	4.76	4.26	3.94	3.71	3.54	3.40	3.30	3.21	3.07	2.94	2.78	2.70	2.62	2.54	2.45	2.35	2.26
24	7.82	5.61	4.72	4.22	3.90	3.67	3.50	3.36	3.26	3.17	3.03	2.89	2.74	2.66	2.58	2.49	2.40	2.31	2.21
25	7.77	5.57	4.68	4.18	3.86	3.63	3.46	3.32	3.22	3.13	2.99	2.85	2.70	2.62	2.54	2.45	2.36	2.27	2.17
26	7.72	5.53	4.64	4.14	3.82	3.59	3.42	3.29	3.18	3.09	2.96	2.82	2.66	2.58	2.50	2.42	2.33	2.23	2.13
27	7.68	5.49	4.60	4.10	3.78	3.56	3.39	3.26	3.15	3.06	2.92	2.78	2.63	2.55	2.47	2.38	2.29	2.20	2.10
28	7.64	5.45	4.57	4.07	3.75	3.53	3.36	3.22	3.12	3.03	2.90	2.75	2.60	2.52	2.44	2.35	2.26	2.17	2.06
29	7.60	5.42	4.54	4.04	3.72	3.50	3.33	3.20	3.09	3.00	2.87	2.72	2.57	2.49	2.41	2.32	2.23	2.14	2.03
30	7.56	5.39	4.51	4.02	3.70	3.47	3.30	3.17	3.07	2.98	2.84	2.70	2.55	2.47	2.39	2.30	2.21	2.11	2.01
40	7.31	5.18	4.31	3.83	3.51	3.29	3.12	2.99	2.89	2.80	2.66	2.52	2.37	2.29	2.20	2.11	2.02	1.92	1.80
60	7.08	4.98	4.12	3.65	3.34	3.12	2.95	2.82	2.72	2.63	2.50	2.35	2.20	2.11	2.03	1.94	1.84	1.73	1.60
120	6.85	4.79	3.95	3.48	3.17	2.96	2.79	2.66	2.56	2.47	2.34	2.19	2.03	1.95	1.86	1.76	1.66	1.53	1.38
∞	6.63	4.60	3.78	3.32	3.02	2.80	2.64	2.51	2.41	2.32	2.18	2.04	1.88	1.79	1.70	1.59	1.47	1.32	1.00

Source: This table is abstracted from "Tables of Percentage Points of the Inverted Beta (F) Distribution," *Biometrika* 33 (1943), by M. Merrington and C. M. Thompson, by permission of the authors and editors of Biometrika.

Table A4 **Percentage Points of the Ratio $s^2(max)/s^2(min)$[a]**

Upper 5% Points

γ \ k	2	3	4	5	6	7
2	39.0	87.5	142	202	266	333
3	15.4	27.8	39.2	50.7	62.0	72.9
4	9.60	15.5	20.6	25.2	29.5	33.6
5	7.15	10.8	13.7	16.3	18.7	20.8
6	5.82	8.38	10.4	12.1	13.7	15.0
7	4.99	6.94	8.44	9.70	10.8	11.8
8	4.43	6.00	7.18	8.12	9.03	9.78
9	4.03	5.34	6.31	7.11	7.80	8.41
10	3.72	4.85	5.67	6.34	6.92	7.42
12	3.28	4.16	4.79	5.30	5.72	6.09
15	2.86	3.54	4.01	4.37	4.68	4.95
20	2.46	2.95	3.29	3.54	3.76	3.94
30	2.07	2.40	2.61	2.78	2.91	3.02
60	1.67	1.85	1.96	2.04	2.11	2.17

Upper 5% Points (con't)

γ \ k	8	9	10	11	12
2	403	475	550	626	704
3	83.5	93.9	104	114	124
4	37.5	41.1	44.6	48.0	51.4
5	22.9	24.7	26.5	28.2	29.9
6	16.3	17.5	18.6	19.7	20.7
7	12.7	13.5	14.3	15.1	15.8
8	10.5	11.1	11.7	12.2	12.7
9	8.95	9.45	9.91	10.3	10.7
10	7.87	8.28	8.66	9.01	9.34
12	6.42	6.72	7.00	7.25	7.48
15	5.19	5.40	5.59	5.77	5.93
20	4.10	4.24	4.37	4.49	4.59
30	3.12	3.21	3.29	3.36	3.39
60	2.22	2.26	2.30	2.33	2.36

ν	k 2	3	4	5	6	7
2	199	448	729	1036	1362	1705
3	47.5	85	120	151	184	216
4	23.2	37	49	59	69	79
5	14.9	22	28	33	38	42
6	11.1	15.5	19.1	22	25	27
7	8.89	12.1	14.6	16.5	18.4	20
8	7.50	9.9	11.7	13.2	14.5	15.8
9	6.54	8.5	9.9	11.1	12.1	13.1
10	5.85	7.4	8.6	9.6	10.4	11.1
12	4.91	6.1	6.9	7.6	8.2	8.7
15	4.07	4.9	5.5	6.0	6.4	6.7
20	3.32	3.8	4.3	4.6	4.9	5.1
30	2.63	3.0	3.3	3.4	3.6	3.7
60	1.96	2.2	2.3	2.4	2.4	2.5

Upper 1% Points (Con't)

ν	k 8	9	10	11	12
2	2063	2432	2813	3204	3605
3	249	281	310	337	361
4	89	97	106	113	120
5	46	50	54	57	60
6	30	32	34	36	37
7	22	23	24	26	27
8	16.9	17.9	18.9	19.8	21
9	13.9	14.7	15.3	16.0	16.6
10	11.8	12.4	12.9	13.4	13.9
12	9.1	9.5	9.9	10.2	10.6
15	7.1	7.3	7.5	7.8	8.0
20	5.3	5.5	5.6	5.8	5.9
30	3.8	3.9	4.0	4.1	4.2
60	2.5	2.6	2.6	2.7	2.7

Source: This table is abstracted from Table 31, *Biometrika Tables for Statisticians*, Vol. 1, 3rd ed., 1966, by permission of the Biometrika Trustees.

[a] s^2(max) is the largest and s^2(min) is the smallest in a set of k independent mean squares, each based on ν degrees of freedom.

Table A5 **Percentage Points of the Chi-Square Distribution**

Probability of a Greater Value

df	0.250	0.100	0.050	0.025	0.010	.005
1	1.32	2.71	3.84	5.02	6.63	7.88
2	2.77	4.61	5.99	7.38	9.21	10.60
3	4.11	6.25	7.81	9.35	11.34	12.84
4	5.39	7.78	9.49	11.14	13.28	14.86
5	6.63	9.24	11.07	12.83	15.09	16.75
6	7.84	10.64	12.59	14.45	16.81	18.55
7	9.04	12.02	14.07	16.01	18.48	20.28
8	10.22	13.36	15.51	17.53	20.09	21.96
9	11.39	14.68	16.92	19.02	21.67	23.59
10	12.55	15.99	18.31	20.48	23.21	25.19
11	13.70	17.28	19.68	21.92	24.72	26.76
12	14.85	18.55	21.03	23.34	26.22	28.30
13	15.98	19.81	22.36	24.74	27.69	29.82
14	17.12	21.06	23.68	26.12	29.14	31.32
15	18.25	22.31	25.00	27.49	30.58	32.80
16	19.37	23.54	26.30	28.85	32.00	34.27
17	20.49	24.77	27.59	30.19	33.41	35.72
18	21.60	25.99	28.87	31.53	34.81	37.16
19	22.72	27.20	30.14	32.85	36.19	38.58
20	23.83	28.41	31.41	34.17	37.57	40.00
22	26.04	30.81	33.92	36.78	40.29	42.80
24	28.24	33.20	36.42	39.36	42.98	45.56
26	30.43	35.56	38.89	41.92	45.64	48.29
28	32.62	37.92	41.34	44.46	48.28	50.99
30	34.80	40.26	43.77	46.98	50.89	53.67
40	45.62	51.80	55.76	59.34	63.69	66.77
50	56.33	63.17	67.50	71.42	76.15	79.49
60	66.98	74.40	79.08	83.30	88.38	91 95
80	88.13	96.58	101.88	106.63	112.33	116.32
100	109.14	118.50	124.34	129.56	135.81	140.17

Source: This table is extracted from Table 8, *Biometrika Tables for Statisticians*, Vol. 1, 3rd ed., 1966, by permission of the Biometrika Trustees.

Table A6 **Percentage Points of the Studentized Range (*Q*): Upper 5 % Points of the Studentized Range. Entries are $q_{.05}$ Where $P(q < q_{.05}) = .95$**

ν	n = 2	3	4	5	6	7	8	9	10	11	12	13	14	15	16	17	18	19	20
1	17.97	26.98	32.82	37.08	40.41	43.12	45.40	47.36	49.07	50.59	51.96	53.20	54.33	55.36	56.32	57.22	58.04	58.83	59.56
2	6.08	8.33	9.80	10.88	11.74	12.44	13.03	13.54	13.99	14.39	14.75	15.08	15.38	15.65	15.91	16.14	16.37	16.57	16.77
3	4.50	5.91	6.82	7.50	8.04	8.48	8.85	9.18	9.46	9.72	9.95	10.15	10.35	10.53	10.69	10.84	10.98	11.11	11.24
4	3.83	5.04	5.78	6.29	6.71	7.05	7.35	7.60	7.83	8.03	8.21	8.37	8.52	8.66	8.79	8.91	9.03	9.13	9.23
5	3.64	4.60	6.22	5.67	6.03	6.33	6.58	6.80	6.99	7.17	7.32	7.47	7.60	7.72	7.83	7.93	8.03	8.12	8.21
6	3.46	4.34	4.90	5.30	5.63	5.90	6.12	6.32	6.49	6.65	6.79	6.92	7.03	7.14	7.24	7.34	7.43	7.51	7.59
7	3.34	4.16	4.68	5.06	5.36	5.61	5.82	6.00	6.16	6.30	6.43	6.55	6.66	6.76	6.85	6.94	7.02	7.10	7.17
8	3.26	4.04	4.53	4.89	5.17	5.40	5.60	5.77	5.92	6.05	6.18	6.29	6.39	6.48	6.57	6.65	6.73	6.80	6.87
9	3.20	3.95	4.41	4.76	5.02	5.24	5.43	5.59	5.74	5.87	5.98	6.09	6.19	6.28	6.36	6.44	6.51	6.58	6.64
10	3.15	3.88	4.33	4.65	4.91	5.12	5.30	5.46	5.60	5.72	5.83	5.93	6.03	6.11	6.19	6.27	6.34	6.40	6.47
11	3.11	3.82	4.26	4.57	4.82	5.03	5.20	5.35	5.49	5.61	5.71	5.81	5.90	5.98	6.06	6.13	6.20	6.27	6.33
12	3.08	3.77	4.20	4.51	4.75	4.95	5.12	5.27	5.39	5.51	5.61	5.71	5.80	5.88	5.95	6.02	6.09	6.15	6.21
13	3.06	3.73	4.15	4.45	4.69	4.88	5.05	5.10	5.32	5.43	5.53	5.63	5.71	5.79	5.86	5.93	5.99	6.05	6.11
14	3.03	3.70	4.11	4.41	4.64	4.83	4.99	5.13	5.25	5.36	5.48	5.55	5.84	5.71	5.79	5.85	5.91	5.97	6.03
15	3.01	3.67	4.08	4.37	4.59	4.78	4.94	5.08	5.20	5.31	5.40	5.49	5.57	5.65	5.72	5.78	5.85	5.90	5.96
16	3.00	3.65	4.05	4.33	4.56	4.74	4.90	5.03	5.15	5.26	5.35	5.44	5.52	5.59	5.66	5.73	5.79	5.84	5.90
17	2.98	3.63	4.02	4.30	4.52	4.70	4.86	4.99	5.11	5.21	5.31	5.39	5.47	5.54	5.61	5.67	5.73	5.79	5.84
18	2.97	3.61	4.00	4.28	4.49	4.67	4.82	4.96	5.07	5.17	5.27	5.35	5.43	5.50	5.57	5.63	5.69	5.74	5.79
19	2.96	3.59	3.98	4.23	4.47	4.65	4.79	4.92	5.04	5.14	5.23	5.31	5.39	5.48	5.53	5.59	5.65	5.70	5.75
20	2.95	3.58	3.96	4.23	4.45	4.62	4.77	4.90	5.01	5.11	5.20	5.28	5.36	5.43	5.49	5.55	5.61	5.66	5.71
24	2.92	3.53	3.90	4.17	4.37	4.54	4.68	4.81	4.92	5.01	5.10	5.18	5.25	5.32	5.38	5.44	5.49	5.55	5.59
30	2.89	3.49	3.85	4.10	4.30	4.46	4.60	4.72	4.82	4.92	5.00	5.08	5.15	5.21	5.27	5.33	5.38	5.43	5.47
40	2.86	3.44	3.79	4.04	4.23	4.39	4.52	4.63	4.73	4.82	4.90	4.98	5.04	5.11	5.16	5.22	5.27	5.31	5.36
60	2.83	3.40	3.74	3.98	4.16	4.31	4.44	4.55	4.65	4.73	4.81	4.88	4.94	5.00	5.06	5.11	5.15	5.20	5.24
120	2.80	3.36	3.68	3.92	4.10	4.24	4.36	4.47	4.58	4.64	4.71	4.78	4.84	4.90	4.95	5.00	5.04	5.09	5.13
∞	2.77	3.31	3.63	3.86	4.03	4.17	4.29	4.39	4.47	4.55	4.62	4.68	4.74	4.80	4.85	4.89	4.93	4.97	5.01

Table A6 (*Continued*): Upper 1 % Points of the Studentized Range. Entries are $q_{.01}$ Where $P(q < q_{.01}) = .99$

ν	n: 2	3	4	5	6	7	8	9	10	11	12	13	14	15	16	17	18	19	20
1	90.03	135.0	164.3	185.6	202.2	215.8	227.2	237.0	245.6	253.2	260.0	266.2	271.8	277.0	281.8	286.3	290.4	294.3	298.0
2	14.04	19.02	22.29	24.72	26.63	28.20	29.53	30.68	31.69	32.59	33.40	34.13	34.81	35.43	36.00	36.53	37.03	37.50	37.95
3	8.26	10.62	12.17	13.33	14.24	15.00	15.64	16.20	16.69	17.13	17.53	17.89	18.22	18.52	18.81	19.07	19.32	19.55	19.77
4	6.51	8.12	9.17	9.96	10.58	11.10	11.55	11.93	12.27	12.57	12.84	13.09	13.32	13.53	13.73	13.91	14.08	14.24	14.40
5	5.70	6.98	7.80	8.42	8.91	9.32	9.67	9.97	10.24	10.48	10.70	10.89	11.08	11.24	11.40	11.55	11.68	11.81	11.93
6	5.24	6.33	7.03	7.56	7.97	8.32	8.61	8.87	9.10	9.30	9.48	9.65	9.81	9.95	10.08	10.21	10.32	10.43	10.54
7	4.95	5.92	6.54	7.01	7.37	7.68	7.94	8.17	8.37	8.55	8.71	8.86	9.00	9.12	9.24	9.35	9.46	9.55	9.65
8	4.75	5.64	6.20	6.62	6.96	7.24	7.47	7.68	7.86	8.03	8.18	8.31	8.44	8.55	8.66	8.76	8.85	8.94	9.03
9	4.60	5.43	5.96	6.35	6.66	6.91	7.13	7.33	7.49	7.65	7.78	7.91	8.03	8.13	8.23	8.33	8.41	8.49	8.57
10	4.48	5.27	5.77	6.14	6.43	6.67	6.87	7.05	7.21	7.36	7.49	7.60	7.71	7.81	7.91	7.99	8.08	8.15	8.23
11	4.39	5.15	5.62	5.97	6.25	6.48	6.67	6.84	6.99	7.13	7.25	7.36	7.46	7.56	7.65	7.73	7.81	7.88	7.95
12	4.32	5.05	5.50	5.84	6.10	6.32	6.51	6.67	6.81	6.94	7.06	7.17	7.26	7.36	7.44	7.52	7.59	7.66	7.73
13	4.26	4.96	5.40	5.73	5.98	6.19	6.37	6.53	6.67	6.79	6.90	7.01	7.10	7.19	7.27	7.35	7.42	7.48	7.55
14	4.21	4.89	5.32	5.63	5.88	6.08	6.26	6.41	6.54	6.66	6.77	6.87	6.96	7.05	7.13	7.20	7.27	7.33	7.39
15	4.17	4.84	5.25	5.56	5.80	5.99	6.16	6.31	6.44	6.55	6.66	6.76	6.84	6.93	7.00	7.07	7.14	7.20	7.26
16	4.13	4.79	5.19	5.49	5.72	5.92	6.08	6.22	6.35	6.46	6.56	6.66	6.74	6.82	6.90	6.97	7.03	7.09	7.15
17	4.10	4.74	5.14	5.43	5.66	5.85	6.01	6.15	6.27	6.38	6.48	6.57	6.66	6.73	6.81	6.87	6.94	7.00	7.05
18	4.07	4.70	5.09	5.38	5.60	5.79	5.94	6.08	6.20	6.31	6.41	6.50	6.58	6.65	6.73	6.79	6.85	6.91	6.97
19	4.05	4.67	5.05	5.33	5.55	5.73	5.89	6.02	6.14	6.25	6.34	6.43	6.51	6.58	6.65	6.72	6.78	6.84	6.89
20	4.02	4.64	5.02	5.29	5.51	5.69	5.84	5.97	6.09	6.19	6.28	6.37	6.45	6.52	6.59	6.65	6.71	6.77	6.82
24	3.96	4.55	4.91	5.17	5.37	5.54	5.69	5.81	5.92	6.02	6.11	6.19	6.26	6.33	6.39	6.45	6.51	6.56	6.61
30	3.89	4.45	4.80	5.05	5.24	5.40	5.54	5.65	5.76	5.85	5.93	6.01	6.08	6.14	6.20	6.26	6.31	6.36	6.41
40	3.82	4.37	4.70	4.93	5.11	5.26	5.39	5.50	5.60	5.69	5.76	5.83	5.90	5.96	6.02	6.07	6.12	6.16	6.21
60	3.76	4.28	4.59	4.82	4.99	5.13	5.25	5.36	5.45	5.53	5.60	5.67	5.73	5.78	5.84	5.89	5.93	5.97	6.01
120	3.70	4.20	4.50	4.71	4.87	5.01	5.12	5.21	5.30	5.37	5.44	5.50	5.56	5.61	5.66	5.71	5.75	5.79	5.83
∞	3.64	4.12	4.40	4.60	4.76	4.88	4.99	5.08	5.16	5.23	5.29	5.35	5.40	5.45	5.49	5.54	5.57	5.61	5.65

Note: Reprinted from *Biometrika Tables for Statisticians*, Vol. 1, 3rd ed. (1966), by permission of the *Biometrika* trustees.

Table A7 **Minimum-Average-Risk *t* Values**

q	f: 6	8	10	12	14	16	18	20	24	30	40	60	120
							(k = 100)						
F = 1.2 (a = .913, b = 2.449)													
2-6	*	*	*	*	*	*	*	*	*	*	*	*	*
8	2.91	2.94	2.96	2.97	2.98	2.99	2.99	2.99	3.00	3.00	3.00	3.00	3.00
10	2.93	2.98	3.01	3.04	3.05	3.06	3.07	3.08	3.09	3.10	3.10	3.11	3.12
12	2.95	3.01	3.05	3.08	3.10	3.12	3.13	3.14	3.16	3.17	3.19	3.20	3.21
14	2.96	3.03	3.08	3.12	3.14	3.16	3.18	3.19	3.21	3.23	3.25	3.27	3.29
16	2.97	3.05	3.11	3.15	3.18	3.20	3.22	3.24	3.26	3.28	3.31	3.33	3.36
20	2.99	3.08	3.14	3.19	3.23	3.26	3.28	3.30	3.33	3.37	3.40	3.44	3.47
40	3.02	3.13	3.22	3.29	3.35	3.39	3.43	3.47	3.52	3.58	3.64	3.72	3.79
100	3.04	3.17	3.28	3.36	3.44	3.50	3.55	3.59	3.67	3.76	3.86	3.98	4.11
∞	3.05	3.20	3.32	3.42	3.50	3.58	3.64	3.70	3.80	3.91	4.06	4.24	4.45
F = 1.4 (a = .845, b = 1.871)													
2-4	*	*	*	*	*	*	*	*	*	*	*	*	*
6	2.85	2.84	2.83	2.82	2.82	2.81	2.80	2.80	2.79	2.78	2.77	2.75	2.74
8	2.88	2.89	2.90	2.90	2.90	2.89	2.89	2.89	2.88	2.88	2.87	2.86	2.85
10	2.90	2.93	2.94	2.95	2.95	2.96	2.96	2.96	2.95	2.95	2.95	2.94	2.93
12	2.92	2.95	2.98	2.99	3.00	3.00	3.01	3.01	3.01	3.01	3.01	3.00	2.99
14	2.93	2.97	3.00	3.02	3.03	3.04	3.04	3.05	3.05	3.06	3.06	3.05	3.05
16	2.94	2.99	3.02	3.04	3.06	3.07	3.08	3.08	3.09	3.09	3.10	3.10	3.09
20	2.95	3.01	3.05	3.08	3.10	3.11	3.12	3.13	3.14	3.15	3.16	3.16	3.16
40	2.98	3.06	3.12	3.16	3.19	3.22	3.24	3.25	3.28	3.30	3.31	3.32	3.32
100	2.99	3.09	3.16	3.22	3.26	3.29	3.32	3.34	3.38	3.41	3.43	3.45	3.42
∞	3.01	3.12	3.20	3.26	3.31	3.35	3.39	3.42	3.46	3.50	3.53	3.54	3.46
F = 1.7 (a = .767, b = 1.558)													
2	*	*	*	*	*	*	*	*	*	*	*	*	*
4	*	*	*	*	*	2.61	2.59	2.58	2.56	2.54	2.52	2.50	2.48
6	2.82	2.79	2.76	2.74	2.72	2.71	2.70	2.69	2.67	2.65	2.63	2.61	2.58
8	2.84	2.83	2.81	2.80	2.78	2.77	2.76	2.75	2.74	2.72	2.70	2.68	2.65
10	2.86	2.86	2.85	2.84	2.83	2.82	2.81	2.80	2.79	2.77	2.75	2.73	2.70
12	2.87	2.88	2.88	2.87	2.86	2.85	2.84	2.84	2.82	2.81	2.79	2.76	2.73
14	2.88	2.90	2.90	2.89	2.89	2.88	2.87	2.86	2.85	2.83	2.81	2.79	2.75
16	2.89	2.91	2.91	2.91	2.90	2.90	2.89	2.89	2.87	2.86	2.84	2.81	2.77
20	2.90	2.93	2.93	2.94	2.93	2.93	2.92	2.92	2.91	2.89	2.87	2.84	2.80
40	2.93	2.97	2.99	3.00	3.00	3.00	3.00	2.99	2.98	2.97	2.94	2.89	2.83
100	2.94	2.99	3.02	3.04	3.05	3.05	3.05	3.05	3.04	3.02	2.98	2.92	2.83
∞	2.95	3.01	3.05	3.07	3.08	3.09	3.09	3.08	3.07	3.05	3.01	2.93	2.81

Table A7 *(Continued)*

q	f: 6	8	10	12	14	16	18	20	24	30	40	60	120
F = 2.0 (a = .707, b = 1.414)													
2	*	*	*	*	*	*	*	*	*	*	*	*	*
4	2.74	2.67	2.63	2.59	2.56	2.54	2.52	2.51	2.49	2.46	2.44	2.41	2.39
6	2.79	2.74	2.70	2.67	2.64	2.62	2.60	2.59	2.57	2.54	2.52	2.49	2.46
8	2.81	2.77	2.74	2.71	2.69	2.67	2.65	2.64	2.62	2.59	2.56	2.53	2.49
10	2.83	2.80	2.77	2.74	2.72	2.70	2.69	2.67	2.65	2.62	2.59	2.56	2.52
12	2.84	2.82	2.79	2.77	2.75	2.73	2.71	2.70	2.67	2.64	2.61	2.57	2.53
14	2.85	2.83	2.81	2.79	2.77	2.75	2.73	2.72	2.69	2.66	2.63	2.59	2.54
16	2.85	2.84	2.82	2.80	2.78	2.76	2.74	2.73	2.70	2.67	2.64	2.59	2.54
20	2.86	2.85	2.84	2.82	2.80	2.78	2.77	2.75	2.72	2.69	2.65	2.61	2.55
40	2.88	2.89	2.88	2.86	2.85	2.83	2.81	2.80	2.77	2.73	2.68	2.62	2.55
100	2.89	2.91	2.90	2.89	2.88	2.86	2.84	2.82	2.79	2.75	2.69	2.62	2.53
∞	2.90	2.92	2.92	2.91	2.90	2.88	2.86	2.85	2.81	2.76	2.69	2.61	2.52
F = 2.4 (a = .645, b = 1.309)													
2	*	*	*	*	*	*	*	*	*	*	*	*	2.18
4	2.71	2.63	2.57	2.53	2.49	2.47	2.44	2.43	2.40	2.37	2.34	2.31	2.28
6	2.75	2.68	2.63	2.58	2.55	2.52	2.50	2.48	2.46	2.42	2.39	2.36	2.32
8	2.77	2.71	2.66	2.62	2.59	2.56	2.54	2.52	2.49	2.45	2.42	2.38	2.34
10	2.79	2.73	2.68	2.64	2.61	2.58	2.56	2.54	2.50	2.47	2.43	2.39	2.34
12	2.79	2.74	2.70	2.66	2.62	2.60	2.57	2.55	2.52	2.48	2.44	2.39	2.35
14	2.80	2.75	2.71	2.67	2.64	2.61	2.58	2.56	2.53	2.49	2.44	2.40	2.35
16	2.81	2.76	2.72	2.68	2.65	2.62	2.59	2.57	2.53	2.49	2.45	2.40	2.34
20	2.82	2.77	2.73	2.69	2.66	2.63	2.60	2.58	2.54	2.50	2.45	2.40	2.34
40	2.83	2.80	2.76	2.72	2.69	2.66	2.63	2.60	2.56	2.51	2.46	2.39	2.33
100	2.84	2.81	2.78	2.74	2.71	2.67	2.64	2.62	2.57	2.51	2.45	2.39	2.32
∞	2.85	2.83	2.79	2.76	2.72	2.68	2.65	2.62	2.57	2.51	2.45	2.38	2.31
F = 3.0 (a = .577, b = 1.225)													
2	*	*	2.41	2.36	2.32	2.29	2.27	2.25	2.22	2.20	2.17	2.14	2.11
4	2.68	2.57	2.50	2.45	2.41	2.38	2.35	2.33	2.30	2.27	2.24	2.20	2.17
6	2.71	2.61	2.54	2.49	2.44	2.41	2.39	2.36	2.33	2.29	2.26	2.22	2.18
8	2.72	2.63	2.56	2.51	2.47	2.43	2.40	2.38	2.34	2.31	2.27	2.22	2.18
10	2.74	2.65	2.58	2.52	2.48	2.44	2.41	2.39	2.35	2.31	2.27	2.22	2.18
12	2.74	2.66	2.59	2.53	2.49	2.45	2.42	2.40	2.36	2.31	2.27	2.22	2.18
14	2.75	2.66	2.60	2.54	2.49	2.46	2.43	2.40	2.36	2.32	2.27	2.22	2.17
16	2.75	2.67	2.60	2.55	2.50	2.46	2.43	2.40	2.36	2.32	2.27	2.22	2.17
20	2.76	2.68	2.61	2.55	2.51	2.47	2.43	2.41	2.36	2.32	2.27	2.22	2.17
40	2.77	2.70	2.63	2.57	2.52	2.48	2.44	2.41	2.37	2.32	2.26	2.21	2.16
100	2.78	2.71	2.64	2.58	2.53	2.49	2.45	2.42	2.37	2.31	2.26	2.21	2.16
∞	2.79	2.71	2.65	2.59	2.53	2.49	2.45	2.42	2.37	2.31	2.26	2.20	2.15

Table A7 (*Continued*)

q	f: 6	8	10	12	14	16	18	20	24	30	40	60	120
F = 4.0 (a = .500, b = 1.155)													
2	2.58	2.44	2.35	2.29	2.25	2.22	2.20	2.18	2.15	2.12	2.09	2.06	2.03
4	2.63	2.50	2.41	2.35	2.30	2.27	2.24	2.22	2.18	2.15	2.12	2.08	2.05
6	2.65	2.52	2.43	2.37	2.32	2.28	2.25	2.23	2.19	2.16	2.12	2.08	2.04
10	2.67	2.55	2.46	2.39	2.34	2.30	2.26	2.24	2.20	2.16	2.12	2.08	2.04
20	2.69	2.57	2.47	2.40	2.35	2.30	2.27	2.24	2.20	2.15	2.11	2.07	2.03
∞	2.71	2.59	2.49	2.42	2.36	2.31	2.27	2.24	2.19	2.15	2.11	2.06	2.02
F = 6.0 (a = .408, b = 1.095)													
2	2.53	2.37	2.27	2.21	2.16	2.13	2.10	2.08	2.05	2.02	1.99	1.96	1.93
4	2.56	2.40	2.30	2.23	2.18	2.14	2.12	2.09	2.06	2.02	1.99	1.96	1.93
6	2.58	2.42	2.31	2.24	2.19	2.15	2.12	2.09	2.06	2.02	1.99	1.95	1.92
10	2.59	2.43	2.32	2.24	2.19	2.15	2.12	2.09	2.06	2.02	1.99	1.95	1.92
20	2.60	2.44	2.32	2.25	2.19	2.15	2.12	2.09	2.05	2.02	1.98	1.95	1.92
∞	2.61	2.44	2.33	2.25	2.19	2.15	2.12	2.09	2.05	2.02	1.98	1.95	1.92
F = 10.0 (a = .316, b = 1.054)													
2	2.48	2.30	2.19	2.12	2.07	2.04	2.01	1.99	1.96	1.93	1.90	1.87	1.85
4	2.49	2.31	2.20	2.13	2.08	2.04	2.01	1.99	1.96	1.93	1.90	1.87	1.84
6	2.50	2.31	2.20	2.13	2.08	2.04	2.01	1.99	1.96	1.93	1.90	1.87	1.84
10–∞	2.51	2.32	2.20	2.13	2.08	2.04	2.01	1.99	1.96	1.93	1.90	1.87	1.84
F = 25.0 (a = .200, b = 1.021)													
2–4	2.40	2.20	2.10	2.03	1.99	1.95	1.93	1.91	1.88	1.86	1.83	1.80	1.78
6–∞	2.41	2.21	2.10	2.03	1.99	1.95	1.93	1.91	1.88	1.86	1.83	1.80	1.78
F = ∞ (a = 0, b = 1)													
2–∞	2.33	2.13	2.03	1.97	1.93	1.90	1.88	1.86	1.84	1.81	1.79	1.76	1.74
(k = 500)													
F = 1.2 (a = .913, b = 2.449)													
2–16	*	*	*	*	*	*	*	*	*	*	*	*	*
20	4.70	4.82	4.89	*	*	*	*	*	*	*	*	*	*
40	4.75	4.91	5.03	5.12	5.20	5.25	5.30	5.34	5.41	5.48	5.55	5.61	5.67
100	4.79	4.98	5.13	5.25	5.34	5.43	5.50	5.56	5.65	5.76	5.89	6.02	6.13
∞	4.81	5.03	5.20	5.34	5.46	5.56	5.65	5.73	5.86	6.02	6.20	6.41	6.56
F = 1.4 (a = .845, b = 1.871)													
2–14	*	*	*	*	*	*	*	*	*	*	*	*	*
16	4.61	4.66	4.68	4.69	4.69	4.69	4.69	4.68	4.67	4.65	4.62	4.58	4.53
20	4.64	4.70	4.73	4.75	4.76	4.77	4.77	4.76	4.76	4.74	4.72	4.68	4.62
40	4.68	4.78	4.85	4.89	4.92	4.94	4.96	4.96	4.97	4.97	4.95	4.90	4.81
∞	4.74	4.88	4.99	5.06	5.12	5.17	5.20	5.23	5.26	5.28	5.26	5.16	4.82

Table A7 ***(Continued)***

q	6	8	10	12	14	16	18	20	24	30	40	60	120
F = 1.7 (a = .767, b = 1.558)													
2-8	*	*	*	*	*	*	*	*	*	*	*	*	*
10	*	*	*	*	*	*	*	*	*	4.08	4.02	3.95	3.87
12	4.50	4.46	4.42	4.38	4.34	4.30	4.27	4.24	4.19	4.14	4.07	3.99	3.90
20	4.55	4.54	4.52	4.49	4.46	4.43	4.40	4.37	4.32	4.26	4.18	4.08	3.95
40	4.59	4.61	4.61	4.60	4.57	4.55	4.52	4.49	4.44	4.36	4.26	4.12	3.93
∞	4.64	4.69	4.71	4.72	4.71	4.69	4.66	4.63	4.57	4.46	4.31	4.07	3.76
F = 2.0 (a = .707, b = 1.414)													
2-6	*	*	*	*	*	*	*	*	*	*	*	*	*
8	*	*	*	*	*	3.98	3.93	3.89	3.83	3.76	3.69	3.60	3.51
10	4.41	4.31	4.22	4.15	4.08	4.03	3.98	3.94	3.88	3.80	3.72	3.63	3.53
20	4.48	4.41	4.34	4.27	4.21	4.16	4.10	4.06	3.98	3.89	3.78	3.65	3.51
40	4.51	4.47	4.41	4.35	4.29	4.23	4.17	4.12	4.03	3.92	3.78	3.62	3.44
∞	4.55	4.53	4.49	4.43	4.37	4.31	4.25	4.19	4.07	3.93	3.75	3.54	3.33
F = 2.4 (a = .645, b = 1.309)													
2-4	*	*	*	*	*	*	*	*	*	*	*	*	*
6	*	*	*	*	3.77	3.71	3.65	3.61	3.54	3.47	3.39	3.30	3.22
8	4.31	4.14	4.01	3.91	3.83	3.76	3.70	3.66	3.58	3.50	3.41	3.32	3.22
10	4.33	4.18	4.05	3.95	3.87	3.79	3.73	3.68	3.60	3.51	3.42	3.31	3.21
20	4.39	4.26	4.14	4.04	3.95	3.87	3.80	3.74	3.64	3.53	3.41	3.28	3.15
∞	4.45	4.35	4.25	4.14	4.03	3.94	3.85	3.78	3.64	3.50	3.34	3.18	3.04
F = 3.0 (a = .577, b = 1.225)													
2	*	*	*	*	*	*	*	*	*	*	*	*	*
4	*	*	*	*	*	3.43	3.38	3.33	3.26	3.19	3.12	3.04	2.97
6	4.19	3.95	3.79	3.66	3.56	3.49	3.43	3.37	3.30	3.21	3.13	3.04	2.95
10	4.24	4.02	3.85	3.72	3.62	3.53	3.46	3.40	3.31	3.21	3.12	3.02	2.92
20	4.28	4.08	3.91	3.77	3.65	3.56	3.48	3.41	3.31	3.20	3.09	2.98	2.87
∞	4.33	4.15	3.97	3.82	3.69	3.57	3.48	3.40	3.28	3.15	3.03	2.92	2.82
F = 4.0 (a = .500, b = 1.155)													
2	*	*	*	*	*	*	*	*	*	*	*	2.81	2.75
4	*	3.74	3.54	3.40	3.30	3.22	3.16	3.11	3.04	2.96	2.89	2.81	2.74
6	4.08	3.78	3.58	3.43	3.32	3.24	3.17	3.12	3.04	2.95	2.87	2.79	2.71
10	4.12	3.83	3.62	3.46	3.34	3.25	3.17	3.11	3.03	2.94	2.85	2.77	2.69
20	4.15	3.86	3.64	3.48	3.35	3.25	3.17	3.10	3.01	2.92	2.83	2.74	2.66
∞	4.19	3.90	3.67	3.49	3.35	3.24	3.15	3.09	2.99	2.89	2.80	2.72	2.65

Column headings 6–120 are values of f.

Table A7 (*Continued*)

q	f: 6	8	10	12	14	16	18	20	24	30	40	60	120
F = 6.0 (a = .408, b = 1.095)													
2	*	*	3.28	3.24	3.04	2.97	2.91	2.87	2.81	2.74	2.68	2.62	2.56
4	3.90	3.54	3.32	3.17	3.06	2.98	2.92	2.87	2.80	2.73	2.66	2.60	2.53
6	3.93	3.57	3.33	3.18	3.06	2.98	2.91	2.86	2.79	2.72	2.65	2.58	2.52
10	3.95	3.59	3.34	3.18	3.06	2.97	2. 1	2.85	2.78	2.71	2.64	2.57	2.51
20	3.97	3.60	3.35	3.18	3.06	2.97	2.	2.84	2.77	2.70	2.63	2.56	2.51
∞	3.99	3.62	3.36	3.18	3.05	2.96	2. 9	2.83	2.76	2.69	2.62	2.56	2.50
F = 10.0 (a = .316, b = 1.054)													
2	3.72	3.33	3.10	2.96	2.86	2.79	2.74	2.70	2.64	2.58	2.52	2.47	2.42
4	3.75	3.35	3.11	2.96	2.86	2.79	2.73	2.69	2.63	2.57	2.51	2.46	2.41
10	3.78	3.36	3.11	2.96	2.85	2.78	2.72	2.68	2.62	2.56	2.50	2.45	2.40
20	3.79	3.36	3.11	2.96	2.85	2.78	2.72	2.68	2.62	2.56	2.50	2.45	2.40
∞	3.80	3.37	3.11	2.95	2.85	2.77	2.72	2.67	2.61	2.56	2.50	2.45	2.40
F = 25.0 (a = .200, b = 1.021)													
2	3.55	3.14	2.92	2.79	2.70	2.64	2.59	2.56	2.51	2.46	2.41	2.36	2.32
10	3.57	3.14	2.92	2.79	2.70	2.64	2.59	2.55	2.50	2.45	2.41	2.36	2.32
∞	3.57	3.14	2.92	2.78	2.70	2.63	2.59	2.55	2.50	2.45	2.41	2.36	2.32
F = ∞ (a = 0, b = 1)													
2-∞	3.39	3.00	2.80	2.69	2.61	2.55	2.51	2.48	2.44	2.39	2.35	2.31	2.27

*No differences are significant

$a = 1/\sqrt{F}$ $\quad b = \sqrt{F/(F-1)}$

Note: Reprinted from A Bayes Rule Rule for the Symmetric Multiple Comparisons Problem, *J. Amer. Stat. Assoc.* 67 (1972), by R. A. Waller and D. B. Duncan, by permission of the managing editor of the American Statistical Association.

Bibliography

Cochran, W. G., and G. M. Cox (1957). *Experimental Designs*, 2nd ed., Wiley, New York.

Cox, D. R. (1958). *Planning Experiments*, Wiley, New York.

John, J. A., and M. H. Quenouille (1977). *Experiments: Design and Analysis*, Macmillan, New York.

LeClerg, E. L., W. H. Leonard, and A. G. Clark (1962). *Field Plot Technique*, 2nd ed., Burgess, Minneapolis.

Little, T. M., and F. J. Hills (1978). *Agricultural Experimentation*, Wiley, New York.

Petersen, R. G. (1985). *Design and Analysis of Experiments*, Marcel Dekker, New York.

Snedecor, G. W., and W. G. Cochran (1980). *Statistical Methods*, 7th ed., Iowa State University Press, Ames.

Steel, R. G. D., and J. H. Torrie (1980). *Principles and Procedures of Statistics*, 2nd ed., McGraw-Hill, New York.

Index